ARIZONA SOILS

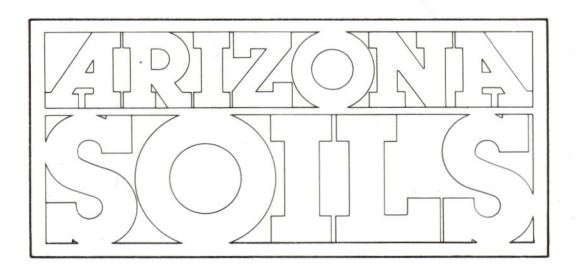

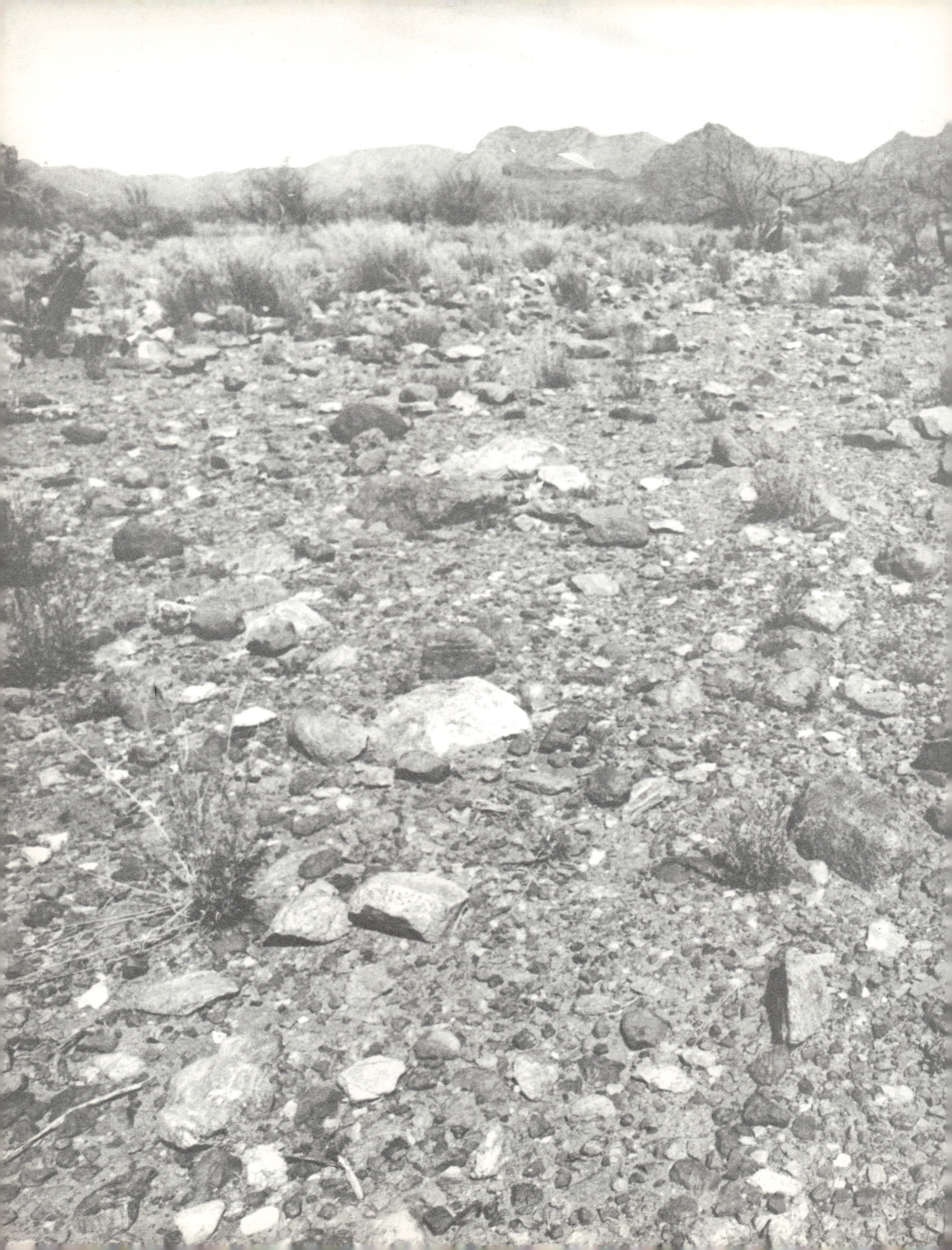

ARIZONA SOILS

DAVID M HENDRICKS

COLLEGE OF AGRICULTURE
UNIVERSITY OF ARIZONA

Editor Richard A. Haney Jr.
Book Design Paul M. Mirocha
Cover Design and Photo by Paul M. Mirocha
Illustrations by Paul M. Mirocha, Billie Jo Lobley, Lori Lieber
Word Processing/Tabular Formatting by Vicki Lee Thomas
Reference List Development by Suzanne N. Taylor
Production Coordination by Stephen D. Godwin
Typesetting by Typecraft, Tucson, Arizona, in 9/11 Mergenthaler Melior
Color Separations by Hollis Phototechnics, Tucson, Arizona
Printing by Fabe Litho Ltd., Tucson, Arizona
Binding by Roswell Bookbinding, Phoenix, Arizona

The University of Arizona
One Hundred Years 1885–1985
A Proud Beginning

A Centennial Publication
of the
College of Agriculture
University of Arizona
Tucson, Arizona
1985

Copyright © 1985
The Arizona Board of Regents
All Rights Reserved
Manufactured in the U.S.A.

Library of Congress Cataloging in Publication Data

Hendricks, David M. 1934-
Arizona soils.

Bibliography: p.
Includes index.
1. Soils—Arizona. I. Title.
S599.A7H46 1985   631.4'9791   85-6724
ISBN 0-932913-02-4 (pbk.)

David M. Hendricks
Department of Soil and Water Science

Contributing Authors

Animals and Soil in Arizona
Paul R. Krausman
R. William Mannan
School of Renewable Natural Resources
University of Arizona

with contributions from
Francis D. Hole
Professor Emeritus, University of Wisconsin

Introduction – Geologic Framework of Arizona
H. Wesley Peirce
Bureau of Geology and Mineral Technology, College of Engineering and Mines
University of Arizona

Appendix E – Descriptions of Soil Series
Steven J. Levine
Department of Soil and Water Science
University of Arizona

with contributions from
Donald F. Post
Department of Soil and Water Science
University of Arizona

Carl W. Guernsey
Y. Harmon Havens
James E. Jay
Merlin L. Richardson
Soil Scientists, Retired,
Soil Conservation Service
U.S. Department of Agriculture

Contents

Preface	xi
Acknowledgments	xiii
Editor's Notes	xv

1 Soils, What They Are and How They Form — 1
- Introduction — 1
- Definition of Soil — 3
- Composition of Soils — 3
- Soil Horizons — 6
- Soil-Forming Factors and Processes — 7
  - Factors — 7
  - Processes — 12

2 Geologic Framework of Arizona — 15
- Introduction — 15
- Basin and Range Province — 17
  - General Nature of the Geology — 17
  - Evolution of Structural Features — 21
  - Geomorphology and Soils — 23
- Colorado Plateau Province — 26
  - General Nature of the Geology — 26
  - Evolution of Structural Features — 26
  - Late Cenozoic Volcanism — 27
  - Geomorphology and Soils — 28

3 Climate of Arizona — 33
- Introduction — 33
  - Temperature — 34
  - Precipitation — 34
- Arizona Climate Zones — 35
- Climatic Zones of the State Soil Map — 38
- Soil Temperature Regimes — 38
- Soil Moisture Regimes — 40
- Water Balance and Thornthwaite's Climatic Classification — 40
- Paleoclimate — 43

4 Natural Vegetation of Arizona — 47
- Introduction — 47
- Bioclimatic Classification of Vegetation — 48
  - Alpine Tundra — 49
  - Spruce-Alpine Fir Forest — 49
  - Montane Conifer Forest — 49
  - Riparian Deciduous Woodland — 50
  - Pinyon-Juniper Woodland — 50
  - Encinal and Mexican Oak-Pine Woodland — 51
  - Interior Chaparral — 51
  - Grasslands — 51
  - Desert Scrub — 52

5 Animals and Soil in Arizona — 55
Introduction — 55
General Role of Animals in the Soil Community — 56
Soil Fauna: Classification — 56
Role of Animals in the Soil Community — 57
    Invertebrates — 57
    Vertebrates — 60
Effects of Soil on Animals — 61
    Animals That Live in the Soil — 61
    Animals That Live on the Surface — 62

6 The United States Soil Classification System and Its Application in Arizona — 63
Introduction — 63
Overview of Soil Taxonomy — 63
    Diagnostic Horizons — 63
    Categories — 64
    Nomenclature — 65
    Soil Individual — 66
Application to Arizona — 68
    Orders and Suborders — 68
    Soil Families — 69

7 Description of the Mapping Units of the Arizona General Soil Map — 71
Basis and Development of the Units — 71
Mapping Unit Descriptions — 72
    HA Hyperthermic Arid Soils — 74
    TA Thermic Arid Soils — 86
    TS Thermic Semiarid Soils — 93
    MA Mesic Arid Soils — 114
    MS Mesic Semiarid Soils — 122
    MH Mesic Subhumid Soils — 134
    FH Frigid Subhumid Soils — 142

Appendices — 151
Appendix A: Geologic Time Scale — 151
Appendix B: Vegetation: Common and Scientific Names — 155
Appendix C: Animals: Common and Scientific Names — 157
Appendix D: Outline of *Soil Taxonomy* Nomenclature System — 159
Appendix E: Descriptions of Soil Series — 161
Appendix F: Glossary — 213

List of References — 225
Index — 235

List of Plates

The unpaginated plate section follows Chapter Five, between pages 62 and 63.

Plate   Title
1. Arizona General Soil Map
2. Landform Features and Physiographic Provinces of Arizona
3. Geology of Arizona
4. Generalized Zones of Elevation in Arizona
5. Average Annual Precipitation in Arizona
6. Winter Precipitation as the Percentage of Annual Precipitation in Arizona
7. Mean January Air Temperature in Arizona
8. Mean July Air Temperature in Arizona
9. Potential Annual Evapotranspiration in Arizona
10. Actual Annual Evapotranspiration in Arizona
11. Vegetation Distribution in Arizona
12. The Four Desert Scrub Vegetative Communities in Arizona
13. Average Annual Runoff in Arizona
14. The Four Major Soil Temperature Regimes in Arizona
15. Horizons That Could Occur in a Soil Profile
16. Actual Soil Profile and Associated Vegetation
17. Biotic Provinces in Arizona

List of Figures

Figure	Title	Page
1	Indian Reservations in Arizona	2
2	Generalized Soil Composition Diagram	3
3	Representative Shapes and Sizes of Sand, Silt and Clay	3
4	Soil Texture Triangle	4
5	Crystal Structure of Muscovite Clay	4
6	Crystal Structure of Montmorillonite Clay	5
7	Crystal Structure of Kaolinite Clay	5
8	Representative Diagrams of Forms of Soil Aggregation	5
9	Form of Water in Soil	6
10	Water in an Unsaturated, Coarse-Textured Soil	6
11	Schematic of Various Factors That Affect Soil	8
12	Representative Illustration of the Evolution of Soil	8
13	Generalized Soil Profile	9
14	Effects of Slope and Aspect on Soils	10
15	Green's Peak	11
16	Depth of Carbonate Accumulation in Soils Relative to Parent Material Texture	11
17	Churning Action of Vertisols	13
18	Selected Recreation Areas in Arizona	14
19	Cross Sections of the Physiographic Provinces of Arizona	18
20	Generalized Topography of Arizona	20
21	Cross Section of a Typical Basin and Range Province Valley; Inset, Classic Schematic of Block Faulting	22
22	Generalized Models of Tilted and Listric Block Faulting	23
23	Typical Cross Sections of Arizona Basin and Range Province Valleys and Associated Soil Mapping Units	24
24	Representations of an Anticline, a Monocline and a Syncline	26
25	Map of Volcanic Rock Distribution near the Colorado Plateau	27
26	Schematic Representation of the Evolution of Typical Landforms in the Navajo Section of the Colorado Plateau	30
27	Schematic of Precipitation Distribution from Atmosphere to Soil	33
28	Climates of Arizona	36
29	Generalized Water Cycle in Hyperthermic, Thermic and Some Mesic Regions	37
30	Generalized Water Cycle in a Frigid Subhumid Region	37
31	Frost-Alternation Days in Arizona	39
32	Generalized Annual Moisture Budgets for Soils near 14 Climate Stations in Arizona	42
33	Altitudinal Distribution of the Major Vegetation Zones of the San Francisco Peaks in Northern Arizona	48
34	Altitudinal Distribution of the Major Vegetation Zones of the Santa Catalina Mountains in Southern Arizona	49
35	Low Elevation Riparian Woodland	50
36	Cross Section of the Grand Canyon and Associated Life Forms	56
37	Representative Soil Animals	58
37a	Termites	58
37b	Harvester Ants	58
37c	Spadefoot Toad	58
37d	Kangaroo Rat	58
37e	Shovel-Nosed Snake	59
38	Soil Microfauna, Mesofauna and Macrofauna Classification	59
39	Diagram of the Hierarchical Soil Classification System	65
40	Diagram Showing Soil Pedon, Polypedon and Other Features Related to a Soilscape	67
41	Explanation of Soil and Rock Symbols in Soilscapes and Profiles	72
42	Representative Hyperthermic Arid Soils Soilscape and Profiles	73
43	Geographic Distribution of Hyperthermic Arid (HA) Soils in Arizona	74
	HA1 Torrifluvents Association	75
	HA2 Casa Grande-Mohall-La Palma Association	76
	HA3 Mohall-Vecont-Pinamt Association	77
	HA4 Gunsight-Rillito-Pinal Association	78
	HA5 Laveen-Rillito Association	79
	HA6 Lithic Camborthids-Rock Outcrop-Lithic Haplargids Association	80
	HA7 Laveen-Carrizo-Antho Association	81
	HA8 Tremant-Coolidge-Mohall Association	82
	HA9 Harqua-Perryville-Gunsight Association	83
	HA10 Superstition-Rositas Association	84
44	Representative Thermic Arid Soils Soilscape and Profiles	85
45	Geographic Distribution of Thermic Arid (TA) Soils in Arizona	86
	TA1 Torriorthents-Camborthids-Rock Outcrop Association	87
	TA2 Anthony-Vinton-Agua Association	88

List of Figures Continued

	TA3	Lithic Torriorthents-Rock Outcrop-Lithic Haplargids Association	89
	TA4	Latene-Anthony-Tes Hermanos Association	90
	TA5	Paleorthids-Calciorthids-Torriorthents Association	91
46	Representative Thermic Semiarid Soils Soilscape and Profiles		92
47	Geographic Distribution of Thermic Semiarid (TS) Soils in Arizona		93
	TS1	Torrifluvents-Torripsamments Association	94
	TS2	Torrifluvents Association	95
	TS3	Tubac-Sonoita-Grabe Association	96
	TS4	White House-Bernardino-Hathaway Association	97
	TS5	Caralampi-Hathaway Association	98
	TS6	Lithic Torriorthents-Lithic Haplustolls-Rock Outcrop Association	99
	TS7	White House-Caralampi Association	100
	TS8	Caralampi-White House Association	101
	TS9	Latene-Nickel-Pinaleno Association	102
	TS10	Chiricahua-Cellar Association	103
	TS11	Gothard-Crot-Stewart Association	104
	TS12	Continental-Latene-Pinaleno Association	105
	TS13	Karro-Gothard Association	106
	TS14	Nickel-Latene-Cave Association	107
	TS15	Bonita-Graham-Rimrock Association	108
	TS16	Penthouse-Latene-Cornville Association	109
	TS17	Signal-Grabe Association	110
	TS18	Graham-Lampshire-House Mountain Association	111
	TS19	Anthony-Sonoita Association	112
48	Representative Mesic Arid Soils Soilscape and Profiles		113
49	Geographic Distribution of Mesic Arid (MA) Soils in Arizona		114
	MA1	Badland-Torriorthents-Torrifluvents Association	115
	MA2	Moenkopie-Shalet-Tours Association	116
	MA3	Sheppard-Fruitland-Rock Outcrop Association	117
	MA4	Tours-Navajo Association	118
	MA5	Torriorthents-Torrifluvents Association	119
	MA6	Fruitland-Camborthids-Torrifluvents Association	120
50	Representative Mesic Semiarid Soils Soilscape and Profiles		121
51	Geographic Distribution of Mesic Semiarid (MS) Soils in Arizona		122
	MS1	Tortugas-Purner-Jacks Association	123
	MS2	Winona-Boysag-Rock Outcrop Association	124
	MS3	Palma-Clovis-Trail Association	125
	MS4	Rudd-Bandera-Cabezon Association	126
	MS5	Roundtop-Boysag Association	127
	MS6	Lithic Torriorthents-Lithic Haplargids-Rock Outcrop Association	128
	MS7	Cabezon-Thunderbird-Springerville Association	129
	MS8	Pastura-Poley-Partri Association	130
	MS9	Lonti-Balon-Lynx Association	131
	MS10	Pastura-Abra-Lynx Association	132
52	Representative Mesic Subhumid Soils Soilscape and Profiles		133
53	Geographic Distribution of Mesic Subhumid (MH) Soils in Arizona		134
	MH1	Casto-Martinez-Canelo Association	135
	MH2	Lithic Haplustolls-Lithic Argiustolls-Rock Outcrop Association	136
	MH3	Showlow-Disterheff-Cibeque Association	137
	MH4	Roundtop-Tortugas-Jacks Association	138
	MH5	Overgaard-Elledge-Telephone Association	139
	MH6	Pachic Argiustolls-Lynx Association	140
54	Representative Frigid Subhumid Soils Soilscape and Profiles		141
55	Geographic Distribution of Frigid Subhumid (FH) Soils in Arizona		142
	FH1	Mirabal-Dandrea-Brolliar Association	143
	FH2	Sponseller-Ess-Gordo Association	144
	FH3	Soldier-Hogg-McVickers Association	145
	FH4	Soldier-Lithic Cryoborolls Association	146
	FH5	Mirabal-Baldy-Rock Outcrop Association	147
	FH6	Eutroboralfs-Mirabal Association	148
	FH7	Cryorthents-Eutroboralfs Association	149
	FH8	Gordo-Tatiyee Association	150
A1	Modern Geologic Time Scale, Appendix A		152
A2	Stratigraphic Column of Northern Arizona, Appendix A		153

List of Tables

Table	Title	Page
1	Comparison of San Francisco Volcanic Field Subdivisions	29
2	Erosional Cycles along the Little Colorado River in Northeastern Arizona	32
3	Computed Coefficients of Variation of Winter, Summer and Yearly Amounts of Precipitation for Four Arizona Climate Stations	34
4	Temperature-Precipitation Zones of the Arizona General Soil Map	38
5	Summary of Site Disturbance Indices for Deer, Elk, Cattle and Horses in the Carson National Forest, New Mexico	62
6	Diagnostic Horizons Common in Arizona Soils	64
7	Names and Important Properties of the Orders	66
8	Soil Order Names and Formative Elements	67
9	Examples of Formative Elements of Suborder Names	67
A	The Classical Subdivision of the Quaternary for North America. Appendix A.	154

Preface

This book presents our knowledge about the nature and distribution of the soils of Arizona and their relation to the diversity of geology, climate, vegetation and fauna of the state. Many soil scientists and investigators in related fields have contributed to the fund of knowledge about Arizona soils in particular and soils in general. Plate 1, the Arizona General Soil Map, was prepared in cooperation with the Soil Conservation Service of the United States Department of Agriculture.* Much of the data used to compile the map were obtained by field soil scientists as part of the National Cooperative Soil Survey. Through basic research other scientists have extended knowledge about how soils formed and the basic properties of soils, which is invaluable in predicting and understanding the distribution of soils.

The nature and distribution of geology, physiography and climate are emphasized since they are important in affecting soil distribution. Vegetation distribution is addressed because of its relationship to the distribution of both climate and soils. Also, vegetation in turn may influence the nature of the soils. A chapter about animals and soils was not planned originally since the author has only limited knowledge and background in zoology and animal ecology. When Dr. Francis Hole of the University of Wisconsin reviewed the first draft of the manuscript he suggested that such a chapter would emphasize how animals might affect soil properties and how soil and vegetation create habitats for animals. Consequently, a chapter about animals and soil was written by Drs. Paul R. Krausman and R. William Mannan of the University of Arizona School of Renewable Natural Resources, with contributions from Dr. Hole. It is emphasized that the descriptions of the relationships between the geology, physiography, climate, vegetation and animals are keyed to the Arizona General Soil Map (Plate 1) and are, for the most part, generalized.

The mapping units of the Arizona General Soil Map (Plate 1) are named mostly by soil series, the lowest category of *Soil Taxonomy*, even though it would have been much more desirable and scientifically sound to have used the nomenclature of a higher category, such as the subgroup. Soil series were used as the main bases in developing the map because many potential users of the map might not be familiar with the *Soil Taxonomy* nomenclature of the higher categories and would be more comfortable with the more familiar soil series.

Soil series were used to describe some mapping units even though insufficient information or only preliminary data were available about the dominant soil series in the area. In several areas soil surveys have been completed and in others are nearing completion since the soil mapping units were developed and named. The more detailed information resulting from these recent soil surveys has revealed that soil series different from those used to describe some mapping units dominate. On the other hand, the soil series used to identify the mapping units of the Arizona General Soil Map (Plate 1) and the more recently

---

*Copies of the original map from which Plate 1 was developed are available at no cost from the State Soil Scientist, 3008 Federal Building, 230 N. 1st Ave., Phoenix, Arizona 85025.

recognized soil series are usually members of the same subgroup, hence the desirability of having used the higher category had it been possible.

The boundary between thermic and hyperthermic soils now is considered to be farther west in western Pima County than shown on the Arizona General Soil Map (Plate 1) and to extend somewhat into southern Maricopa County and southwestern Pinal County. This temperature boundary still is only approximate. It is anticipated that as additional knowledge about some Arizona soils is obtained that additional descriptive shifts will occur.

Dave Hendricks
April 1985

Acknowledgments

The contributions and suggestions made by those who reviewed all or portions of this book are deeply appreciated. My thanks to you for your time and effort.

Reviewers

Wilford R. Gardner, Department of Soil and Water Science
Lloyd W. Gay, School of Renewable Natural Resources
Francis D. Hole, Professor of Soil Science Emeritus, University of Wisconsin
Allan D. Matthias, Department of Soil and Water Science
Steven P. McLaughlin, Office of Arid Lands Studies
H. Wesley Peirce, Arizona Bureau of Geology and Mineral Technology
Gary Peterson, Professor of Soil Science, Pennsylvania State University

I thank my colleagues named below from the University of Arizona, the Offices of the State Climatologist and the State Soil Scientist, and the U.S. Department of Agriculture for their help, suggestions, support and encouragement.

Anne G. Bailey
Anthony J. Brazel
B. P. Cardon
Anna L. Elias-Cesnik
L. T. Clark
Karen L. Coonce
Larry A. Crowder
K. James DeCook
L. W. "Pete" Dewhirst
Monika C. Escher
Daniel J. Farrell
Kenneth E. Foster
Edward G. Frisch
Lay James Gibson
Hector J. Gonzales
Douglas H. Graham
Barbara S. Hutchinson
Charles F. Hutchinson
Jack D. Johnson
George A. Kew
Lorraine B. Kingdon
Joseph R. LaVoie
Floyd McCormick Jr.
Edgar J. McCullough
Thomas G. McGarvin
William L. Nutting
Patricia Paylore
Susan A. Pearce
Douglas Pease
Ian L. Pepper
Janice E. Raffler
Richard W. Reeves
Kenneth Renard
Sol D. Resnick
Stephen J. Reynolds
Shirley Roberts
Patricia A. St. Germaine
Susan M. Scalero
Robert B. Scarborough
Barbara P. Sears
Robert L. Smith
Allyn G. Spence
Mary R. Traynor
Mercy A. Valencia
Emily E. Whitehead
N. Gene Wright

Illustration Credits

Figures

1 Paul Mirocha
2 Lori Lieber
3 Billie Jo Lobley
4 Lori Lieber
5 Lori Lieber
6 Lori Lieber
7 Lori Lieber
8 Billie Jo Lobley
9 Billie Jo Lobley
10 Billie Jo Lobley
11 Lori Lieber
12 Paul Mirocha
13 Paul Mirocha
14 Lori Lieber
15 David Hendricks
16 Lori Lieber
17 Paul Mirocha
18 Paul Mirocha
19 Bureau of Geology and Mineral Technology
20 Paul Mirocha
21 Paul Mirocha/Robert Scarborough
22 Paul Mirocha
23 Billie Jo Lobley
24 Paul Mirocha
25 Paul Mirocha
26 Paul Mirocha
27 Lori Lieber
28 Paul Mirocha
29 Billie Jo Lobley
30 Billie Jo Lobley
31 Paul Mirocha
32 Paul Mirocha
33 Billie Jo Lobley
34 Billie Jo Lobley
35 Billie Jo Lobley
36 Robin Mouat
37a Paul Mirocha/Vicki Lee Thomas
37b Paul Krausman
37c Paul Mirocha
37d Arizona-Sonora Desert Museum
37e Norman Smith
38 Paul Mirocha
39 Lori Lieber
40 Paul Mirocha
41 Paul Mirocha
42 Lori Lieber
43 Paul Mirocha
44 Lori Lieber
45 Paul Mirocha
46 Lori Lieber
47 Paul Mirocha
48 Lori Lieber
49 Paul Mirocha
50 Lori Lieber
51 Paul Mirocha
52 Lori Lieber
53 Paul Mirocha
54 Lori Lieber
A1 Paul Mirocha
A2 Paul Mirocha

Plates

1 J.E. Jay et al/Paul Mirocha
2 Paul Mirocha
3 Paul Mirocha
4 Paul Mirocha
5 Paul Mirocha/Billie Jo Lobley
6 Paul Mirocha/Billie Jo Lobley
7 Paul Mirocha/Billie Jo Lobley
8 Paul Mirocha/Billie Jo Lobley
9 Paul Mirocha/Billie Jo Lobley
10 Paul Mirocha/Billie Jo Lobley
11 Paul Mirocha
12 Paul Mirocha
13 Paul Mirocha/Billie Jo Lobley
14 Paul Mirocha/Billie Jo Lobley
15 Lori Lieber/Paul Mirocha
16 David Hendricks/Paul Mirocha
17 Paul Mirocha

Editor's Notes

These notes are explanations of what may appear to the reader to be inconsistencies of style, spelling and cartography. I hope that I have caught them all. Numbers of four digits or more in the text and tables have a comma between the third and fourth digit; they do not in the illustrations and plates. This is because some of the more complex illustrations and plates that have many numbers were produced from existing negatives from other publications that used a different style. We promised not to alter any negatives we borrowed, so sacrificed consistency to economy.

Names of mammals used in the text and appendix are after the *1973 Checklist of North American Mammals North of Mexico* (Jones, Carter and Genoway, 1973). Bird names and spellings, common and scientific, are after *The Auk* (1973; Parkes, 1978). Common and scientific names of reptiles and amphibians are after *The Vertebrates of Arizona* (Lowe, 1972). Plant common names vary greatly. Scientific names for those plants, listed in Appendix B, are after *A Catalogue of the Flora of Arizona* (Lehr, 1978).

Readers may note that a species of sage is identified as being Mojave sage while a species of yucca is known as Mohave yucca. Perhaps an entry in the 1935 edition of *Arizona Place Names*, by Will C. Barnes, explains this apparent inconsistency.

Mohave County

In northwest corner of State along Colorado (R)iver. One of the four original counties. After local Indians. "An Indian word meaning 'three mountains' from their proximity to the 'Needles.'" Hodge.

An Act of the 13th Legislative Assembly of Arizona contained a clause to the effect that: "The county seat of Mohave county shall be at Mineral Park or some place located on the Atlantic and Pacific (R)ailway within said county."

The Howell code originally spelled this word "Mojave," but according to authorities of that day, "due to an ignorant clerk," the legislative act above quoted spelled it "Mohave." Word now seems to be uniformly spelled Mohave, excepting the post office in California which has always used the "j."

Similarly Dice (1943) spelled the name of one of his biotic provinces in Arizona "Navahonian." Once again to Barnes (1935).

Spelling of this word is not well defined. Congressional appropriation bills for the Navajo Indians spell it Navajo.

Authorities differ on it. Hodge says Navajo, while others write it Navaho. Broadly speaking, scientists and ethnologists have decided on the spelling(s) Navaho and Mohave as most satisfactory and (they are) gradually coming into use. As long as Congress uses a certain form, that is legal for the particular place or thing.

Apparently Congress won out, but not before some places or things were named, hence the Navahonian Biotic Province.

The volcanic formations near Flagstaff in Coconino County have been variously known as the San Francisco Mountains, San Francisco Mountain and San Francisco Peaks. We elected to use the name San Francisco Peaks since the primary formations are known as the Agassiz, Fremont and Humphreys peaks.

La Paz County does not appear on the Arizona General Soil Map (Plate 1) in accordance with an agreement between the author, editor and the state soil scientist, whose permission was necessary for us to reproduce the map using colors that are keyed to the colors selected for the other color plates. The original map was published in 1975. All other state maps show La Paz County, which became operational January 1, 1983 following a vote by Yuma County citizens on September 7, 1982. The tilde (˜) is not used in spelling proper names in Spanish, again, because none was used in the text of the original soils map.

ARIZONA SOILS

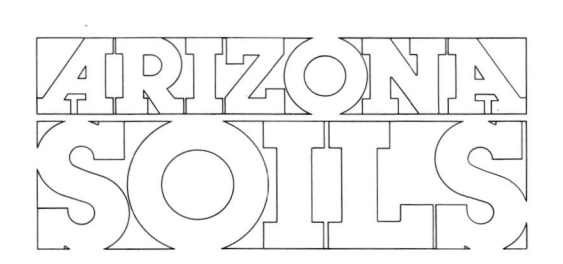

Twenty-four images transmitted to Earth by Landsat between August 24, 1972 and October 12, 1973 were used to make the Arizona Satellite Image Map 1972-1973 shown here. Landsat orbits at an altitude of about 920 kilometers (570 miles). The map was published by the U.S. Geological Survey in cooperation with NASA.

1

Soils, What They Are and How They Form

Introduction

Arizona, the sixth largest state, has an area of 295,146 square kilometers (113,956 square miles) or 29,515,111 hectares (72,931,840 acres). The state soil resources are the outermost surface of the land and range in depth from several centimeters (a few inches) to 1 or 2 meters (3 to 7 feet). The soils are among the most important state natural resources. Soils provide sustenance for animals and man and support buildings and highways, while contributing to the economies of our cities, the productivity of our farms and rangelands, and the vitality of our wildlife and wilderness areas.

Put differently by Jacks and Whyte (1939), "Below that thin layer comprising the delicate organism known as the soil is a planet as lifeless as the moon." Either statement acknowledges an awareness of the necessity of soil in our lives. It was important too between 12,000 and 15,000 years ago to the first humans who entered what is now called Arizona. They must have been aware of the support soil gave to the various kinds of life forms that they encountered in their travels, of the interactions between soil and the life forms.

To these people the most important life forms were faunal and it probably was pursuit of animals that brought them to this part of the world, perhaps from Asia. Earliest Arizonans moved with the animals that sustained them, leaving an occasional glimpse of their presence here in prehistoric campsites and the remains of elephants, camels, horses, sloths and bison that they killed.

As the climate changed and as the big game became less plentiful, early Arizonans sustained themselves by hunting the smaller species and gathering food from plants. These hunter-gatherers lived here for perhaps 8,000 years and are known to researchers as the Cochise people because the first evidence of them was found in Cochise County.

The region experienced climatic shifts over time, however, to produce a rich ecological system for sustaining ancient animals, and early man. Many plants, like paloverde and mesquite, produced abundant protein sources for food consumption as well as nitrogen that aided general plant growth in the system. This environmental richness before the time of Christ was extended by the adoption of farming practices.

Evidence of the first farmers in the region has been found in the rich alluvial valleys of Cochise County dating from around 2,000 BC. At this time, corn or maize was first domesticated as a reliable food source. Ancient corn had separately sheathed kernels in a husk, attached to a small cob. Acceptance and cultivation of this ancestor of corn transformed human social and economic life from hunting and gathering to farming and trading. The simultaneous introduction of beans, chili and squash, along with corn, provided complete nutrition for early man. With this storable source of food available in sufficient, and later surplus, quantities, the small bands of humans became larger in numbers, more elaborate in organization and more concentrated in definite centers. Later, trading food for specialized finished goods, like fabrics and pottery, led to development of sophisticated societies.

Archaeologists have noted that distinctions in geographic regions began to emerge over time. The Hohokam, Mogollon and Anasazi became unique cultural entities, with identifiable traits. The Hohokam were in the riverine valleys. The Mogollon were in areas of the mountains and the Mogollon

Rim. The Anasazi were in the far north on the high plateau. All these cultures had economic and social distinctions and attained their highest levels of development in the period from 900 AD to 1450 AD, a wetter period. However, the Hohokam, Mogollon and Anasazi did share numerous traits.

All three cultures seem to have been stimulated by the introduction of new crops, agricultural practices and social customs that probably came from highly developed cultures in Mexico (University of Arizona Press, 1972). This stimulation apparently led to trade, as evidenced by discovery of pottery designs traditional to northern peoples in southern sites and by traditional legends of south-to-north movements.

The agricultural practices adopted by these people and their successors had to be refined in each region and adapted to the constraints of soils and water. Mountain and plateau people were essentially dryland farmers, depending on rainfall and soil moisture. Arable lands were limited by the hilly and mountainous terrain. Desert-dweller domains were limited by the sources of water, whether flowing rivers or runoff from rains. Desert dwellers practiced irrigation farming in river valleys and water harvesting to augment dryland farming.

The Hohokam built large irrigation canals that drew water from flowing streams. Several hundred miles of farm engineering systems can be seen today in the Salt and Gila river valleys. Some match modern systems in size. There is evidence that pioneers restored Hohokam irrigation ditches by patching and cleaning them and then used them to irrigate their crops. Certainly, no Indian achievements north of central Mexico in pre-Conquest times surpassed the Hohokam canal system. They stand as examples of sound planning, enormous expenditures of effort and evident intercommunity organization. The Hohokam were master farmers, producing corn, beans and squash, as well as a unique contribution, cotton.

Outward from the Hohokam agriculture centers were other desert dwellers, living on areas now overlapping the current Papago Reservation. Here, people did not have access to live streams. Water-harvesting techniques were perfected to divert runoff from rainfall. One water-harvesting gathering or diversion ditch ran in a westerly direction for nearly 16 km (10 miles) from the base of Baboquivari Peak. It cut across numerous small natural drainage washes on the gentle piedmont slope, collecting rain and runoff, directing it to fields in fertile lowland ground (University of Arizona Press, 1972).

The Mogollon people lived in high areas that ranged from the Mogollon Rim and White Mountains southward into Mexico. Remains of their pueblo-style dwellings are found surprisingly close to Hohokam sites, often within close walking distances, but on high ground. While water may not have been a problem for the Mogollon, arable land was. Trade with the Hohokam can be seen in Hohokam sites. The Mogollon evidently specialized in pottery production, trading with the Hohokam, who could grow surpluses of food and fiber. Among the earliest people to leave evidence of cultivated corn, the Mogollon became a part of this regional mosaic of cultures (University of Arizona Press, 1972).

Farther north were the Anasazi, inhabiting the high plateau region, where running streams were scarce. Populations were concentrated into large centers at Mesa Verde, Chaco Canyon, Hopi and Black Mesa. In the White Mountains, the Anasazi and Mogollon evidently overlapped. The diversified environment and terrain of the Anasazi region demanded development of various methods for cultivating limited arable soils to produce food to support these large communities. Techniques included intensive cropping of carefully cultivated arable soils and water harvesting of rain and floodwater.

During the first Spanish explorations of the Southwest, between 1535 and 1604 AD, the descendants of the Hohokam, Mogollon and Anasazi offered their visitors food, not the gold Spanish tales predicted. Food included, corn, beans and squash, also found in archaeological sites dated from earlier times. Most of the tribes met by the Spanish are still represented in Arizona (See Figure 1).

The study of soils is now the occupation of the soil scientist. An important aspect of soil science is the study of the soil as a natural phenomenon on the surface of the Earth. The soil scientist, then, is interested in the appearance of the soil, its mode of formation, its physical, chemical and biological processes and composition, and its classification, distribution and use.

A modern soil scientist, referring to Plate 1, might tell us for instance that the Hohokam probably farmed in HA1, HA2,

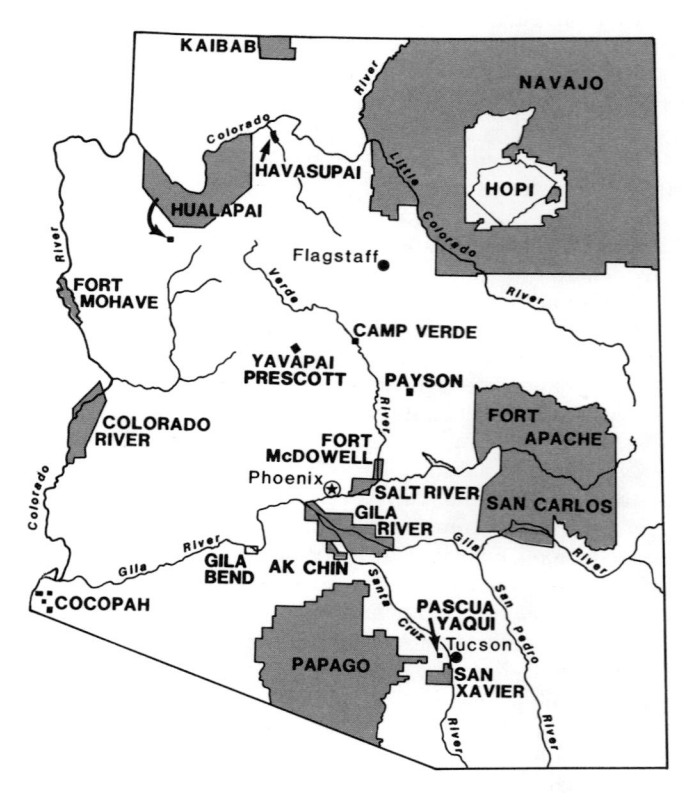

FIGURE 1. Indian Reservations in Arizona

HA3, HA5 and HA8 as well as in some areas in TS2, TS3 and TS19; that the Mogollon probably farmed only in TS9 and TS12 in the northeastern area of the Thermic Semiarid Soils; and that the Anasazi probably farmed mostly in MA5 and MA6 and some in MA1, MA2, MA3 and MA4.

Soil science is an integrative science. It uses many branches of scientific knowledge, including certain aspects of physics, chemistry, biology and geology as well as agriculture, forestry, history, geography and archaeology. From all of these, information is obtained that can be synthesized into a scientific discipline and natural philosophy separate from, and yet closely related to, many other branches of natural science.

Soil science can be viewed as a pure science that investigates the basic processes involved in the formation of soils and that produces soil maps and classifications. The scientific knowledge obtained is extremely important for those who must solve practical problems in agriculture, horticulture, range management, forestry and plan the use of the land.

Definition of Soil

The term soil generally is used to describe the material on the thin skin of the Earth's crust and that has been under the influence of certain physical and biological processes. Soil is considered here primarily as a natural unit in a pedological or ecological sense rather than in an engineering sense.

Review of detailed soil maps from many different regions of the world reveals a pattern of the interplay of climate with other soil forming factors. Soils, then can be considered to be in a state of dynamic equilibrium or slow evolution with their environments.

Soil also can be envisaged as an open system through which various hydrological, biological and geochemical cycles operate, a process-response system in which exists a close relationship between soil properties and the inputs and outputs of mass and energy.

In addition to these concepts, soil is considered in this book according to the definition given in *Soil Taxonomy* (Soil Survey Staff, 1975): "Soil is the collection of natural bodies on the earth's surface, in places modified or even made by man of earthy materials, containing living matter and supporting or capable of supporting plants out-of-doors. Its upper limit is air or shallow water. At its margins it grades to deep water or to barren areas of rock and ice.... Commonly soil grades at its lower margin to hard rock or to earthy materials virtually devoid of roots, animals or marks of other biological activity."

Composition of Soils

Soils have four main constituents: mineral and organic matter, air and water (Figure 2). Mineral matter includes all minerals inherited from the parent material as well as those formed by recombination from substances in the soil solution. Organic matter is derived mostly from decaying plant material broken down and decomposed by the actions of animals and microorganisms living in the soil. Normally, both air and water fill the voids in soil.

Minerals are the major constituent in Arizona soils and are derived from the parent material by weathering. Mineral particles range in size from 2.0 mm to less than 0.002 mm (0.079 to 0.000079 in). These particles constitute the fine earth of soil and are the bases upon which soil texture is determined according to the relative amounts of the various particles in the soil. Soil minerals are sand, particles that range in size from 0.05 to 2.0 mm (0.002 to 0.079 in) in diameter; silt, particles having diameters of 0.002 to 0.05 mm (0.000079 to 0.002 in); and clay, particles less than 0.002 mm (0.000079 in) in diameter (Figure 3).

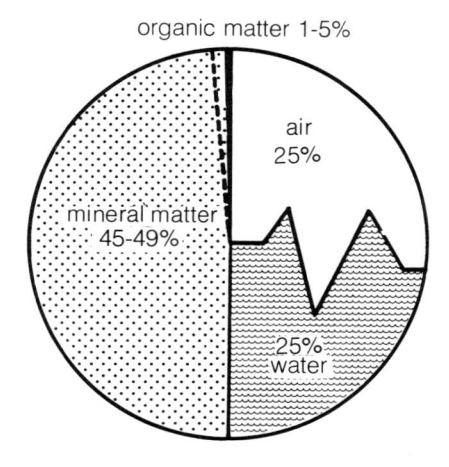

FIGURE 2. Generalized Soil Composition Diagram

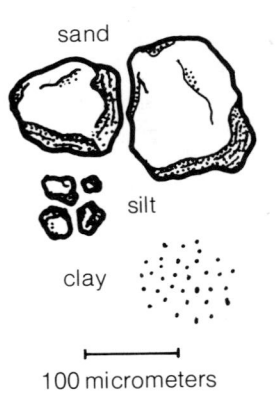

FIGURE 3. Representative Shapes and Sizes of Sand, Silt and Clay

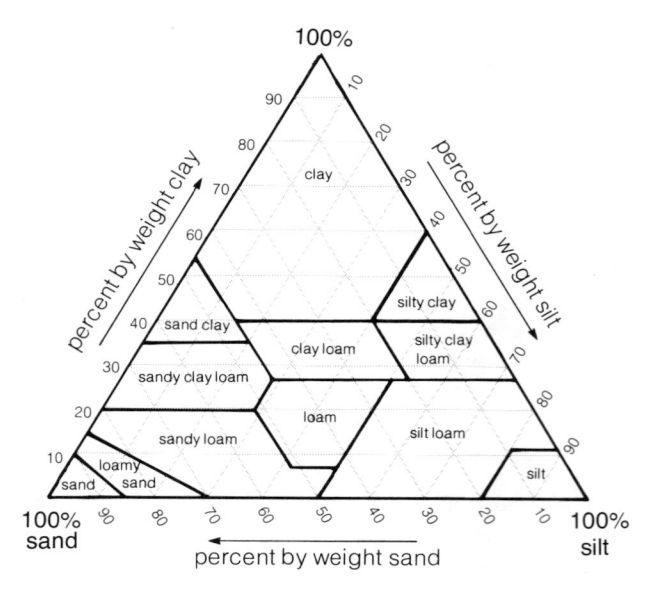

FIGURE 4. Soil Texture Triangle

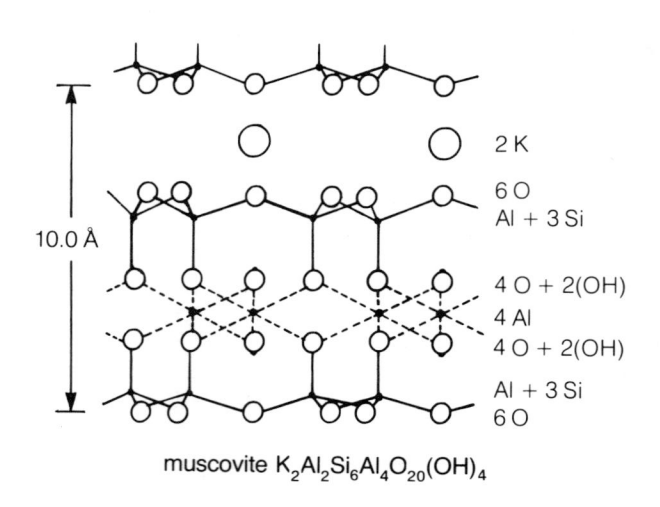

FIGURE 5. Crystal Structure of Muscovite Clay (after P. W. Birkland, 1974)

The textural class of a soil can be determined by using the soil textural triangle (Figure 4). A soil with a loam texture will have the proportions of clay, silt and sand to place it in the "loam area" of the triangle. With practice, the texture of a soil can be approximated in the field by moistening and then working a sample between the fingers and thumb. Based upon how the soil feels, the porportions of clay, silt and sand is approximated. Larger particles such as gravel and other rock fragments also occur in soil, but are considered inert, contributing only their bulk and physical presence to the soil.

Minerals can be inherited directly from the parent material with little or no change or they may be formed from minerals within the parent material during weathering and soil formation. Primary minerals such as quartz, feldspars, various mafic minerals, muscovite and biotite commonly are inherited from the parent material, whereas clay minerals form as weathering products in the soil. Generally, the proportion of primary minerals decreases and the proportion of clay minerals increases as the soil particle size becomes smaller. Thus, the sand- and silt-size particles are dominated by primary minerals and the clay by clay minerals.

Clays are the most important mineral constituents of soils. Because of their small size, clay minerals have large specific surface areas (surface area per unit mass) compared with the larger silt and sand particles. Most chemical and physical reactions and interactions occur on the surfaces of soil particles. Therefore, whether chemical or physical phenomena are considered, the small clay particles with their large specific surface area are most important in determining the fundamental properties of soil.

Clay minerals are characterized by a layered crystalline structure. The clay minerals mica and montmorillonite dominate the clay in most soils in Arizona. A crystal unit of mica, such as muscovite, has one silica sheet on each side of an alumina sheet. Adjacent crystal units are held together by potassium bridges so that the space between the units is not readily accessible for surface reactions (Figure 5). The crystal structure of montmorillonite is similar to that of mica but the adjacent crystal units are bound together by very weak linkages (Figure 6). The entire surface of the montmorillonite crystal, therefore, is accessible for reactions. Distances between the crystal units are determined by the amount of water present. As amounts of water vary, shrinking and swelling occurs with drying and wetting. Kaolinite is a common component in many of the soils in Arizona, although usually in relatively small amounts. The structure of kaolinite differs from that of muscovite and montmorillonite; it has only one silica sheet bonded to one alumina sheet (Figure 7). Kaolinite particles consist of these crystal units stacked on each other and are bonded so that the interlayer positions are not accessible for surface reactions.

A very important and intriguing property of clay minerals is their ability to adsorb and hold bases such as calcium, magnesium, sodium and potassium, and the acid element hydrogen. Clay minerals, because of their chemical composition and structure, possess a net negative charge. The bases, also known as cations, are positively charged when in the soil solution and are attracted to and held on the surfaces of the negatively charged clay particles. The adsorbed cations are held tightly enough to retard their movement from the soil by leaching, yet loosely enough to be replaced by other cations, a process called "cation exchange." The absolute quantity of cations that may be held in a soil by the clay fraction in exchangeable form is the "cation exchange capacity." One of the most significant features of the cation exchange capacity of a soil is that it provides temporary storage of large quantities of plant nutrients such as calcium, magnesium and potassium. When ammonium sulfate fertilizer is added to a soil it dissolves in the soil solution

montmorillonite for example, $X_{0.8}(Al_{0.3}Si_{7.7})(Al_{2.6}Fe'''_{0.9}Mg_{0.5})O_{20}(OH)_4 \cdot nH_2O$

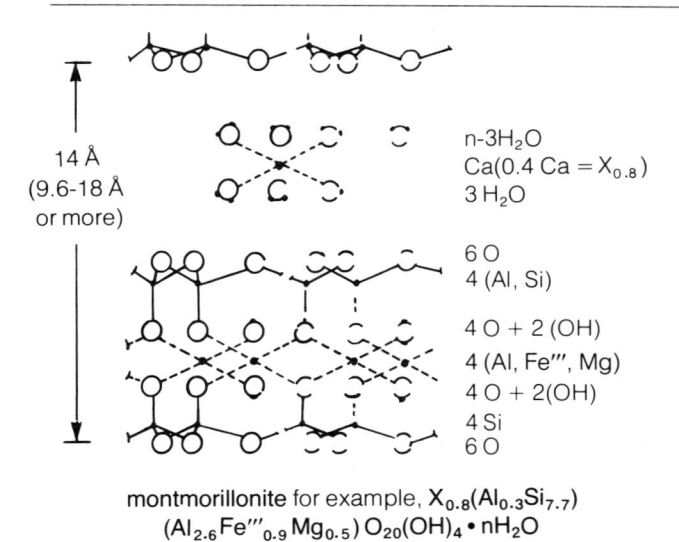

FIGURE 6. *Crystal Structure of Montmorillonite Clay (after P. W. Birkland, 1974)*

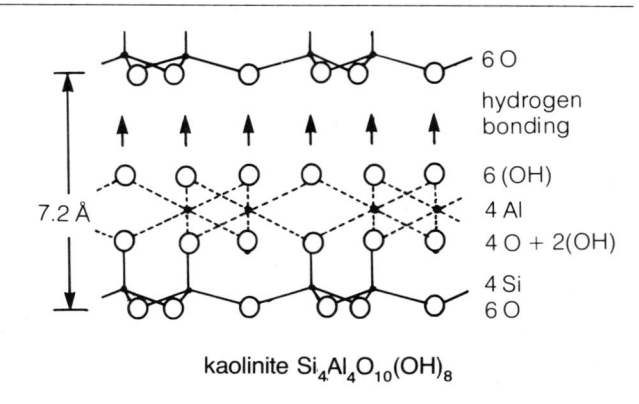

FIGURE 7. *Crystal Structure of Kaolinite Clay (after P. W. Birkland, 1974)*

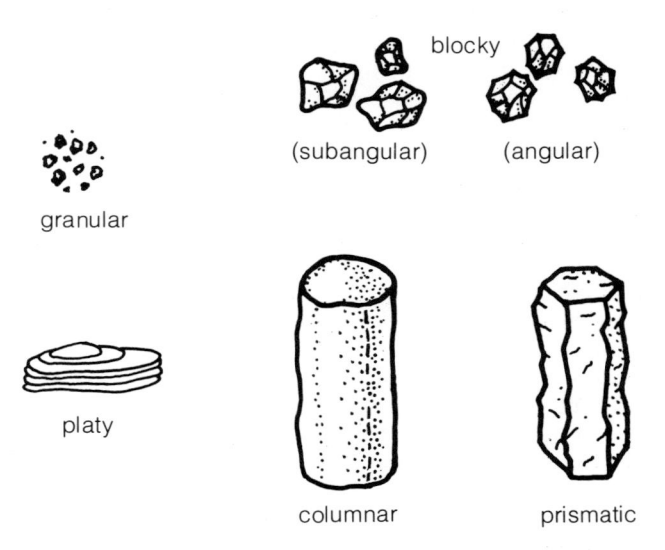

FIGURE 8. *Representative Diagrams of Forms of Soil Aggregation*

forming the ammonium cation and the negatively charged sulfate ion. The ammonium then may replace some of the exchangeable cations and be held on the clay, silt and sand surfaces until used by plants. This reaction occurs most readily on clay particles, to a lesser extent on silt and to a much lesser extent on sand.

Arizona soils generally contain relatively small amounts of organic matter due to limited plant growth and rapid decomposition of dead plant matter. Most of the soils contain less than 1 percent organic matter, although a few high-elevation meadow soils in the White Mountains contain more than 10 percent. Organic matter present in soils is intimately mixed with the mineral matter of the surface horizons and it may lie on the soil surface. Surface organic matter accumulations are quite limited in the deserts of the state but are very common in the forested soils. This surface organic matter may be freshly deposited leaf and other plant debris, as well as matter in different degrees of decomposition. Organic matter also adsorbs and holds ions in exchangeable forms similar to the clay minerals.

Soil structure is an important physical characteristic of any soil. It is produced by the aggregation of particles of sand, silt and clay into larger units called peds. Formation of structure is enhanced by organic matter, clay, $CaCO_3$ (calcite), iron oxides and other substances that can bind particles together. The five classes of soil structure are based on the geometrical shape of the peds (Figure 8): platy, granular, blocky, prismatic and columnar structures. Platy and granular aggregates usually are on the soil surface or in the surface horizon. Blocky, prismatic and columnar aggregates usually occur in the B horizon, with sodium affecting the columnar structure. Soils that lack structures are called structureless. These include soils composed of loose sand grains and those in which the particles form a coherent mass, but do not break out into distinct peds.

Air and water fill voids of the soil. Voids may result from the way soil particles in the form of single grains or peds are packed together. They are called packing voids and normally are connected. Voids other than packing voids may exist that are not normally interconnected with other voids. These voids may have irregularly shaped walls or walls consisting of smooth, simple curves; those that approach being spherical are called vesicles. Vesicles are quite common just below the desert pavement of soils in western Arizona. Channels are voids with a cylindrical form larger than those resulting from normal packing. Chambers are similar to vesicles except that they are interconnected with channels. Other voids may be planer in form, such as those associated with cracks and fissures.

The atmosphere extends into the soil through voids. Soil atmosphere is a natural continuation of the atmosphere. Although soil air is similar in some respects to atmospheric air, it differs in others. Soil air often is saturated with water vapor, has higher levels of carbon dioxide and may have less oxygen. The levels of gases present depend upon the activity of microorganisms in the soil and the gaseous exchange between soil and atmosphere. Additions of leaf litter greatly stimulate microbial activity, resulting in oxygen depletion and carbon dioxide increase. Exchange of oxygen and carbon dioxide with the atmosphere is by diffusion, a process that is hindered if soil voids are small and few. If the voids are filled with water, oxygen cannot easily diffuse in and oxygen that is present is used up, producing anaerobic conditions. The absence of oxygen and the soil chemical changes that result from it combine to make the soil useful only to water-adapted vegetation.

Air and water in the soil are almost complementary, because if the soil is saturated with water, air is displaced. In a saturated soil nearly all voids are filled with water. If the soil drains and all water in the larger voids is removed, the water lost is gravitational water (Figure 9). About two days after flooding or a heavy rain, a freely drained soil loses its gravitational water and is at field capacity. In this state, considerable amounts of water remain in finer voids because of capillary attraction.

If the soil continues to lose moisture from capillary water reserves, plants will be unable to obtain enough water for transpiration. Wilting occurs and the plant does not recover. This is the permanent wilting point. Although by convention the permanent wilting point is considered to correspond to a soil moisture tension of 15 atmospheres, the point is not actually fixed and varies according to the soil and plant involved. Some desert plants, for example, can obtain water from the soil against strong capillary forces. The amount of water held in soil between field capacity and permanent wilting point is the available water capacity. It varies according to soil texture and structure (Figure 10).

Additional water can be obtained from soil in the laboratory. Soil can be brought to an air-dry state, but it is rather variable, depending upon the relative humidity of the atmosphere where the soil is kept. Water held by soil at temperatures up to oven dry, 105 C (221 F), is hygroscopic water and is mostly unavailable to plants.

Water moves in soils through the soil voids by saturated flow. Rain or irrigation water infiltrates at the soil surface and continues downward by gravity and by capillary forces. In unsaturated soils movement is restricted, taking place slowly in response to capillary forces. Water also can move as vapor from a warmed soil layer into a cooler layer where condensation occurs.

Soil water dissolves soluble constituents and as such will constitute the soil solution which is the medium whereby plants are supplied with nutrients. Inorganic salts dissociate into ions in solution, many of which are attracted to and adsorbed onto clay and humus particle surfaces in exchangeable form. An equilibrium may be approached between exchange position numbers and ions still in the soil solution. The concentration of hydrogen ions in solution is defined by the pH scale. Neutrality is pH 7; values below pH 7 are acid, above pH 7 alkaline. Most Arizona soils are between pH 6 and 8. A few sodic soils have pH values exceeding 9 and some leached soils in the higher elevations may have pH values as low as 4.

Soil Horizons

A unique feature of most soils is the horizontal layering that generally is seen in a vertical cut from the soil surface to the underlying rock or unconsolidated rock material. The layers are soil horizons and taken together they constitute the soil profile. Each horizon may differ from its neighbors in color and depth and in physical and chemical properties. Horizons are designated according to soil profile position and the processes that created them. Soil horizon notations currently used in the United States by the National Cooperative Soil Survey (Soil Survey Staff, 1981) are described below.

The capital letters O, A, E, B, C and R represent master soil horizons.

O horizons are dominated by organic material, except limnic layers that are organic. Some are saturated with water for long periods of time or were once saturated but are now artificially drained. Others have never been saturated. They may be subdivided into:

Oa, highly decomposed organic material;
Oe, organic material of intermediate decomposition; and
Oi, slightly decomposed organic material.

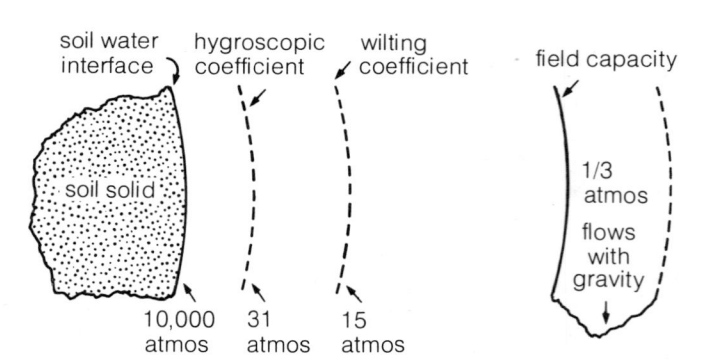

FIGURE 9. Form of Water in Soil

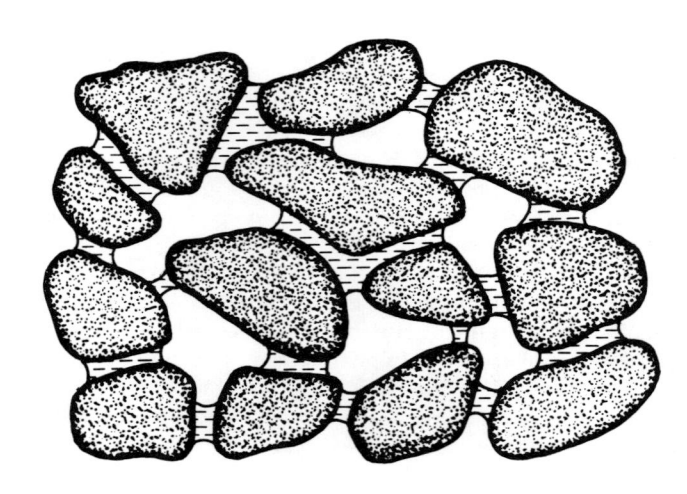

FIGURE 10. Water in an Unsaturated, Coarse-Textured Soil

A horizons are mineral horizons that have formed at the surface or below an O horizon. They are characterized by an accumulation of humified organic matter intimately mixed with the mineral fraction and not dominated by properties characteristic of E or B horizons, or have properties resulting from cultivation, pasturing or similar kinds of disturbances. In the latter case the horizon is designated Ap.

E horizons are subsurface mineral horizons that are lighter in color because of removal of organic matter, iron and aluminum oxide minerals, silicate clay or some combination of these that leaves a concentration of sand and silt particles of quartz and other resistant minerals.

B horizons form below an A, E or O horizon and are dominated by the obliteration of all or much of the original rock structure and by illuvial concentration of silicate clay, iron, aluminum, humus, carbonates, gypsum or silica, above or in combination; evidence of removal of carbonates; residual concentration of sesquioxides; coatings of sesquioxides that make the horizon conspicuously lower in value, higher in chroma or redder in hue than overlying and underlying horizons without apparent illuviation of iron; alteration that forms silicate clay or liberates oxides or both and that forms granular, blocky or prismatic structure if volume changes accompany changes in moisture content; or any combination of the aforementioned. Subdivisions of B horizons that are common in Arizona soils include:

- Bb, buried horizon;
- Bk, accumulation of carbonates;
- Bkm, cemented or indurated with carbonate;
- Bn, accumulation of sodium;
- Bq, accumulation of silica;
- Bqm, cemented or indurated with silica;
- Bt, accumulation of silicate clay;
- Bw, development of color or structure;
- By, accumulation of gypsum; and
- Bz, accumulation of salts more soluble than gypsum.

C horizons or layers, excluding hard bedrock, are little affected by pedogenic processes and lack properties of O, A, E or B horizons. Most are mineral layers, but limnic layers, whether organic or inorganic, are included. The material of C layers may be either like or unlike that from which the solum presumably formed. A C horizon may have been modified even if there is no evidence of pedogenesis. Subdivisions of C horizons in Arizona soils include:

- Ck, accumulation of carbonates;
- Cq, accumulation of silica;
- Cr, weathered or soft bedrock;
- Cy, accumulation of gypsum; and
- Cz, accumulation of salts more soluble than gypsum.

R horizons are used to designate hard or very hard bedrock.

AB is a horizon transitional between A and B horizons dominated by properties characteristic of an overlying A horizon but having some subordinate properties of an underlying B horizon.

BA is a horizon transitional between A and B horizons dominated by properties characteristic of an underlying B horizon but having subordinate properties of an overlying A horizon.

EB is a horizon transitional between E and B horizons dominated by properties characteristic of an overlying E horizon but having some subordinate properties of an underlying B horizon.

BE is a horizon transitional between E and B horizons dominated by properties characteristic of an underlying B horizon but having some subordinate properties of an overlying E horizon.

BC is a horizon transitional between B and C horizons dominated by properties of an overlying B horizon but having some subordinate properties of an underlying C horizon.

A/B, B/A, E/B, B/E, A/C and B/C are horizons in which distinct parts have recognizable properties of two kinds of horizons. The two capital letters are separated by a virgule (/). The first letter is that of the horizon that makes up the greater volume.

Limnic layers are composed of materials deposited or formed in freshwater such as coprogenus and diatomatious earth and marl.

A hypothetical soil profile with many of the different kinds of horizons in Arizona is shown in Plate 15. No given soil would contain all these horizons.

Soil-forming Factors and Processes

Horizons in soil reflect the combined effects of soil-forming factors and soil-forming processes. Climate, organisms (mostly vegetation), parent material, topography and time are the five soil-forming factors (Figure 11). The sequence of changes that the parent material goes through in forming a specific soil are soil-forming processes. Additions to, removals from and vertical transfers and transformations within the soil are the four basic kinds of soil-forming processes. The relative importance and specific nature of these processes vary according to the effects on the soil of the soil-forming factors, which also affect the evolution of the soil and its profile (Figures 12 and 13).

Factors

The importance of the soil-forming factors has been recognized for about 100 years. More recently, Jenny (1941, 1961 and 1980) extended the concept and attempted to give them a rigorous mathematical treatment. Jenny's approach is to show genetic or geographical relationships among soils using a "fundamental" equation of state,

$$s = f\,(cl,\,o,\,r,\,p,\,t),$$

where s is any soil property, f indicates a functional relationship, cl is climate, o is biological activity (organisms), r is topography or relief, p is the soil parent material and t is the time during which the soil forms. Jenny also attempted to

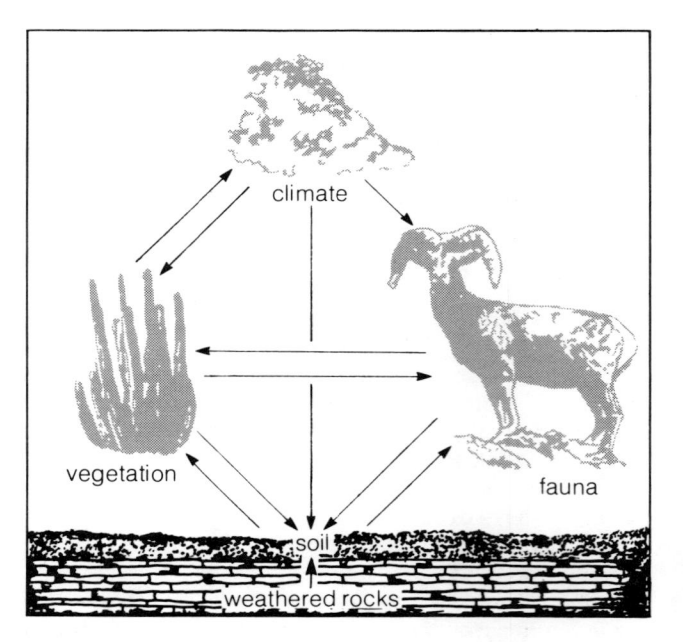

FIGURE 11. Schematic of Various Factors That Affect Soil

evaluate the influence of each factor on different soil properties by keeping the other factors constant.

For example, to assess the role that precipitation (*cl*) has on a soil property, the soil property is measured under different precipitation conditions under the same or similar temperature conditions, topography and parent materials, and using soils of similar age. Although practical difficulties are encountered in applying this equation, some very useful generalizations have been obtained. Because of the importance of these soil-forming factors, their roles in the formation of Arizona soils will be discussed briefly.

Climate. Climate includes type and amount of precipitation, temperature, humidity, evapotranspiration, duration of sunshine and a number of other variables.

Precipitation and temperature are the main parameters of climate and are important in influencing the nature of the soil that forms. Precipitation is important because of its effect on the soil-moisture regime. Much of the precipitation that falls in Arizona is lost to runoff and evaporation from vegetation and soil surfaces. Still more of the moisture that enters the soil is taken up by vegetation through roots and transpired from plant leaf surfaces, further reducing soil moisture levels. In low precipitation areas, not enough water is added

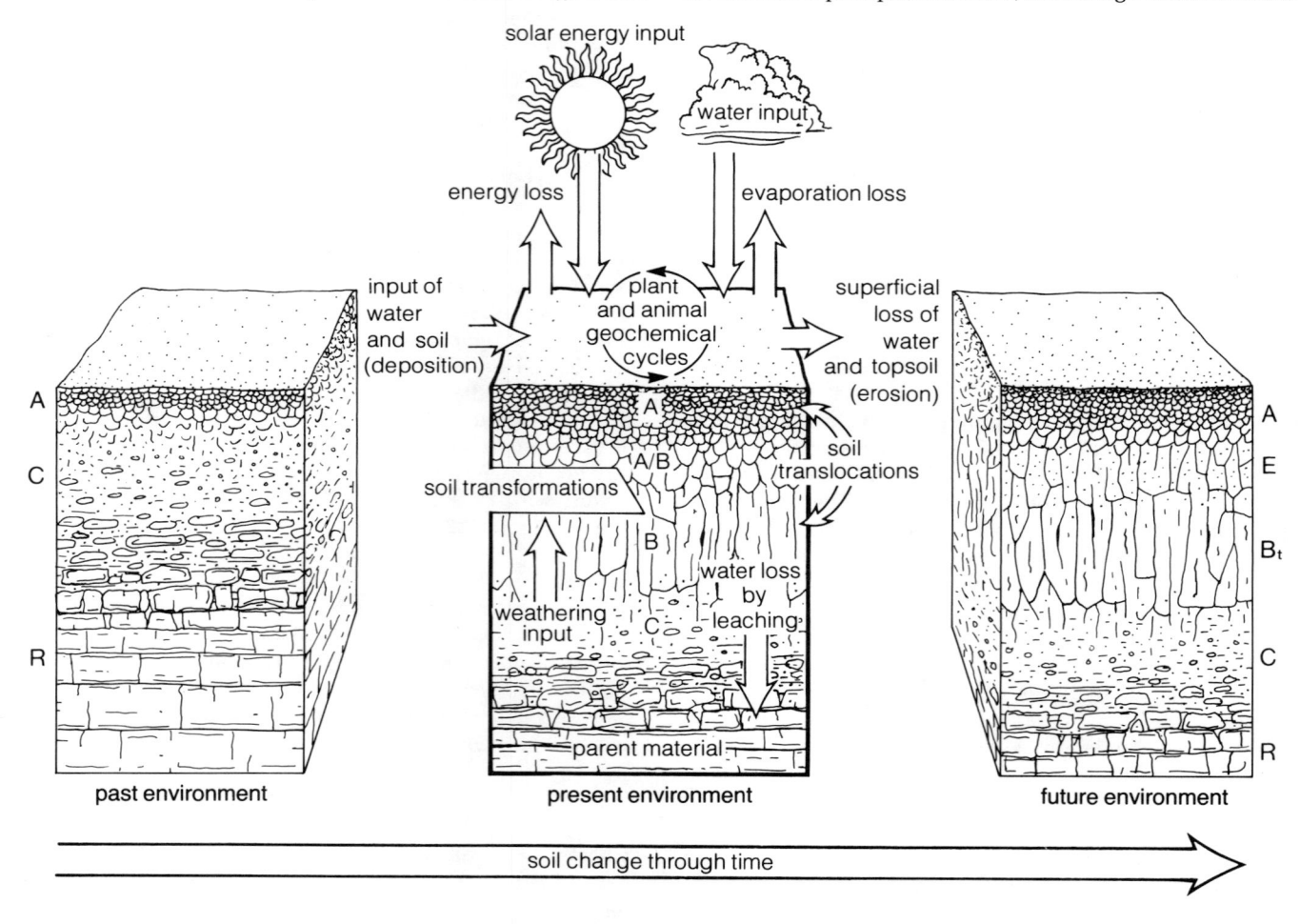

FIGURE 12. Representative Illustration of the Evolution of Soil (after E. M. Bridges, 1978)

to the soil for through leaching. Water that enters the soil penetrates to such a limited depth that soluble constituents are not removed from the soil profile. Thus, most of Arizona that does not receive much precipitation contains soils that have soluble salts and carbonates, and generally are neutral to alkaline in reaction (pH). The higher elevations in Arizona receive more precipitation than the lower, and more water percolates through the soil to remove soluble substances. Soils in these areas generally are leached of soluble salts, lack carbonates and are more acid.

Lack of moisture in the lower, arid regions of the state also limits vegetative production and inhibits soil organic matter buildup. Soils in the higher, more humid regions, on the other hand, are higher in organic matter because of greater vegetative growth.

Temperature is important in soil formation because it influences the effectiveness of precipitation. Arizona has two principal rainy seasons, December through March and July through August. Higher temperatures associated with summer rains cause much more moisture loss by evaporation and transpiration than is lost during the winter season; therefore, soil moisture levels and leaching rates are much less than during the winter season. Soils at the higher elevations in Arizona that receive more precipitation also are cooler. Thus, the amount of water available to moisten soils is greater and leaching intensities are enhanced further.

Temperature also influences biological activity since greater vegetative production occurs in soils with higher temperatures. But in Arizona limited moisture restricts vegetative growth in the warmer areas and accelerates decomposition of the organic material available to the soil. Decomposition of soil organic matter in the more humid, cooler and higher elevation soils is slower and the buildup of organic matter is higher.

The combined effect on soil properties of increasing precipitation and decreasing temperatures as elevations increase has been studied in many areas of the world. Two such studies were made in Arizona by Martin and Fletcher (1943) in the Pinaleno Mountains near Safford, and by Whittaker et al (1968) in the Santa Catalina Mountains near Tucson. The latter authors determined that soil organic matter increased by about 3.23 percent per 1,000 m (3,300 ft) increase in elevation up to 2,000 m (6,600 ft). A higher rate of increase was observed in the coniferous forests above 2,000 m (6,600 ft). Surface soil pH in the Santa Catalina Mountains decreased with increasing elevations by 1.29 units per 1,000 m (3,300 ft), according to Whittaker et al (1968).

Organisms. Organisms, particularly vegetation, long have been recognized as important factors in soil genesis. Vegetation is controlled or strongly influenced by climate, however, which makes a separate evaluation of its role in soil formation difficult. Jenny (1958, 1980) and Crocker (1960) have described theoretical methods of dealing with this problem.

Plants may influence soil properties in a number of ways. The amount and depth of organic matter is related to the type of vegetation growing on the soil. Forest soils tend to have organic matter concentrated at or near the surface because most organic matter added to the surface is leaf litter. Grass vegetation, on the other hand, generally contributes annually much more organic matter to the soil and to greater depths below the surface; the total amount of organic matter throughout grassland soils often is higher than that in forests. Forest soils generally are more acid and more strongly leached than grassland soils. Although these differences may be attributed mostly to the climatic conditions associated with the vegetation, the differing chemical nature of organic matter formed during the decomposition of the two kinds of plant materials also may contribute.

Zinke (1962) and Zinke and Crocker (1962) described and applied a method of assessing the effects of individual trees on soil properties. Since the life of a tree is rather short from a soil-formation point of view, only soil properties that change rapidly in response to the living tree would be affected.

Soils near the tree trunk were compared with those under the tree canopy and in adjacent open areas or under a neighboring tree. Soils near the trunk are influenced by bark litter and rain diverted by the tree as stem flow. Soils under the canopy are influenced by leaf litter and canopy drip. Soils in the open are much less affected by the tree. The general trend for most tree species observed by Zinke (1962) was for soil pH to increase radially from the trunk outward. Similarly, nitrogen content, exchangeable cations and cation exchange capacities were low near the trunk, increased to a maximum under the canopy some distance from the trunk and then generally declined farther outward.

The approach of Zinke (1962) was applied by Tiedemann and Klemmedson (1972, 1973, 1977) and Barth and Klemmedson (1978, 1982) to mesquite and paloverde on the Santa Rita Experimental Range south of Tucson. Tiedemann and Klemmedson (1972, 1973, 1977) found more nitrogen, potassium, sulfur, soluble salts and organic matter in soils under mesquite trees than in open spaces between trees. Mesquite trees redistribute nutrients that are absorbed through roots to a zone mostly beneath the canopy where most leaves and other plant parts accumulate.

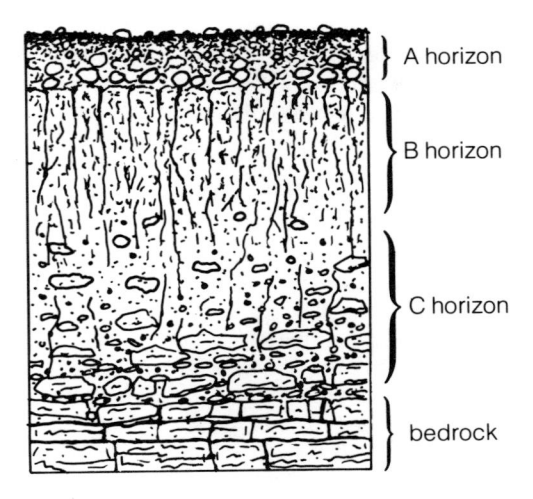

FIGURE 13. Generalized Soil Profile (after D. Hillel, 1982)

Barth and Klemmedson (1978) observed that the surface organic litter derived from both velvet mesquite and paloverde decreased in weight as distance from the plants' center increased, whereas litter originating from understory vegetation displayed weak spatial patterns in dry weight. They also found that the percent of soil nitrogen and organic carbon decreased with horizontal distance from the plants' center and with depth, to 60 cm (24 in).

Soil pH tended to increase under velvet mesquite but to decrease with horizontal distance from the center of paloverde plants. Soil pH values increased with depth under both velvet mesquite and paloverde.

These distribution patterns of soil nitrogen, organic carbon and pH are most marked in the surface soil (0 to 5 cm, or 0 to 2 in). Barth and Klemmedson (1982) further found no apparent increase in soil nitrogen and organic carbon with increased size of paloverde plants, whereas under velvet mesquite they estimated that the soil accumulated nitrogen at the rate of 11.2 g/m² (.036 oz/ft²) per meter (39 in) of height and organic carbon at the rate of 0.11 kg/m² (3.6 oz/ft²) per meter (39 in) of height.

Topography (or Relief). Soil properties commonly vary laterally with topography. One cause is the steepness of slope and its influence on surface-water runoff and erosion. Steeper slopes generally have rapid runoff and less moisture entering the soil. Additional moisture may be received by soils in low-lying areas from upslope runoff.

Erosion potential is greater on steep slopes. Surface soil material is removed more rapidly from them than from more gentle slopes. Consequently, soils on steep slopes often are shallow and lack horizon development. The eroded material usually is deposited as alluvium in low-lying areas, along stream channels or in floodplains. This alluvium then serves as parent material for soils that form in those areas.

Low areas sometimes have high water tables and the soils are saturated with water. Saturated soils may lack oxygen, and chemical reactions favored in anaerobic conditions occur, such as reduction of iron and manganese. High amounts of soluble salts accumulate readily in saturated soils because water containing dissolved salts moves toward the soil surface by capillary rise. When this water evaporates from the soil the soluble salts that it carried remain in the soil. Over time, fairly large accumulations of soluble salts accrue by this process and the soil will support growth only of salt-tolerant plants.

Another aspect of topography that affects soil properties is hillslope orientation, mostly because of its influence on the microclimate. Soil temperature regimes are not alike on slopes of different exposure because of disparate solar radiation, soil moisture, vegetative cover and wind direction and velocity (Figure 14). Southern sun-facing slopes usually receive more heat from solar radiation and are drier than northern slopes. Different soils, therefore, form on sun-facing slopes than do on shady slopes. Many hills and low mountains in the intermediate to higher elevations in Arizona have distinctly different vegetation on north- and south-facing slopes. South-facing slopes commonly are covered with grass, the north-facing slopes with woodland or forests.

shaded slope	sun-facing slope
colder soils	warmer soils
wetter soils	drier soils
restricted soil fauna	varied soil fauna
surface accumulation of acid organic matter	organic matter incorporated

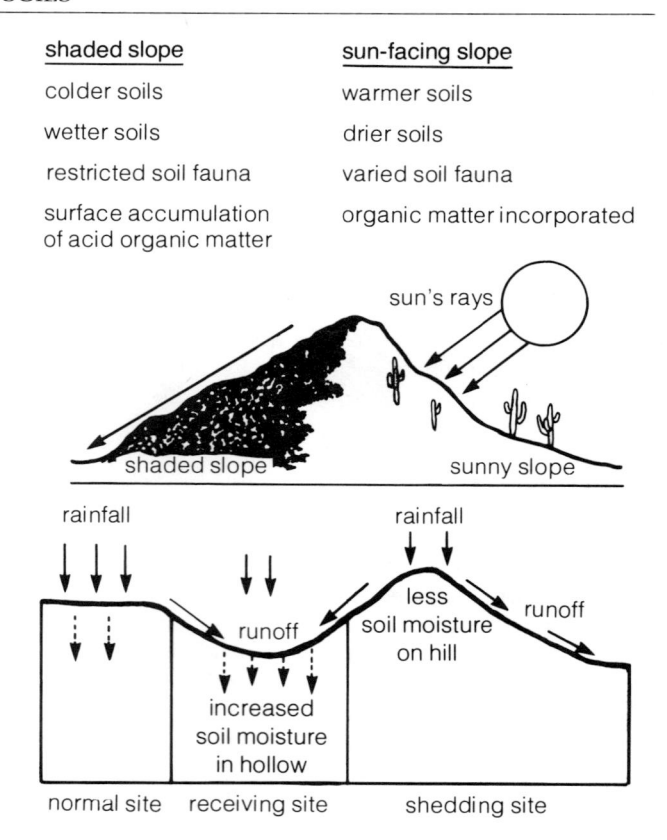

FIGURE 14. *Effects of Slope and Aspect on Soils (after E. M. Bridges, 1978)*

The soils also are different. For example, Green's Peak, a cinder cone in southern Apache County, has an Engelmann spruce forest—with some admixture of quaking aspen—on the north-facing slope and a grassland on the south-facing slope (Figure 15). Hendricks and Davis (1979) reported that forested soils are more acid, higher in clay and free iron oxides content, and have lower nitrogen content than grassland soils. Although slopes that face the wind receive more precipitation than those that do not, the wind also has a drying effect and may carry fallen snow off the slope.

Parent Material. Jenny (1941) defined parent material as being the initial state of soil systems. When parent material is exposed to environment at a site, soil-forming processes begin.

Parent material exerts the greatest influence in arid regions and during the early stages of soil formation. In humid regions and in older, more extensively weathered and evolved soils parent material differences tend to be less evident.

Parent material influences development of many soil properties to varying degrees. Soil constituents such as sand- and silt-sized particles may be inherited directly from the parent material with little if any change. Other constituents such as clay minerals evolve through chemical changes in material components. Parent material influences may or may not be obvious. For example, coarse-grained parent materials often produce coarse-textured soils. Coarse-grained parent

FIGURE 15. Green's Peak

materials that weather easily to form clay, on the other hand, tend to produce fine-textured soils. Thus, coarse-grained granitic rocks composed mostly of minerals not susceptible to chemical weathering generally produce coarse-textured soils high in sand. Basaltic rocks that contain mostly minerals quite susceptible to chemical weathering yield fine-textured soils high in clay content. Soils formed on granitic rocks also generally are deeper than soils formed from basalt because granitic parent materials are more susceptible to physical disintegration than basaltic parent materials in which smaller mineral grains are tightly interlocked.

Unconsolidated parent material such as stream-deposited sediments in Arizona valleys may influence soil formation through the ability of its texture to retain or otherwise affect moisture movement. Sediment texture influences the rate and depth of leaching, which, in turn affects many soil properties, including the zone of carbonate accumulation as illustrated in Figure 16.

Because of chemical makeup some parent materials produce soils that have higher inherent fertility than others and that support greater vegetative growth. Fertility and resultant vegetation produce soils richer in organic matter. Welch and Klemmedson (1973), for example, reported that in the Colorado Plateau ponderosa pine zone near Flagstaff basic volcanic rocks (andesite and basalt) produced soils in both tree and in grass ecosystems that contained more nitrogen than soils derived from acid volcanic rocks (rhyolite).

Time. The length of time that a soil has been forming dictates the degree of horizon expression and other pedogenic soil features that evolve given a combination of

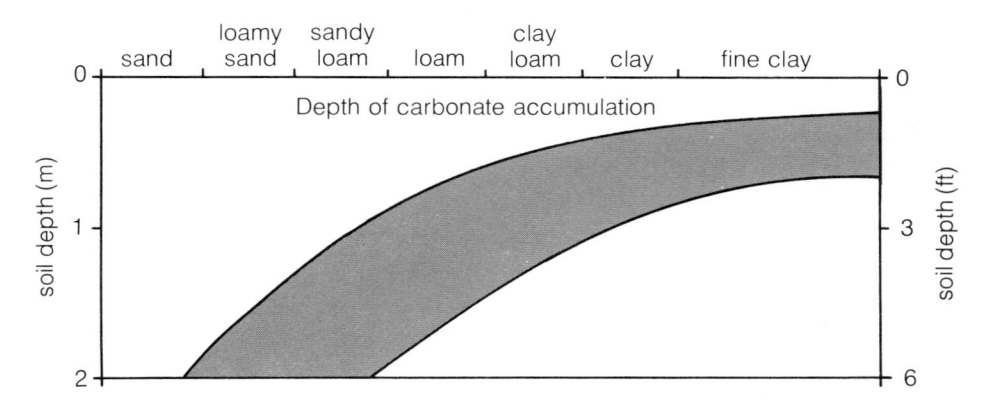

FIGURE 16. Depth of Carbonate Accumulation in Soils Relative to Parent Material Texture

climate, organisms, topography (or relief) and parent material. The time necessary for various features to evolve depends not only on the four other soil-forming factors, but on the nature of the particular feature that is formed. Formation of zones of soluble salt accumulation and A horizons by organic matter buildup happens relatively rapidly, less than 100 to 1,000 years (Yaalon, 1971). Other features, such as argillic (clay enriched) and petrocalcic (lime-cemented pan) horizons, form quite slowly, 1,000 to 10,000 years or more.

Most soil properties develop more rapidly during the initial stages of soil formation. Decreasing rates of development follow until a steady-state eventually is reached. Steady-state is a condition in which no apparent additional change in a soil property occurs because of time. Thus, rates of soil formation initially are rapid and then slow until it is zero, as demonstrated by chronosequence studies reviewed by Stevens and Walker (1970). However, a more recent review by Bockheim (1980) cast some doubts about soils actually reaching a steady-state with environments.

Several approaches have been made to assess the role of time in soil formation: 1) by determining the relative stage in soil formation from a time zero, the estimated time when parent material was first exposed to weathering; 2) by determining the rate of formation of a unit depth of soil or soil horizon; 3) by referring to the age of the slope or landform on which the soil formed; and 4) by absolute dating (radiometric dating) of a part of the soil profile (Bunting, 1965).

Processes

Many possible physical, chemical and biological reactions and interactions occur in soils that are categorized as soil-forming processes. These numerous processes are grouped into four basic kinds of changes: additions to, removals (losses) from, vertical transfers and transformations within the soil (Simonson, 1959, 1978). Vertical transfers and transformations are grouped once again under the heading of Translocations in this discussion. Acting under the influence of and in conjunction with the soil-forming factors, these processes produce the variety of soils that exist.

Additions. Organic matter that accumulates in soils includes the remains of organisms, plants and animals, but mostly plants. When plant debris falls onto the soil surface it may accumulate to form a surface litter layer or it may be mixed into the surface mineral horizons. Organic matter also may be added directly to the soil from roots and by animals.

Dissolved substances and solid particles are added to soil by precipitation. Solid particles also are added by aeolian action. Aeolian materials can have a strong influence on soil properties. Slow carbonate accretion in soil often is attributed to aeolian actions. Over a long period of time carbonates build up to relatively high levels in some soils, although the soil parent material lacks carbonates and is low in other calcium-containing minerals. The buildup of soluble salts in some other Arizona soils by this mechanism also is recognized (Nettleton et al, 1975).

Soils in floodplains and that are subject to flooding receive fresh deposits of alluvium from the floodwaters. Floodplain soils are subject to more frequent flooding and lack horizons because not enough time passes for their development before another flood deposits new material on the soil surface. The soils of floodplains, then, are characterized by limited pedogenic horizon development, but do have layers of material deposited by different flooding events.

Losses. Precipitation sometimes exceeds evapotranspiration at higher elevations in Arizona. Where this occurs downward percolating water dissolves readily soluble salts that later are carried from the soil by underground water into an aquifer or stream channel. Over a long time, percolating water removes materials that are less soluble, such as calcium carbonate minerals. The soils at higher elevations on the Kaibab Plateau north of the Grand Canyon, for instance, are free of calcium carbonate even though they formed from Kaibab limestone. The calcium carbonate originally in Kaibab limestone leached away and left noncarbonate impurities. These and added aeolian materials constituted the parent material of the soils. Leaching processes also may remove exchangeable bases such as calcium, magnesium and potassium and leave behind hydrogen or aluminum. Thus, a major effect of leaching is gradually to make the soil more acid.

Losses of soil material by wind and water erosion are common. Wind erosion is especially active in the more arid regions of Arizona where the vegetation is sparse and much of the surface is exposed. In these areas, wind may remove smaller soil particles and leave desert pavement, a stone or gravel layer on the soil surface. Desert pavement protects additional small particles underneath it from further wind erosion. Soils formed in some floodplains from finer alluvium and those formed from fine-grained sedimentary rocks that do not contain gravel may continue to be subject to wind erosion unless vegetation is established, either naturally or by man.

Water erodes soil material by raindrop splash and surface runoff, often called sheet erosion. Sheet erosion removes smaller soil particles, organic matter and soluble constituents from the soil surface. Rill and gully erosion results when surface runoff concentrates in surface depressions and forms well-defined channels, or rills. As runoff continues to erode the rills, gullies form and water removes soils from lower horizons as well as from the surface. Soils unprotected by vegetation or by desert pavement are most susceptible to water erosion. The high-intensity rains characteristic of the summer thunderstorms in Arizona are most effective in producing voluminous, rapid runoff that carries away massive amounts of soil.

Translocations. The movement of dissolved materials and very small particles from one part of the soil to another is very important in the formation of soil horizons. The soluble salts are removed from other soils and/or weathering rocks, especially salt-rich sediments, and carried by groundwater. Dissolved salts introduced to soils that have high water tables by groundwater are drawn upwards into the soil by capillary action and deposited as the water evaporates. As a result, these soils contain concentrations of salt at or near the surface that, in some cases, can develop as a surface encrustation.

Dissolved materials are carried downward in well-drained soils by percolating water, unless changing chemical conditions or dehydration cause them to precipitate. This is the principal mechanism by which carbonate-enriched horizons form in soils. Calcium carbonate in the upper part of the soil dissolves in water and is carried downward by it to be precipitated at a lower depth. In addition, calcium released by chemical weathering of noncarbonate minerals also percolates down through the soil to be precipitated as carbonates.

The zone where the carbonates are deposited represents the average depth of penetration of the percolating water. Given enough time, (usually more than 10,000 years) this process may form a horizon completely plugged, cemented and indurated with carbonates. Most soils in the arid and semiarid regions of Arizona have been affected by this process. At the higher elevations in Arizona, leaching removes carbonates from the profiles of the more humid soils.

Soil material also translocates as small particles in suspension from the surface horizons by water percolating to lower horizons in the soil. This process involves dispersion of clay and other colloidal soil materials and their movement downward with soil water, and finally by their deposition in lower horizons. For dispersion to take place, the upper horizons must be free of carbonates and low in soluble salts, and have a pH above about 5. Once dispersed, clay particles move through the larger voids by noncapillary or gravitational water, according to most theories. The suspended clay particles may be flocculated by calcium and magnesium from carbonates or by soluble salts in lower horizons. The suspended clay also is deposited when percolating water reaches a drier subsoil, causing the noncapillary flow to cease. When this happens, clay is deposited on the walls of the larger voids as water withdraws.

One of the most important results of clay translocation is development of a Bt horizon enriched in clay, a horizon referred to as being argillic. Clay content in the argillic horizon can be increased considerably compared with the amount remaining in overlying A or E horizons. Thus, a common feature of soils that undergo clay translocation is development of films or coatings of clay lining walls of larger voids and on ped surfaces. These coatings in argillic horizons sometimes are destroyed in Arizona soils by the shrinking and swelling of the soil when drying and wetting cycles occur.

Most soils swell when wet and shrink as they dry. In some soils the magnitude of change with wetting and drying may be quite small, whereas in other soils it may be fairly large, for instance deep, wide cracks may extend about 1 m (3.3 ft) into some soils during dry seasons. Soil material from the surface often falls into the cracks. When the soil becomes wet and swells, the cracks close exerting pressures upward because of the added excess material that fell into the lower horizons (Figure 17). Soil material then is forced toward the surface. With time, these soils *invert* themselves. Material originally in the subsoil moves up and becomes the surface soil and the surface soil becomes the subsoil. Soils that exhibit this phenomenon are Vertisols. Arizona Vertisols formed mostly from basalt in the central and east-central part of the state at elevations ranging from 1,215 to 1,970 m (4,000 to 6,500 ft). A common feature of Vertisols in Arizona is the large accumulation of basalt stones on the soil surface. Soil body stone content is low, suggesting that the stones originally in the soil worked up to the surface over time. Here they accumulate since they are too large to fall into cracks during the dry season.

Nutrient elements absorbed by plant roots at some depth within the soil move upward through the plant in its tissue. When the plant dies or sheds leaves, these nutrient elements are returned to the soil. This process, nutrient cycling, transfers elements from various horizons within the soil to the surface. Some large trees extend roots several meters into the soil and, sometimes, into cracks in underlying rocks where they remove nutrients later deposited on the surface.

Soil animals play an important role in moving solid particles from one part of the soil to another. These animals' activities, whether on a small scale such as those of insects, worms and other invertebrates or on a larger scale by burrowing snakes, gophers and mice, mix soil material from one horizon with that of another.

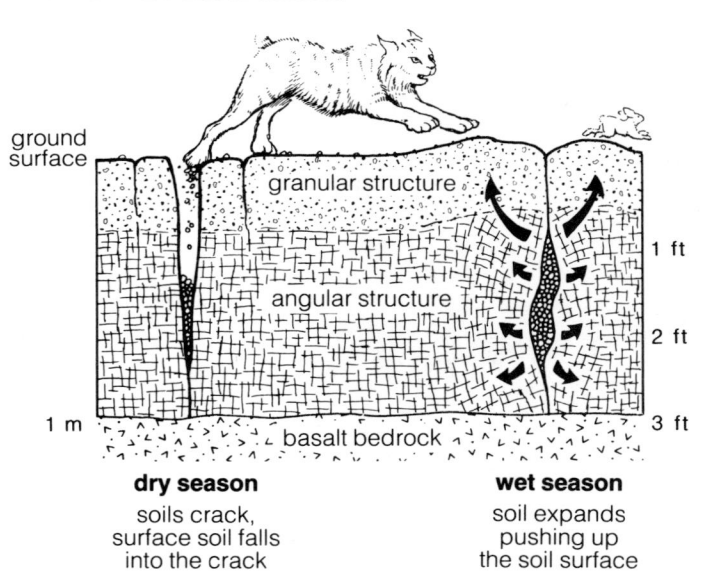

FIGURE 17. Churning Action of Vertisols

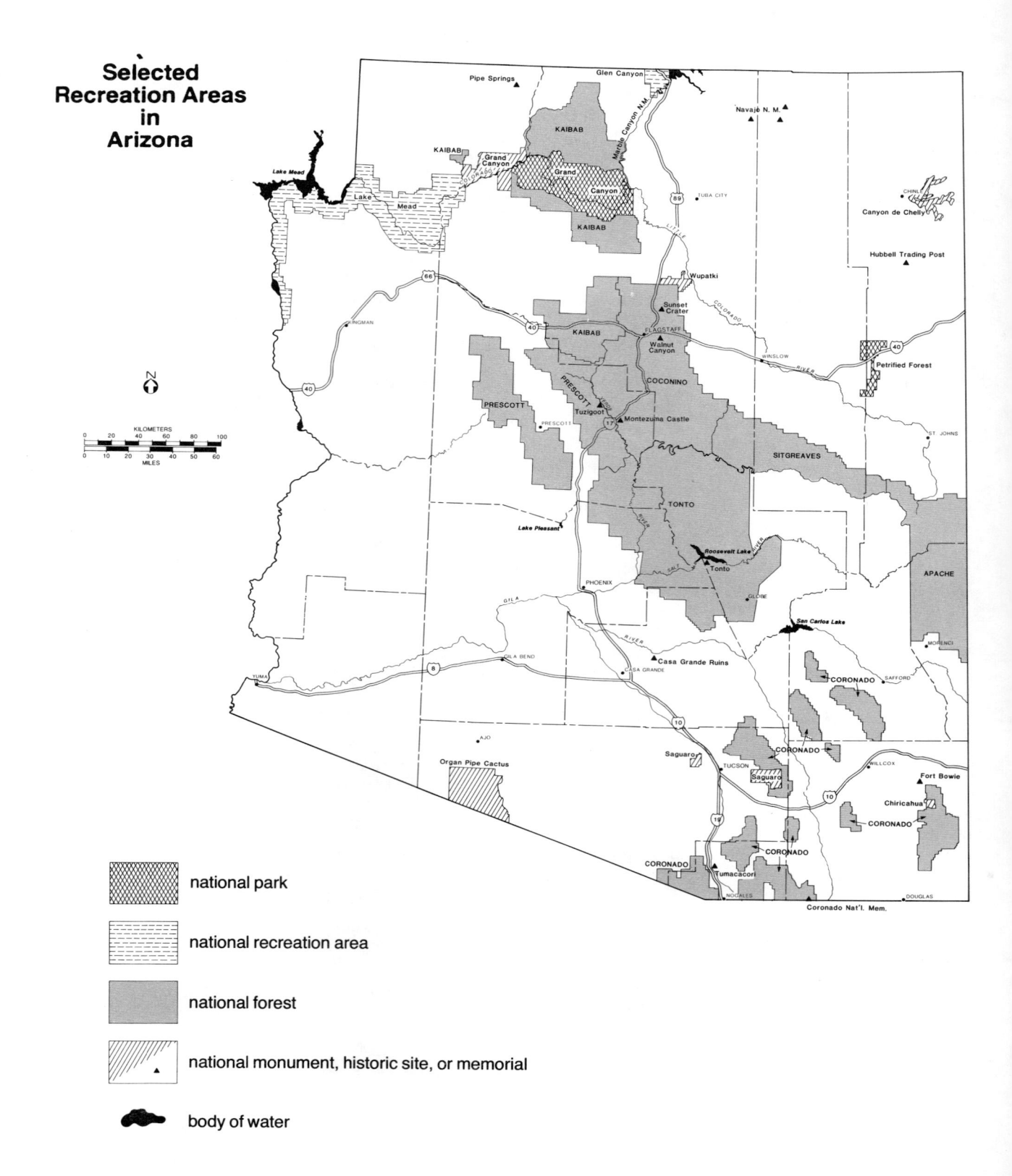

FIGURE 18. Selected Recreation Areas in Arizona (after M. E. Hecht and R. W. Reeves, 1981)

2

Geologic Framework of Arizona

Introduction

Arizona, shaped by a variety of geologic events and processes acting over at least 1.8 billion years of Earth history, is unusual in many respects. The state's great floral and faunal diversity is directly attributable to the diversity of habitats created by the interplay between "things geologic" and climate.

Much of Arizona's world-renowned scenery is geologic. The Grand Canyon is one of the world's wonders, while the Petrified Forest National Park, southeast of Holbrook in Apache and Navajo counties, contains the most spectacular display of fossil wood in the world. *Arizona Highways* magazine has made famous the red rocks around Sedona and in Monument Valley and Canyon de Chelly. In fact, Arizona has 16 national monuments, more than any other state (Figure 18). Some are geologic features such as Sunset Crater, just northeast of Flagstaff. It is a red volcanic cone that erupted about 1065 AD and rose 245 m (800 ft) out of lava and cinders. A short distance west of Winslow lies Meteor Crater, which is about 1,215 m (4,000 ft) across and ranges in depth to 180 m (600 ft).

In recent years phenomena called "metamorphic core complexes" have been recognized by geologists. A series of these complexes begins in southeastern Arizona and curves through the western states into Oregon. They are an unusual kind of mountain peculiar to this region and begin with the Rincon and Santa Catalina mountains north and east of Tucson. Metamorphic core complexes, formed before the late Cenozoic Basin and Range disturbance, have become the focus of intense study by geologists not only from Arizona, but from other parts of the world.

Before the written word or petroglyphs, Earth history was recorded only by rocks, either by information left in them or by events that affected them. Arizona is rich in both kinds of records. The fossil record in Arizona is bountiful and, of course, the Grand Canyon is an outstanding record of an event, erosion, that affects rocks.

Arizona's unique contributions to the life-on-Earth record are many. Rocks of the Chinle Formation in the Petrified Forest have yielded important reptilian evolutionary records. Other of the state's fossils show changes that occurred in bones and skeletal structure as mammals evolved from reptilian life.

One of the better faunal life records of the last few million years in North America is preserved in the San Pedro Valley in southeastern Arizona. Through application of paleomagnetic dating techniques, the fossils of San Pedro Valley have provided "tight" age zones to scientists studying evolution. The fossils are those of precursors of fauna now living in Arizona as well as those of rhinoceros, llama, camel, tapir, giant ground sloth and mammoth.

In 1981 the Kayenta Formation in northeastern Arizona, approximately 180 million years old, gave up the oldest known mammalian fossil found thus far in the Western Hemisphere. The fossil jawbone is about 1 cm (0.4 in) long, similar in size to that of a mouse. It was part of a find in a quarry on the Navajo Indian Reservation that also included the discovery of the oldest turtle skeleton known in North America. The "mouse-like" jawbone appears to belong to a previously unknown and unnamed group of small mammals similar in age to some discovered previously in England, Wales and China.

Earlier life in Arizona was not unlike that in other parts of the world. During the Paleozoic Era, about 550 million to 200 million years ago, inhabitants included marine animals such as brachiopods, mollusks, corals, sponges and trilobites. Fossil teeth and plates of bony armor of primitive fish offer evidence of vertebrate life during Devonian time, about 400 million years ago, when most of what we know as Arizona was under water. By the end of the Paleozoic Era reptiles and amphibians had appeared, but this part of the record in Arizona is scanty.

Just as geologic events in Arizona preserved prehistory, so too did they create a wealth of mineral deposits. The state, more precisely Tucson, is the hub of copper mining in the United States, contributing more than 60 percent of the copper mined in the nation. In fact, the amount of copper in Arizona is so unusual that it has been called a planetary resource. These large deposits are known as "porphyry coppers." Nowhere else in the world are deposits of this kind so well known, concentrated and studied. Economic geologists come from all points on the globe to study Arizona's porphyry copper deposits.

The industrial mineral zeolite, variety chabazite, was first mined in North America from a unique deposit in southeastern Arizona. It is the only known chabazite mine in the northern Western Hemisphere. Zeolite is often called the "environmental mineral" because of its ability to adsorb pollutants in natural gas and in effluents and other wastes.

Arizona also has the thickest, youngest known bedded salt (NaCl) deposits in North America, if not the world. One deposit in the Red Lake Basin north of Kingman in Mohave County is estimated to contain more than 400 km^3 (100 mi^3) with thicknesses approaching 3,035 m (10,000 ft). West of Phoenix lies the Luke salt body, also thought to be about 3,035 m (10,000 ft) thick.

Some scientists believe that the Picacho Basin near Eloy in south-central Arizona contains the thickest anhydrite sequence in the world. Anhydrite (CaSO$_4$) is a gypsum-like mineral. The sequence consists of about 90 percent anhydrite and 10 percent interbedded shale and is slightly less than 1,820 m (6,000 ft) thick. Anhydrite is one of the evaporite minerals that form, under certain circumstances, when large volumes of water evaporate.

The Pinta Dome east of Holbrook on the Colorado Plateau contains the only knowm helium field in the world where non-fuel gas has been commercially extracted. Normally, helium is a minor by-product in certain natural gas fields. But the Pinta Dome contains only helium and nitrogen.

Yet another kind of deposit that abounds in the southern and western portions of Arizona is water, groundwater, that results from geologic action. Arizona is divided into two primary provinces, the Plateau in the northeast and the Basin and Range in the southwestern half of the state. Today, about 67 percent of the water used in Arizona is pumped or otherwise produced from underground reservoirs in the Basin and Range Province. Perhaps the most practical way to illustrate the impact on the state of this bipartite subdivision is by assessing its effect on human activity. More than 94 percent of Arizona's population, value of mineral production, agricultural acreage and volume of water produced are in the Basin and Range Province. What caused this disparate condition? Geologic events, of course.

Episodes of block faulting, the Basin and Range disturbance, occurred between 13 million and 6 million years ago to create southwestern Arizona's mountains and valleys. Topographic relief was created and was attacked by natural physical and chemical processes. The mountains eroded and the components were transported and deposited in the adjacent lowlands or basins. Water, the prime agent of sediment transport, did its surface work, then seeped into the loose materials to become groundwater. This mountain and valley geomorphic couple promoted the slow accumulation of sediment and storage of vast volumes of subsurface water. The regional surfaces constructed on these sedimentary materials were more or less planar, gentle slopes capable of supporting vegetation and associated organisms.

Undrained valley centers tended to become playas. Because of the later integration of the Colorado-Gila rivers drainage system through the valleys, most playas in Arizona were eroded by down-cutting processes. Only two large playas remain, Willcox in Cochise County and Red Lake in Mohave County. Floodplains and associated earth materials evolved in those valleys that had axial streams. Water infiltrated over a lengthy time and tended to fill the subsurface water storage capacity of basin-fill materials.

Once water percolates into the basin-fill materials, it occupies small spaces between the sedimentary particles and becomes an important part of the surficial foundation support system. That is, the presence of water keeps the particles from packing together as tightly as they might if it were absent. In some parts of Arizona more water is withdrawn from basins that is recharged to them. This practice not only depletes the groundwater supply, but also disturbs the natural balance of forces within the basin fill materials and causes the surface to subside and crack. Pumping groundwater in excess of recharge in the Eloy area of Pinal County has resulted in the subsidence of the community by at least 2 m (7 ft) since 1948 (Laney, Raymond and Winikka, 1978).

The title of this chapter, "Geologic Framework of Arizona," alludes to the basic three-dimensional characteristics of Arizona that have been acquired during at least 1.8 billion years. It is this history of geologic events that created the diversity of soil parent materials and topography that in turn produced the diversity of soils and the complex nature of their distribution in Arizona.

Arizona has two primary physiographic provinces, the Basin and Range in the south and west and the Colorado Plateau in the north and east (Plate 2). Different geologic structural activity created the elements that distinguish the two provinces (Plate 3). In the Basin and Range rock formations have been intensively deformed and occur as numerous, relatively elevated and depressed blocks, while in the Colorado Plateau they have undergone only moderate deformation (Figure 19). These structural and topographical differences are important in relation to the nature and distribution of soils in the two provinces (Figure 20).

Between the two primary provinces is the Transition Zone, a zone defined by Wilson and Moore (1959) as being characterized by canyons and large structural troughs. And within the Basin and Range Province are two subdivisions, the desert and the mountain regions, although the boundary between them is not sharply defined. The generalized zones of elevation in Arizona are illustrated in Plate 4.

These are some of the facts about Arizona, a "thing geologic." The rest of this chapter is devoted to Arizona's geologic framework, its formation and how its diversity has influenced soil genesis.

Basin and Range Province

General Nature of the Geology

The Basin and Range Province is characterized by numerous mountain ranges that rise abruptly from broad, plain-like valleys or basins. Altitudes of these mountain masses range from about 90 m (300 ft) to more than 3,035 m (10,000 ft) above sea level. They range in length from a few kilometers to 100 km (60 mi) or more. Width of the mountains is from less than 2 km (1 mi) to more than 25 km (15 mi) (Fenneman, 1931). Ranges and associated basins in Arizona generally trend north to northeast and have through-flowing drainage. The only large, closed, dry basins in Arizona are Willcox Lake or Playa in Cochise County and Red Lake in Mohave County.

The variety of rocks exposed in the Basin and Range Province mountains is shown in Figure 19. Age of the rock ranges from Precambrian to Quaternary. The three major rock classes of igneous, metamorphic and sedimentary are well represented. Precambrian and Tertiary granitic rocks are quite common. Abundant volcanic rocks cover the spectrum from rhyolitic to basaltic (acidic to basic) and range in age from Mesozoic to Quaternary. Composition of the older volcanics is mostly intermediate to silicic. Welded ash flows (tuffs) or ignimbrites are particularly widespread. Younger volcanic rocks in the Basin and Range Province are mostly basaltic. Metamorphic rocks include gneiss and schist, all mostly Precambrian and Mesozoic. Limestone, sandstone, quartzite and shale are sedimentary rocks mostly Paleozoic, Mesozoic and Cenozoic in age.

Sediments filling the intermontane basins contain gravels, sands, silts, clays, marl, gypsum and salt that represent combinations of fluvial, lacustrine, colluvial and alluvial fan deposits. Excepting areas in the lower Colorado River valley, basin fill is the product of continental sedimentation rather than of marine. The fill also has lesser amounts of interbedded volcanic rocks.

The fill in the various basins generally is quite deep (Eberly and Stanley, 1978; Oppenheimer and Sumner, 1980, 1981). A number of basins exceed 2,425 m (8,000 ft) and a few 3,400 m (11,200 ft) in depth to bedrock as depicted by Oppenheimer and Sumner (1980).

Eberly and Stanley (1978) divided Cenozoic sediments in southwestern Arizona into two unconformity-bounded units: an older Unit I, Eocene to late Miocene in age; and a younger Unit II, late Miocene to Holocene in age. The boundary between the two units is a widespread unconformity surface produced by a period of subsidence, block faulting and erosion that began in late Miocene time, 13 million to 12 million years ago. Sedimentation that formed Unit I occurred in broad, interior depressions. Unit II sediments were deposited in troughs and grabens created by the late Miocene block-faulting episode, the Basin and Range disturbance. Scarborough and Peirce (1978) recommended that the term "basin fill" be restricted to material deposited in basins created by the Basin and Range disturbance (Unit II). Such fill deposits range up to about 3,035 m (10,000 ft). Scarborough and Peirce (1978) also excluded deposits formed by relatively modern integrated stream systems. These deposits are generally coarser-grained and probably no older than Pleistocene. Unit II sediments correspond to these basin fill materials.

Surfaces that extend from basin centers to mountain fronts may appear to be quite flat, but actually rise steadily and with increasing steepness from the axial trough toward the mountains where they may have slopes of 8 or 9 degrees (Fenneman, 1931). Typically, the plains meet mountain fronts at sharp angles referred to as nickpoints (Rahn, 1966). The upper slopes often are fans or fan terraces that may be no more than a thin mantle of alluvium over planed-rock surfaces (pediments) near mountain fronts. These surfaces are most distinct where ephemeral streams emerge from mountains, but lose their identities as they coalesce toward valley centers, producing a single, broad slope. The term piedmont slope is used to describe that part of the intermontane basin that comprises all of the constructional and erosional, major and component landforms from the basin floor to the mountain front and on into alluvium-filled mountain valleys.

Melton (1965) believed that alluvial fan deposits flank almost every mountain range in southern Arizona. These coarse-grained alluvial deposits, characteristically composed of gravels, cobbles and boulders, are most prevalent around mountain ranges 1,200 m (3,960 ft) or higher, with relief greater than 460 m (1,520 ft). Melton (1965) further suggested that major fans composed mostly of bouldery alluvium are not now forming in southern Arizona because climate and relief and a lack of tectonic activity do not favor their formation. Nevertheless, young alluvial deposits are on piedmont slopes in the Arizona Basin and Range Province (Refer to Figure 21). These deposits are associated with modern, or Holocene, stream channels that emerge from adjacent mountain ranges and extend across piedmont slopes which usually contain complex patterns of young and old alluvium (Hendricks, 1974). Lateral migration of streams, hence of the areas of deposition, happens occasionally and younger alluvium is deposited on older alluvium. Active alluvial fans, some of which may be ephemeral, are often at the foot of an older, stream-entrenched alluvial surface where the stream enters a floodplain. Alluvium of these active fans may be composed partially of material derived from older alluvium.

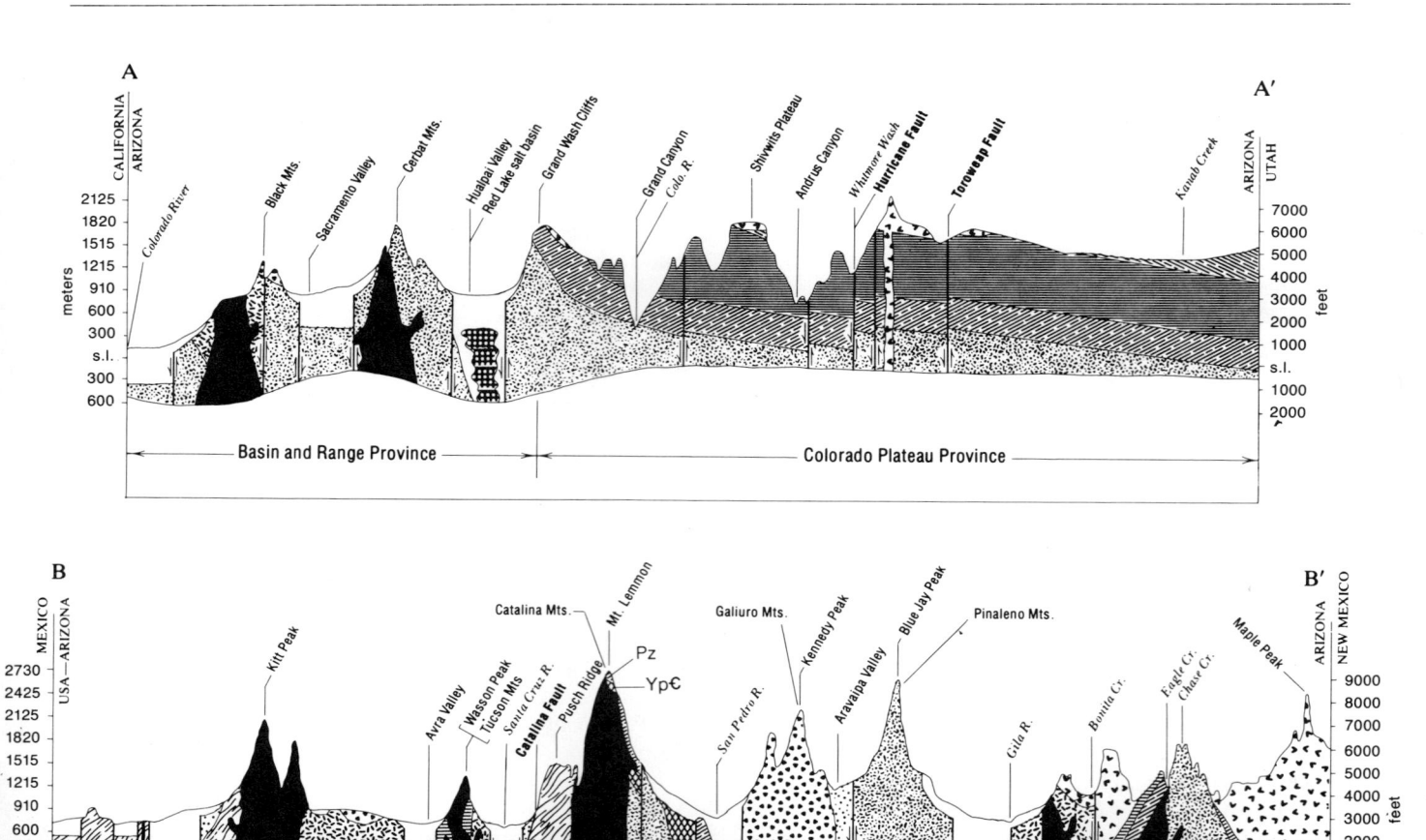

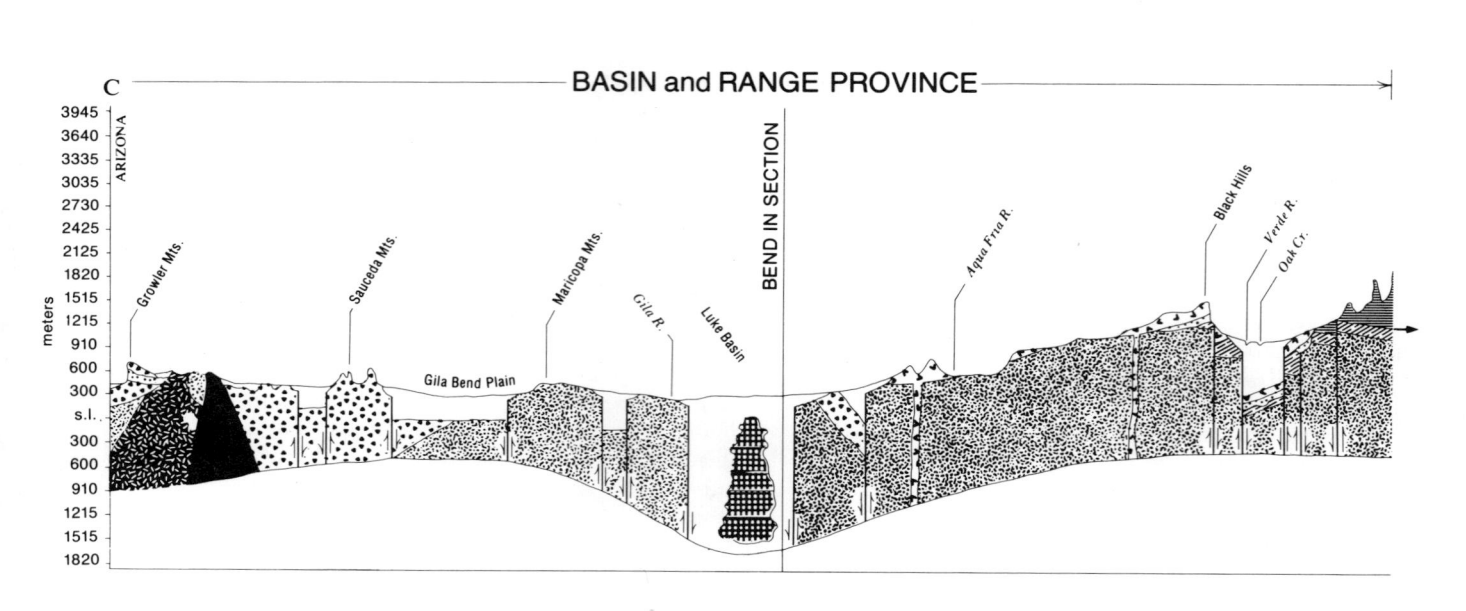

FIGURE 19. Cross Sections of the Physiographic Provinces of Arizona (Compiled by H. W. Peirce)

GEOLOGIC FRAMEWORK OF ARIZONA

Sedimentary and Volcanic Rocks

QTs — Quaternary and upper Tertiary (Pliocene) sedimentary rocks, mostly unconsolidated; includes scarce lava and silicic tuff

QTh / QTa — Evaporites
- QTh-Halite
- QTa-Anhydrite
- Gypsum

QTv — Quaternary and upper Tertiary volcanic rocks, mostly basaltic in composition

Ts — Middle Tertiary (Miocene and Oligocene) sedimentary rocks; locally include lava and tuff

Tv — Middle Tertiary volcanic rocks of silicic to basaltic composition; includes related intrusive rocks

Cenozoic

Ks — Cretaceous sedimentary rocks

TMzv — Lower Tertiary to Triassic volcanic rocks; includes some sedimentary rocks

JŦs — Jurassic and Triassic sedimentary rocks

Mesozoic

PⅠPs — Permian and Pennsylvanian sedimentary rocks; shown only on Colorado Plateau

MЄ / Pz — Mississippian through Cambrian sedimentary rocks on Colorado Plateau; all Paleozoic sedimentary rocks in Basin and Range Province

Paleozoic

YpЄ — Younger Precambrian sedimentary rocks and intrusive diabase

OpЄ — Older Precambrian rocks of all types, including schist, gneiss, and fine- to coarse-grained igneous intrusive rocks

Precambrian

Other Metamorphic and Intrusive Igneous Rocks

Tku — Tertiary and Upper Cretaceous intrusive igneous rocks

TMz — Post-Paleozoic gneiss and schist

KŦ — Mid-Cretaceous to Triassic intrusive igneous rocks

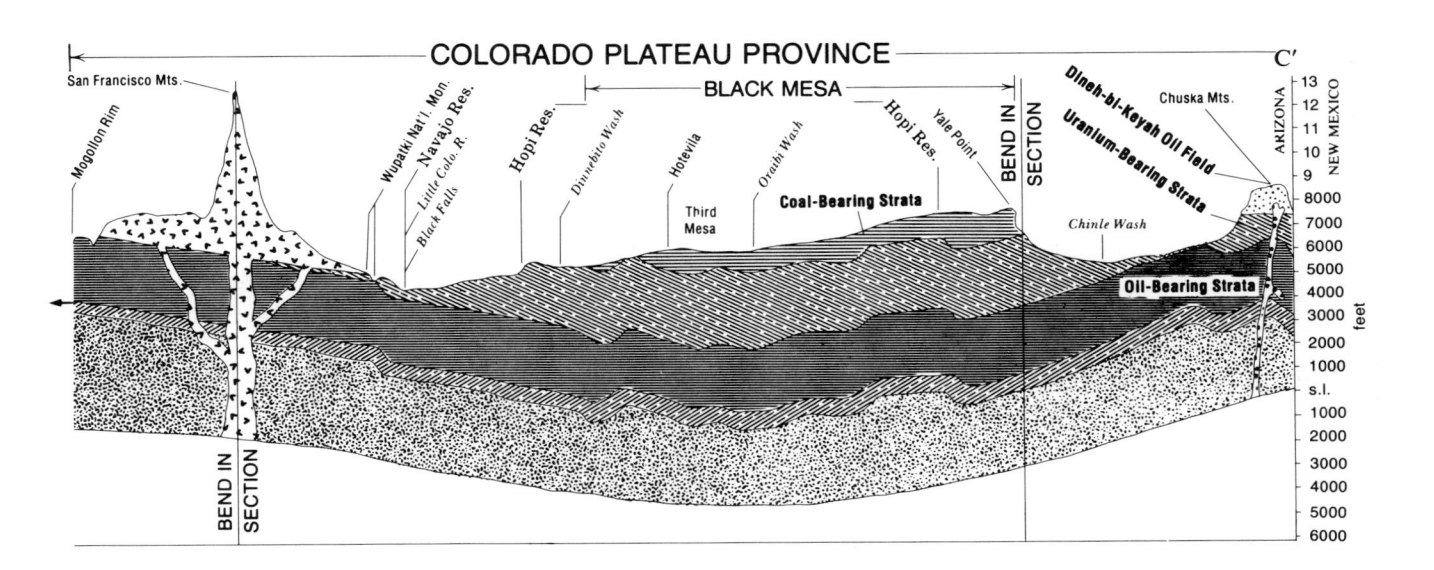

Cross-section C–C′: COLORADO PLATEAU PROVINCE

Labels along section: San Francisco Mts., Mogollon Rim, Wupatki Nat'l. Mon., Navajo Res., Little Colo. R., Black Falls, Hopi Res., Dinnebito Wash, Hotevila, Third Mesa, Oraibi Wash, Hopi Res., Yale Point, BEND IN SECTION, Coal-Bearing Strata, Chinle Wash, Dineh-bi-Keyah Oil Field, Uranium-Bearing Strata, Chuska Mts., ARIZONA | NEW MEXICO, Oil-Bearing Strata, BEND IN SECTION.

Elevation scale (feet): 13000, 12000, 11000, 10000, 9000, 8000, 7000, 6000, 5000, 4000, 3000, 2000, 1000, s.l., 1000, 2000, 3000, 4000, 5000, 6000

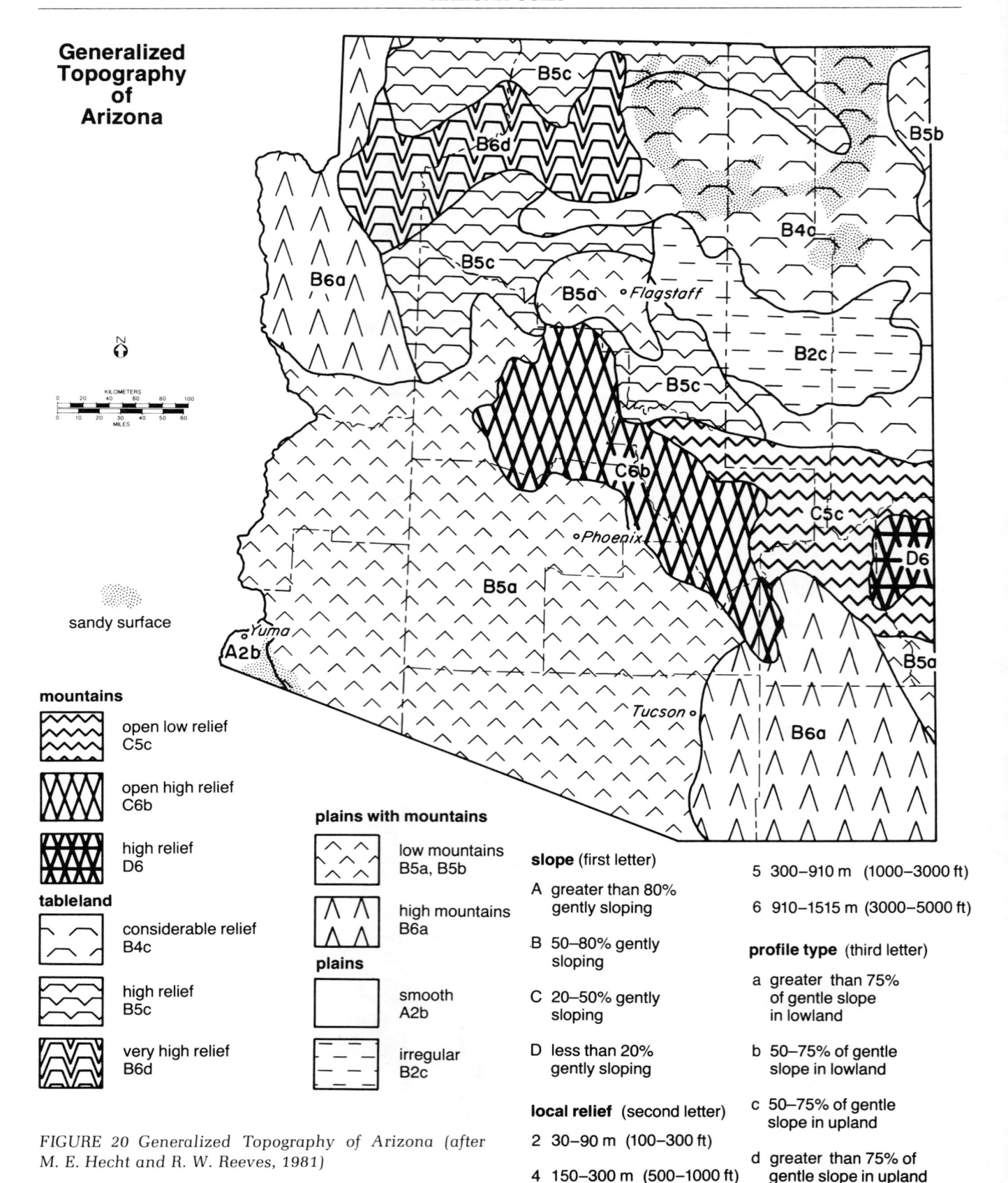

FIGURE 20 Generalized Topography of Arizona (after M. E. Hecht and R. W. Reeves, 1981)

Valley centers consist of a wash or, in some valleys such as the Santa Cruz and Gila, a through-flowing ephemeral stream. Unlike streams that emerge from mountain fronts onto piedmont slopes, these streams flow approximately parallel to mountain fronts in the valley axial troughs. Floodplains may be associated with these axial streams. Width of floodplain alluvium may be as much as a few hundred meters (several hundred feet), depending upon the drainage system size. The Holocene alluvium that forms the floodplain is separated in time from the underlying deposits of fill by hundreds of thousands of years (Martin, 1963). According to Martin (1963), most floodplain alluvium is quite shallow, considerably less than 30 m (100 ft).

Before the late 19th century, many floodplains had extensive fresh water marshes (cienegas) and shallow semipermanent streams, often without distinct channels (Bryan, 1925). Channel cutting began in many floodplains late in the 19th century and most cienegas were drained (Hastings, 1959; Cooke, 1974; Cooke and Reeves, 1976).

Planed-rock surfaces, with or without shallow alluvial covers, are a feature that often characterizes arid and semiarid landscapes. These surfaces, believed to be created by water erosion, are around mountain ranges near the margins of intermontane depositional areas. McGee (1897) applied the term "pediment" to these features in southwestern Arizona. Hadley (1967) stated that of all the landscape features in nature, pediments received inordinate attention in geologic literature. He suggested that this is because pediments constitute much of the arid and semiarid landscapes on Earth and because no general agreement exists about the processes that formed them. Pediments are quite common throughout the Arizona Basin and Range Province but are best and most extensively developed in the state's southwest.

Many basin-fill deposits are dissected by wide terrace-like features along major drainages. These broad benches, called pediment terraces, slope from the foot of the bordering mountain range toward the basin axis and are thought to have formed in poorly consolidated basin fill deposits from pedimentation processes (Mammerickx, 1964; Royse and Barsch, 1971; Barsch and Royse, 1971). A pediment terrace resembles an alluvial fan but has only a thin veneer of coarse, angular gravel on the erosion surface (Royse and Barsch, 1971). But some pediment terraces may be remnants of old alluvial fans; these are sometimes called fan terraces. Several pediment terraces (or fan terraces) normally occur along trunk streams and their major tributaries. Different periods of erosion attributed to tectonism and/or climatic change are associated with creating pediment terraces at different elevations above a stream (Barsch and Royse, 1972). Reported examples of these pediment terraces include those along the Gila River near Safford (Gelderman, 1970) and Duncan (Morrison, 1965), along Tonto Creek in the Tonto Basin (Royse and Barsch, 1971), along Canada del Oro northeast of Tucson (McFadden, 1978, 1981) and in the Sonoita Creek Basin (Menges, 1981; Menges and McFadden, 1981).

Fluvial terraces also are in intermontane basins. These features are the product of geomorphic processes involving changes in sediment load and stream competence. Fluvial terraces formed from alluviation followed by marked stream channel downcutting. They are similar to pediment terraces but are narrower, have lower gradients, commonly below 0.5 degrees, and have a much thicker gravel cap (Royse and Barsch, 1971). Péwé (1978) described the sequence of fluvial terraces associated with the Salt River in the Phoenix Basin.

Evolution of Structural Features

Development of Basin and Range Province features follows a pattern resulting from Miocene block faulting as first suggested by Gilbert (1875) and later elaborated upon by Davis (1903). The relatively uplifted fault blocks (horsts) eroded to form mountains and pediments separated by nickpoints. The relatively downfaulted blocks (grabens) filled in with sediment and piedmont slopes developed across the pediment and alluvial surfaces. Both a conceptual and a more realistic although still generalized cross section of a Basin and Range valley typical of Arizona are illustrated in Figure 21.

Actually, this ideal development of piedmont slopes, pediments and mountains from the initial structure is complicated for several reasons (Thornbury, 1965; Kesel, 1977; Scarborough and Peirce, 1978). Some normal faults consist of many fault slices. Erosion of horsts and filling of grabens proceeded as faulting occurred rather than as a distinct step following faulting. Volcanism obscured much of the block-fault landscape as many volcanoes were produced by magma that emerged along normal faults. Block tilting of different scales is evident in the Basin and Range Province, from blocks at least as large as entire mountain ranges to relatively small blocks less than 90 m (300 ft) across (Stewart, 1980). But this block faulting may be from a faulting event that occurred before the late Miocene Basin and Range faulting.

In addition to the classical model shown in Figure 21, two other generalized models of basin-range structure were proposed (Stewart, 1980) (Figure 22). The tilted-block model depicts the blocks as tilting as well as downdropping. It is related to fragmentation of an upper crystal slab into buoyant blocks. The listric-fault model is related to downward flattening faults that bottom along a sliding surface (listric fault) approximately parallel to the Earth's surface. Listric faulting in Arizona is considered to have occurred during the mid-Miocene and not during the Basin and Range deformation.

The orogeny most responsible for the configuration of the Basin and Range Province began 13 million to 12 million years ago in Arizona and ended 10 million to 6 million years ago in southwest Arizona (Shafiqullah et al, 1980). Menges and McFadden (1981) believed that this deformation lasted until as recently as 6 million to 3 million years ago in southeast Arizona. Basin and Range deformation was followed by some high-angle, normal faulting but on a much reduced, gradually diminished and more localized scale (Morrison, Menges and Lepley, 1981). This post-Basin and Range tectonism (neotectonism) is considered to have mostly terminated by the Quaternary in southwest Arizona but to have continued in places in the southeast and northwest, although infrequently, into the late Pleistocene and perhaps into the early Holocene (Morrison, Menges and Lepley, 1981). Morrison, Menges and Lepley (1981) also believed that most Basin and Range Province neotectonic faults are in the intermontane basins and a few in the mountain blocks. They also considered that most displacements happened more than 100,000 years ago and were single events. Calvo and Pearthree (1982) provided field evidence that two faulting events occurred during the past 250,000 years on the western piedmont slope of the Santa Rita Mountains south of Tucson. Their evidence indicated that the most recent event happened about 100,000 years ago.

The modern concept of plate tectonics and continental drift also is used to explain the origin of the physiographic features of the Basin and Range Province. This concept emerged about 20 years ago and provided significant insight into understanding the origins and relationships between the Earth's major surficial structural features. The basic tenets underlying the theory are discussed by a number of authors, including Dietz (1961), Vine (1966), Ewing and Ewing (1967), Bird and Isache (1972), Marvin (1973) and Tarling and Runcorn (1972).

In brief, the concept of plate tectonics suggests that most active tectonic features of the Earth's surface are related to motions between a small number of semirigid plates in the lithosphere. Where plates move apart, new volcanic crust forms. These areas are called rises and are associated with sea-floor spreading. Trenches develop where plates converge and one plate is underthrust beneath the other in a process called subduction. As the underthrust plate descends, it melts to become magma.

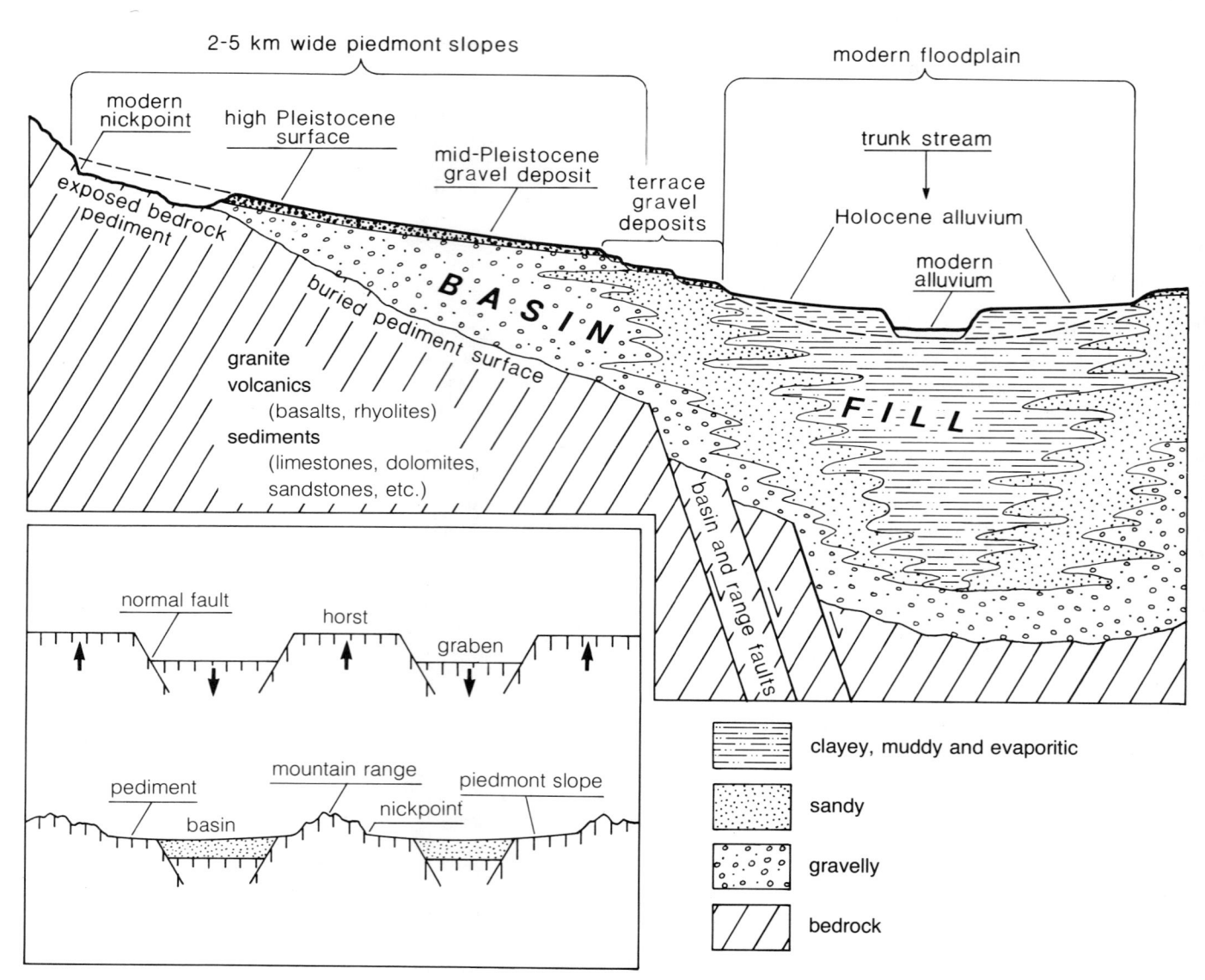

FIGURE 21. Cross Section of a Typical Basin and Range Province Valley; Inset, Classic Schematic of Block Faulting (cross section by R. B. Scarborough; inset after P. H. Rahn, 1966)

Two deformational environments suggested by modern plate tectonics existed in the western United States and contributed to the structural features of the Basin and Range Province (Atwater, 1970). Compressional shearing occurred during pre-Miocene time as the North American Plate converged with and overode the East Pacific Rise (Damon, 1971). The resultant folding and faulting was particularly severe in the Basin and Range Province. A transition from compressional to tensional environments in the Basin and Range Province occurred in mid-Cenozoic (Miocene) time. Intense, normal faulting, often referred to as "basin and range faulting," and crustal extension are associated with this later tensional environment. This Basin and Range deformation may have resulted from the direct contact of the American and Western Pacific plates along the San Andreas Fault, a right-lateral transform fault system (Christiansen and Lipman, 1972). The mechanisms that caused change in the tectonic setting in the mid-Cenozoic are obscure or, at least, not agreed upon (Huntoon, 1974a; Stewart, 1978). Considerable volcanism also occurred during the Miocene.

Correlated with the two-stage tectonic Cenozoic history in the western United States are two distinct types of volcanism (Lipman et al, 1972; Christiansen and Lipman, 1972). Volcanic rocks that formed in pre-Miocene and Miocene times during compressional shearing are mostly intermediate (andesite and rhyodacite) in composition. Those formed during tensional shearing in late Cenozoic time are primarily basaltic with lesser amounts being rhyolitic.

Coney (1978) described six phases of the tectonic evolution of southeastern Arizona during 200 million years. Scarborough and Peirce (1978), however, divided the Cenozoic history of Arizona into four relatively well-defined geologic events that indicate a tectonic history more complex than might be inferred from the two tectonic stages commonly associated with the Cenozoic of the Arizona Basin and Range Province. According to Scarborough and Peirce (1978), the events are: 1) Eocene quiescence, 2) massive late Oligocene-early Miocene, calc-alkaline volcanism and plutonism and associated sedimentation, 3) the Basin and Range disturbance and associated late Miocene-Pliocene basin filling, and 4) maximum filling of basins in middle Pleistocene followed by stream downcutting, development of terraces and valley unloading by erosion along the major rivers of the region.

Geomorphology and Soils

The nature of Arizona Basin and Range Province soils generally is related to the geomorphic surface with which they are associated. Sometimes a mapping unit or soil association on the Arizona General Soil Map (Plate 1) relates to a particular geomorphic surface, but other times a mapping unit consists of soils on two or more geomorphic surfaces. Generalized cross sections of typical valleys in the Arizona Basin and Range Province showing the geomorphic surfaces and examples of associated mapping units are illustrated in Figure 23. These geomorphic surface-soil relationships correspond to similar ones described by Gile (1975, 1977) and Gile and Hawley (1972) in the Chihuahuan Desert in southern New Mexico and by Shlemon (1978) in the southeastern Mohave Desert in California and Arizona. Peterson (1981) prepared detailed descriptions and classifications of various Basin and Range Province landforms that are especially pertinent to conditions in Nevada.

Mountain soils characteristically are shallow, rocky and gravelly. They are derived from various kinds of rocks and usually moderately to steeply sloping. Soil profile development is variable, depending in general, upon erosional surface stability and the kind of parent rock. Mountain soils in the Basin and Range Province generally can be grouped by the kind of parent rock: soils formed on granitic and schistose rocks and soils formed on volcanic rocks.

On exposed rock surfaces granitic and related rocks tend to produce, upon weathering, fine-grained gravel and sand. The main process involved is granular disintegration where the individual grains of the rock break apart with little or no chemical alteration. This weathered material may be readily transported by wind and/or water movement, particularly on steep slopes. The remaining shallow, weathered materials, soils, usually have little profile development and are composed of coarse sand, gravel and rock fragments at the surface. Immediately below this surface the soil is gravelly sandy loam or gravelly loam. Arizona General Soil Map (Plate 1) unit TS10, Chiricahua-Cellar Association, contains soils formed mostly from granitic rock. The Cellar soils are shallower, lack significant profile development and are on steeper slopes than the deeper, better-developed Chiricahua soils.

Effects of the weathering processes on volcanic rocks differ markedly from granitic rocks. Volcanic rocks show an abrupt transition from fresh rock to a usually fine-grained soil with little, if any, intermediate saprolite. This process suggests that the rocks weather from the outside inward. Weathered products removed from the rock are incorporated into the

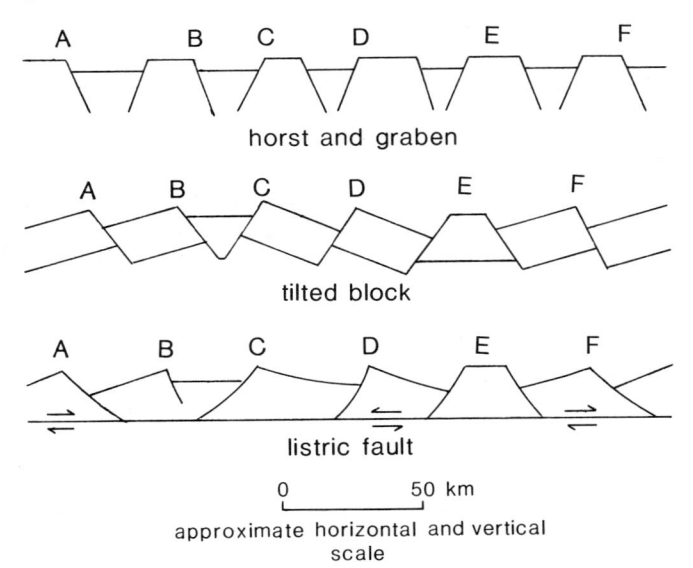

FIGURE 22. *Generalized Models of Tilted and Listric Block Faulting (after J. H. Stewart, 1980)*

soil as they form. A sharp weathering front on basalt and similar rocks is quite common in humid regions as well where a complete transition from fresh rock to saprolite extends for less than 0.2 cm (0.08 in) (Cady, 1960). Weathering rinds similar in appearance to those described by Colman and Pierce (1981) are common. Volcanic rocks exposed at the surface will tend to have a coating of desert varnish, especially in western Arizona. Profile development varies considerably in soils formed on volcanic rocks, ranging from essentially bare rock surfaces to soils with strongly developed B horizons, and is governed mostly by the soil-material removal rate. TS18, Graham-Lampshire-House Mountain Association, contains soils formed on volcanic rocks that are shallow, less than 50 cm (20 in). Graham soils show strong development of a Bt horizon, Lampshire and House Mountain soils show limited soil profile development.

Granitic, schistose and volcanic soils usually are not delineated on general soil maps because of complex lithology and intricate distribution pattern of the soils. HA6, Lithic Camborthids-Rock Outcrop-Lithic Haplargids Association, and TA3, Lithic Torriorthents-Rock Outcrop-Lithic Haplargids Association, are examples of mapping units that contain soils associated with complex lithology (See Plates 1 and 3 and Figure 19).

Rock pediments that lack an alluvial cover are included with mountain soils and are not delineated. Soils on pediments generally are shallow but in some cases have strongly developed profiles.

Several mountain ranges in southeastern Arizona have elevations of more than 2,730 m (9,000 ft). Soils on these mountains differ from soils on lower-elevation mountains because of the cooler and moister climatic conditions at the higher elevations. High mountain residual soils generally are more acid and higher in organic matter than those of the lower elevations (Martin and Fletcher, 1943; Whittaker et al, 1968). Soils of mapping unit MH2, Lithic Haplustolls-Lithic Argiustolls-Rock Outcrop Association, are at intermediate elevations while soils of mapping unit FH5, Mirabal-Baldy-Rock Outcrop Association, are at the high elevations of the Santa Catalina and Pinaleno mountains.

Soils on old alluvial surfaces characteristically exhibit considerable profile development since they have been exposed to weathering and soil formation for a long time. These surfaces include pediment terraces, fan terraces and fluvial terraces. The highly developed profiles are identified by increased clay accumulations in B or argillic horizons and formation of zones of accumulation of calcium carbonate, or calcic horizon development. In advanced stages of soil development, calcic horizons become cemented into hard indurated layers, the petrocalcic horizons.

It is generally accepted that soils with argillic horizons are old soils and that in arid and semiarid regions they may have formed under a more humid climate (Smith, 1965; Gile and Hawley, 1968; Nettleton et al, 1975). Two schools of thought prevail concerning genesis of argillic soil horizons in arid and semiarid regions. Nikiforoff (1937) postulated that B horizons in desert soils resulted from *in situ* formation of clay by weathering. The difference in texture between A and B horizons he attributed to differential weathering because the B horizon remained moist longer to permit more extensive weathering. Green (1966), Oertel and Giles (1967) and Oertel (1968) concluded that the marked particle-size differentiation with depth in soils is due entirely to differential weathering in more humid soils as well. Their interpretation was that clay is destroyed (broken down into ionic components

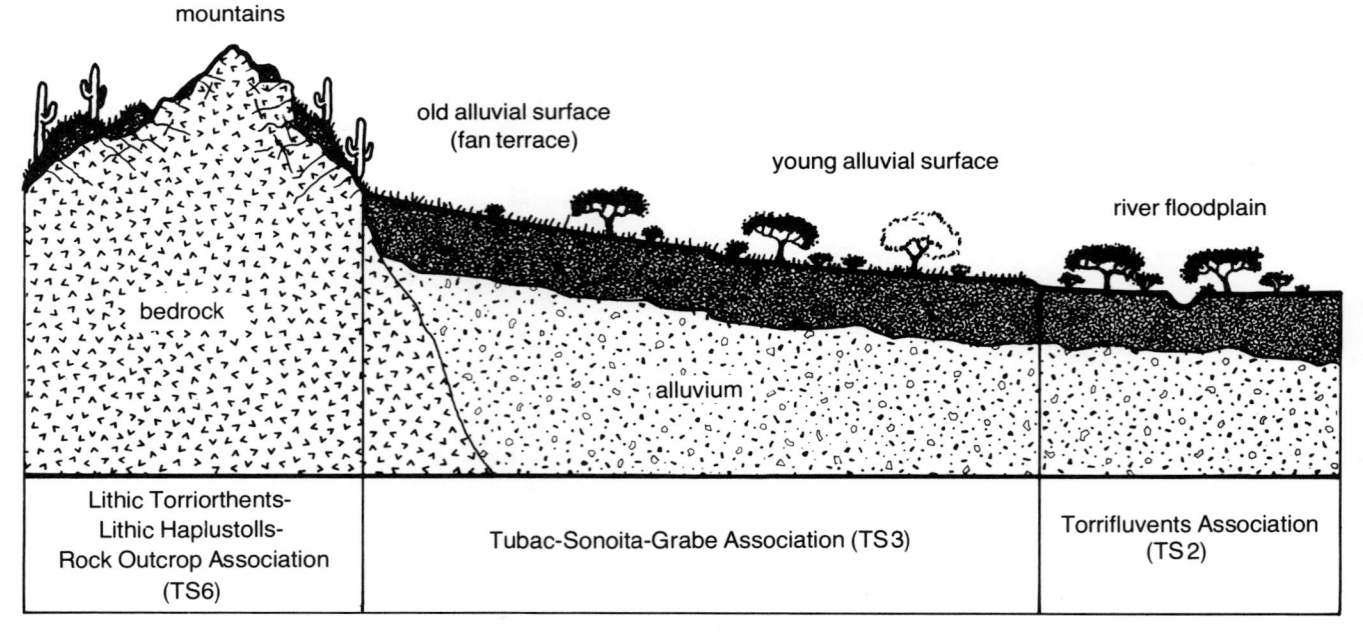

FIGURE 23a. *Typical Cross Sections of Arizona Basin and Range Province Valleys and Associated Soil Mapping Units (D. M. Hendricks)*

that are leached out) by more extensive chemical weathering in A horizons, resulting in finer-textured B horizons relative to the A horizons. The prevailing opinion, however, is that the higher clay content of B horizons is due to illuviation of clay in suspension from the overlying horizons (Soil Survey Staff, 1975). Nevertheless, it is conceded that the proportion of illuviated clay is relatively small in comparison with clay inherited from the parent material and/or formed *in situ* in the B horizon by weathering. The quantitative data of Brewer (1968) and Hill (1970) support this conclusion. Still, the illuviated clay can account for the textural differences between the argillic and overlying eluvial horizons. Mapping units that consist of soils with strongly developed argillic horizons include HA3, Mohall-Vecont-Pinamt Association, and TS7, White House-Caralampi Association.

Soils with calcic and petrocalcic horizons often are associated with calcareous parent materials, although aeolian deposits of carbonates may be an important factor (Ruhe, 1967). Soils with strongly developed petrocalcic horizons date from the late- or mid-Pleistocene (Gile, Peterson and Grossman, 1966; Gardner, 1972; Machette, 1978). Formation in southern New Mexico of such horizons occurs more rapidly in gravelly alluvium on late-Pleistocene surfaces than in non-gravelly alluvium on mid-Pleistocene surfaces according to Gile, Peterson and Grossman (1966). Calcic horizons take less time to form, occur in soils on younger surfaces and are believed to precede formation of petrocalcic horizons. A mapping unit dominated by soils with well-developed calcic horizons is HA5, Laveen-Rillito Association. TA5, Paleorthids-Calciorthids-Torriorthents Association, and TS14, Nickel-Latene-Cave Association, contain soils with both calcic and petrocalcic horizons. Several mapping units contain some soils with argillic horizons and other soils with calcic and/or petrocalcic horizons, including HA9, Harqua-Perryville-Gunsight Association, TS5, Caralampi-Hathaway Association, and TS12, Continental-Latene-Pinaleno Association.

Developed soils with hard indurated layers in which silica is the dominant cementing agent (duripans) also are in some Basin and Range Province soils. Although these soils are not as prevalent as soils with petrocalcic horizons, their extent is now considered to be greater than it was thought to be a few years ago. The reason is that distinguishing between a duripan and a petrocalcic horizon is sometimes difficult, especially in the field, and some horizons that are identified as petrocalcic could be duripans.

A duripan forms when silica is released into solution by weathering in one part of the soil profile and accumulates in another part of the same profile or, sometimes, adjacent profiles. Flach et al (1969) and Flach, Nettleton and Nelson (1974) suggested that the silica is derived primarily from volcanic ash and related pyroclastic parent materials that contain constituents with very low resistance to chemical weathering. Duripans also could form in sodic soils such as the Stewart soils of the TS11, Gothard-Crot-Stewart Association, mapping unit. Sodic soils generally have a high pH, sometimes as high as 9.0 to 9.5. The solubility of silica differs little throughout the pH 4.0 to 8.0 range of most soils. Sodic soils with a pH above about 8.5 experience very rapid increases in the solubility of silica and would favor mobilization of silica that may promote formation of a duripan (Krauskopf, 1956).

Soils derived from young alluvium on piedmont slopes and active fans generally lack profile development. These soils tend to be coarse-textured sandy loam or gravelly sandy loam and are well represented by Antho and Anthony soils.

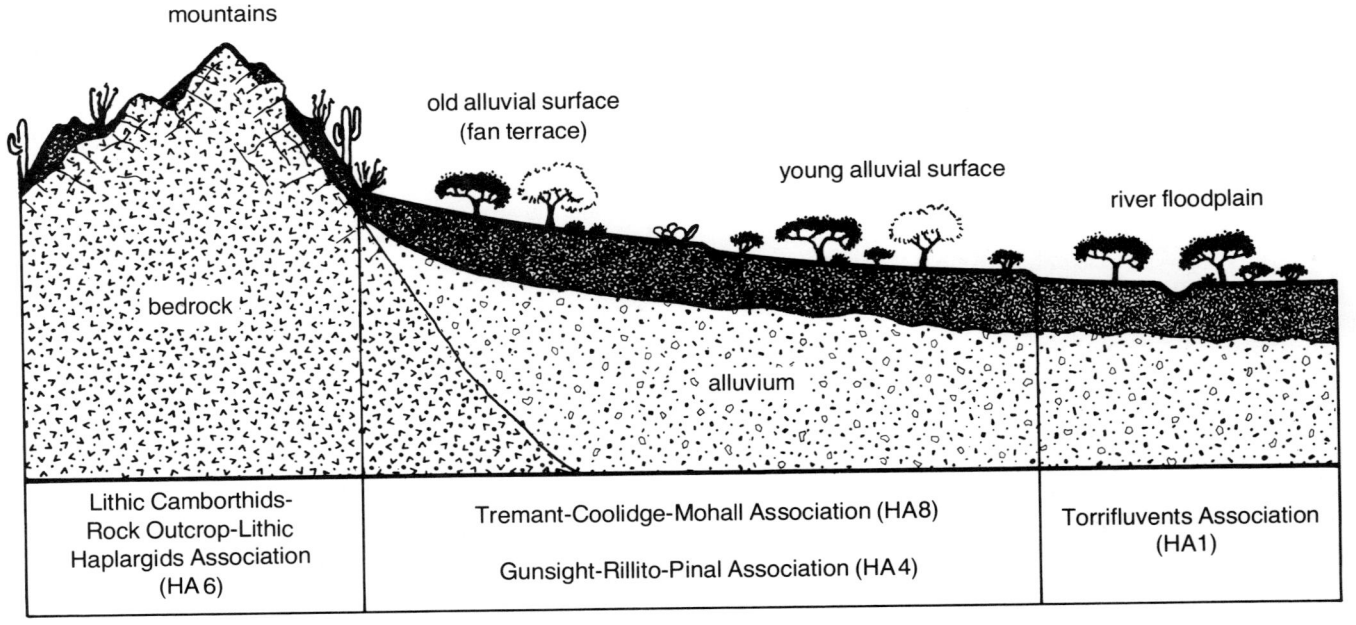

FIGURE 23b. *Typical Cross Sections of Arizona Basin and Range Province Valleys and Associated Soil Mapping Units (D. M. Hendricks)*

The Antho soil is a component of HA1, Torrifluvents Association, and HA7, Laveen-Carrizo-Antho Association. In HA1 the Antho soil is on the lower piedmont slope near the floodplain and sometimes forms from fan alluvium extending below the older alluvial surface into the floodplain. The Anthony soil is a component of TA2, Anthony-Vinton-Agua Association, TA4, Latene-Anthony-Tres Hermanos Association, and TS19, Anthony-Sonoita Association. Most of these associations also include other soils formed in older alluvium reflecting the complex pattern of distribution of young and old alluvium on piedmont slopes.

Floodplains also are composed of young alluvium and contain soils somewhat similar to those formed in young alluvium on piedmont slopes, except they generally are finer textured and some have a relatively high organic-matter content of more than 1 percent. The percent organic matter is higher than would be expected in view of the prevailing climate and may be the result of two factors.

The first factor is that many of the soils might be relict since they formed under cienega, or meadow-like, conditions that would have favored organic-matter buildup. Except in a few areas, these conditions no longer exist, but the associated soil properties persist. The second factor involves the stratification of the soil parent material characterized by the irregular depth distribution of soil organic matter. This indicates that organic matter was deposited, at least in part, with the parent material by periodic floodwaters. Organic matter in the parent material, then, would be from soils in upstream source areas.

Long-term irrigation also may increase soil organic matter in arid regions (Soil Survey Staff, 1975). But the extent, if any, of organic matter buildup from irrigating floodplain soils in Arizona is not known. It is known, however, that soils that have never been irrigated may have a relatively high organic-matter content. Soil mapping units composed mostly of floodplain soils include HA1, Torrifluvents Association, and TS2, Torrifluvents Association.

Colorado Plateau Province

General Nature of the Geology
The Colorado Plateau has a thick sequence of flat to gently dipping sedimentary rocks eroded into majestic plateaus and dissected by deep canyons. Unlike the Basin and Range Province, Colorado Plateau relief is more the result of deep canyons cut into moderately flat terrain than of mountains and valleys created mostly by deformation. Volcanic mountains exist within the province, but block-fault structural mountain ranges do not. Hunt (1967) described the topography as being analogous to a stack of saucers, tilted toward the northeast into Utah and Colorado where the plateau meets the Rocky Mountains. The younger Tertiary rocks crop out in basins on the north and east sides of the plateau and the older Paleozoic rocks crop out along the southern rim overlooking the Basin and Range Province.

The Colorado Plateau Province has six sections (Fenneman, 1931), four in Arizona (Plate 2). The southwestern plateau is the Grand Canyon Section, the highest part of the province. The oldest exposed rocks are complexly deformed Precambrian formations overlain by 1,210 to 1,520 m (4,000 to 5,000 ft) of Paleozoic formations in the Grand Canyon. Precambrian rocks also are exposed along the extreme southwestern edge of the plateau. Permian rocks predominate at the surface except where covered by volcanic rocks. About a third of the Grand Canyon Section is covered by lavas from several volcanic fields, the largest being the San Francisco field near Flagstaff. The Grand Canyon is the dominant topographic feature. North of the Grand Canyon are several high plateaus bounded by fault or fault-line scarps.

The Datil Section is in the southeastern Colorado Plateau in Arizona and northwestern New Mexico. Most of the Datil Section in Arizona is the White Mountain volcanic field.

North of the Grand Canyon and Datil sections is the Navajo Section, a structural depression most of which is in Arizona and New Mexico. This section is a somewhat poorly defined area of broad plateaus and valleys. The valleys tend to be wide and open rather than canyon-like. Volcanic vents and associated flows and pyroclastic materials also are common.

The Canyon Land Section lies north of the Navajo Section mostly in Utah and western Colorado, with only a small area in Arizona. As the name implies, canyons are the dominant features of this section. The portion of the section that extends into Arizona includes Marble Canyon, just northeast of the Grand Canyon.

Evolution of Structural Features
The Colorado Plateau is structurally unique in the western United States because it is only moderately deformed compared with the more intensely deformed regions that surround it. Geological evidence of the structural evolution of the Colorado Plateau in Arizona, particularly relative to the origin of the Grand Canyon, is summarized by a number of authors, including McKee et al (1967), Hunt (1969), Lucchitta (1972) and Breed and Roat (1974).

Monoclines appear to be the most distinctive structural features of the plateau; most of the deformation occurred along them (Kelley, 1955) (Figure 24). Colorado Plateau monoclines and related structures developed principally from late Cretaceous to early Tertiary (Laramide) time. Although the Colorado Plateau has only limited deformation

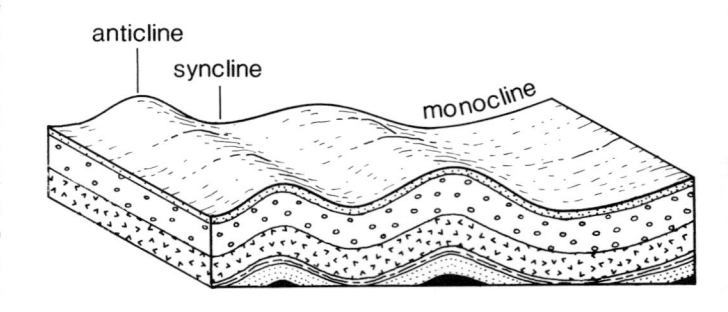

FIGURE 24. Representations of an Anticline, a Monocline and a Syncline

some parallelism between the structural features of the plateau and the adjacent Basin and Range Province in northwest Arizona has been noted by Lucchitta (1974), indicating that both were subject to the same stress field. Lucchitta (1974) further suggested that the much lower degree of deformation on the Colorado Plateau is due to a more competent crust. A related factor is that the sialic crust beneath the Colorado Plateau is much thicker than it is under the Basin and Range Province (Pakiser, 1963; Keller, Braile and Morgan, 1979). Zones of weakness, some very old, in the Colorado Plateau basement rocks are thought to have localized deformation. Many of the zones have been active repeatedly, commonly with reverse movement. Thus, the compressional stress field associated with the Laramide orogeny to the west resulted in reverse faulting in the basement and corresponding folding in the plateau sedimentary cover. The later extensional stress field associated with the Basin and Range deformation resulted in normal faulting along many of the same faults. Thompson and Zoback (1979), on the other hand, provided geophysical evidence that the modern stress field in the middle of the Colorado Plateau is different than that of the Basin and Range Province.

Between late Cretaceous and middle Eocene time, a period of roughly 40 million years, the Colorado Plateau and adjacent Basin and Range Province experienced regional uplift (Huntoon, 1974b). It is suggested that the uplift resulted from western North America overriding the eastern flank of the East Pacific Rise. The uplift of this large region apparently stopped during the middle Eocene, leaving the region about 1.6 km (1 mi) above sea level (Damon, 1971). Plateau definition was enhanced in late Cenozoic time (Pliocene and late Miocene) when it became more obviously separated structurally from the Basin and Range Province in southwest Utah (Rowley et al, 1978) and in Arizona. Peirce, Damon and Shafiqullah (1979) pointed out that in central and western Arizona the southern border of the plateau, represented by the escarpment zone known as the Mogollon Rim, is primarily an Oligocene erosional feature and not a major tectonic boundary associated with late Cenozoic differential plateau uplift as suggested by McKee and McKee (1972).

Late Cenozoic Volcanism

Cenozoic volcanic rocks are an important feature of the Colorado Plateau, occurring for the most part near its margin (Figure 25). These volcanic rocks are large central-type volcanoes, such as the San Francisco Peaks near Flagstaff, and

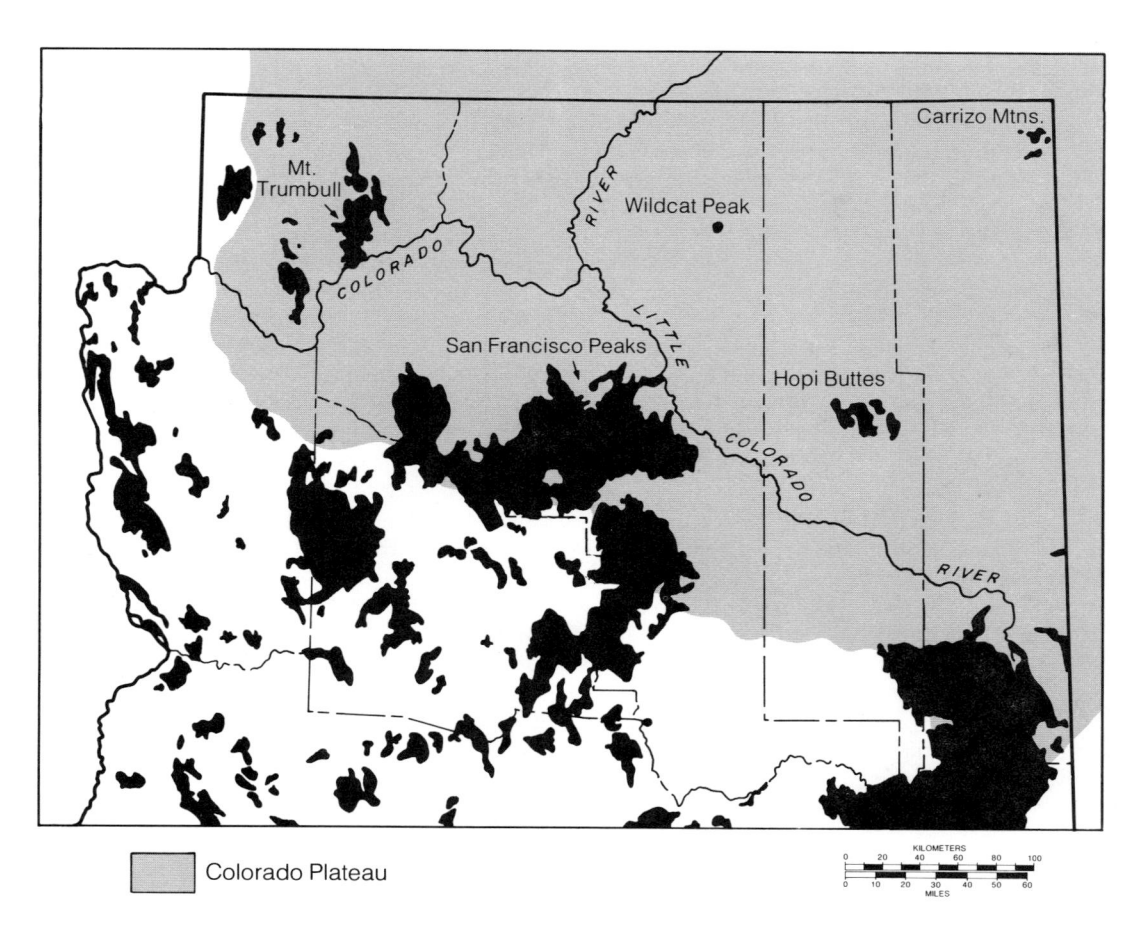

FIGURE 25. Map of Volcanic Rock Distribution near the Colorado Plateau (after C. B. Hunt, 1956)

extensive sheets of lavas and pyroclastic rocks, including numerous volcanic fields.

The San Francisco volcanic field near Flagstaff covers more than 5,200 sq km (2,000 sq mi) of the southern Colorado Plateau. It is composed of late Tertiary and Quaternary volcanic rocks and contains hundreds of basaltic lava flows, cinder cones and minor intermediate to silicic volcanic rocks. The San Francisco Peaks are the most prominent feature of the field.

The geochronology of the San Francisco volcanic field has been studied fairly extensively beginning with Robinson (1913) who divided the rocks into three periods of eruption: 1) basaltic volcanics, probably of the late Pliocene, 2) rhyolitic to andesitic volcanics, probably of the early Pleistocene, and 3) basaltic volcanism, probably of the latter part of the Quaternary. Colton (1937, 1967) divided the basalts of the field into five stages based on weathering and erosion of the cinder cones and flows. Cooley (1962) expanded Colton's (1937) classification by relating the basalt stages to erosional surfaces and alluvial deposits in the Little Colorado drainage basin. More recently Moore, Ulrich and Wolfe (1974) and Moore, Wolfe and Ulrich (1976) defined five episodes of basaltic volcanism based primarily on: 1) stratigraphic and physiographic relations, 2) weathering and erosion, 3) potassium-argon and tree-ring age determinations, and, in part, 4) chemical and 5) petrographic data. The different subdivisions of the San Francisco volcanic field are compared in Table 1.

The Mormon Mountain volcanic field is the extension to the south of the San Francisco field and covers about 2,600 sq km (1,000 sq mi). Beus, Rush and Smouse (1966) recognized three, or possibly four, stages of volcanic activity in the Beaver Creek area of the field. The oldest stage is bedded tuffs and tuff breccias, while the younger two stages are basalt and cinders. Scholtz (1969) suggested that the basalt stages are late Pliocene to Pleistocene in age.

The Mount Floyd volcanic field is the western extension of the San Francisco field and covers about 2,600 sq km (1,000 sq mi) overlain with lavas and ash beds. This field is separated from the San Francisco field by a break between the cities of Williams and Ashfork. The rocks are considered to be Tertiary-Pleistocene in age.

The Uinkaret volcanic field is north of the Grand Canyon. It caps the Uinkaret Plateau and includes basaltic lava flows that spilled into the Grand Canyon and dammed the Colorado River on several occasions (Hamblin, 1974). The extent of the field is less than 2,600 sq km (1,000 sq mi). Koons (1945) classified the flows of the Uinkaret Plateau by methods similar to those Colton (1937) used in the San Francisco volcanic field. More recently Hamblin and Best (1970) and Hamblin (1974) modified the classification of Koons (1945). They based their classification not only on weathering and erosion but also on the nature of the surface upon which the lavas were deposited. According to this classification four major periods of volcanic activity in the Uinkaret field can be recognized.

The White Mountain volcanic field in the Datil Section resembles the San Francisco volcanic field in a number of ways. Mount Baldy and Mount Ord, remnants of the Mount Baldy Volcano of late Tertiary age, are in the White Mountain volcanic field and are composed of latite and quartz latite (Merrill and Péwé, 1977). Cinder cones and lava flows of basaltic composition surround Mount Baldy and Mount Ord and constitute most of the White Mountain volcanic field. These basaltic rocks range widely in age as indicated by a variety of such surface features as degree of preservation of original flow features and weathering and soil formation (Merrill and Péwé, 1977). Thus, these basalts probably erupted intermittently beginning in early Pliocene to late Miocene time through the Pleistocene, perhaps into the Holocene. Merrill and Péwé (1977) also correlate the basaltic lava flows in the White Mountain volcanic field with the Cedar Ranch, Woodhouse and Tappan age groups of the San Francisco volcanic field (See Table 1).

The Hopi Buttes volcanic field is in the southern part of Black Mesa Basin. Igneous rocks in the Hopi Buttes were intruded as dikes and sills, and erupted pyroclastics, lava flows and lava domes. About 200 volcanoes erupted during the Pliocene (Sutton, 1974). Depending on the materials that filled volcano vents after eruption, three topographic forms evolved that characterize the Hopi Buttes field: 1) prominent necks, or plugs, that rise above the landscape as narrow, nearly circular and steep-sided buttes surrounded by talus slopes, 2) lava-capped mesas that resulted from the erosion of lava domes on flows that overlie maar craters, and 3) maar craters that have no protective covering of lava, massive breccia or agglomerate.

Geomorphology and Soils

Grand Canyon Section. The evolution of the Grand Canyon has been the subject of numerous studies since Powell (1875) first explored the area, but controversy continues concerning its formation (Breed, 1969, 1970; Hunt, 1974). Most studies focused on the history of the course of the Colorado River. Less attention was and is given to processes that formed and are enlarging the canyon. Ford et al (1974), however, described several processes that are removing and transporting material from the canyon sides including slab-failure (sudden collapse of a cliff face cut in solid rock), rock avalanche (collapse of a rock face due to failure of a random pattern of microjoints), rock fall (fall of single small blocks from a cliff face), talus slides (sudden downward movement of loose talus), granular disintegration (fallen rocks broken into sand grains suitable for removal by runoff of rainfall) and mudflows (mixtures of rock particles of varying sizes and water that move as viscous masses). In addition to these processes small-scale wastage landslides (rapid creep erosion) (Huntoon, 1975) and dissolution of calcium carbonate from limestone formations (Lange, 1956) are considered important. Wind also undoubtedly has played a role. Cole and Mayer (1982) estimated that the rate of cliff retreat of the Mississippian Redwall Limestone in the eastern Grand Canyon to average 0.45 m (1.5 ft) each 1,000 years. They based their calculations on the distances between dated packrat middens and the cliff faces.

Grand Canyon soils are mostly shallow to only moderately deep and rock outcrop is quite extensive because of relatively rapid material removal from the canyon surfaces. The Grand Canyon is in mapping unit TA1, Torriorthents-Camborthids-Rock Outcrop Association. In addition to rock outcrops and associated soils TA1 contains scattered areas of talus deposits as well as areas of alluvium along the Colorado River (Museum of Northern Arizona, 1976) that consists of recent floodplain and older terrace materials. Howard and Dolan (1981) described the fluvial deposits along the Colorado River as being three intergrading components: 1) tributary, alluvial fan bouldery deposits, 2) cobble and gravel bars, and 3) fine-grained (sandy) terraces. The nature of the soils on these deposits in Plate 1 mapping unit TA1, Torriorthents-Camborthids-Rock Outcrop Association, has not been determined.

Navajo Section. Landforms in the Navajo Section of the Colorado Plateau are characterized and affected by alternating resistant and weak rock strata (Cooley et al, 1969). Resistant rocks form ledges, cliffs, mesas and benches that are separated by slopes, valleys and badlands carved in the weak shaley beds. Three types of canyons are common, depending upon the resistance of the rocks in which they form. Vertical walls are typical in canyons eroded in resistant rocks such as Navajo and De Chelly sandstones. A box or modified box is carved if the canyon is rimmed by a resistant bed that has a gentle dip and a floor of soft sediments. If cut into moderately resistant rocks, a V-shaped canyon with steeply inclined walls forms. All the wide valleys and many extensive slopes have formed principally in the softer sediments such as Chinle, Fruitland, Kirkland and Bidahochi formations and in the Mancos Shale. The influence that alternating resistant

TABLE 1.
Comparison of San Francisco Volcanic Field Subdivisions

Age			Million Years	Colton (1937, 1967)	Cooley (1962)		Moore et al (1974)
Quaternary	Holocene			Stage V	Stage V	Younger Basalts	Sunset Age Group
			0.01	Stage IV	Stage IV		Merriam Age Group
	Pleistocene		0.20	Stage III	Stage III		Tappan Age Group
				Stage II	Stage II		
			1.00		Stage IA		Woodhouse Age Group
			2.00	Stage I	Stage IB		
Tertiary	Pliocene				Transition Basalts		Basalt of Cedar Ranch Age Group
					Older Basalts		

29

and soft rock formations can have on geomorphic feature evolution is illustrated in Figure 26 (Stokes, 1973). Scarp recession appears to be important in the evolution of these landforms. Schumm and Chorley (1966) noted that the rapid weathering and removal of talus from the feet of cliffs promotes the relatively rapid retreat of these scarps.

Colorado River downcutting was accompanied by development of the Colorado River drainage system on the Colorado Plateau. Remnants of old surfaces and terraces along the Little Colorado River in northeastern Arizona are attributed to four major erosional cycles initiated by tectonic events followed by minor perturbations associated with Pleistocene climatic fluctuations (Childs, 1948; Cooley and Akers, 1961; Cooley et al, 1969; Rice, 1976) (Table 2). Holocene terraces also line the Little Colorado River and its tributary washes (Cooley et al, 1969).

Despite great age differences no apparent soil chronosequences are present on these surfaces in northeastern Arizona because of wind erosion and deposition. The results of wind activity are particularly evident in this region. Wind has been a powerful influence since at least Miocene time as described and illustrated by Cooley et al (1969) and emphasized by Stokes (1973). Depositional features include forming and older arrested dunes, some of which overlie terrace deposits of Pliocene to Pleistocene age. Stokes (1973) also suggested the possibility that the amount of material removed by wind may have been roughly comparable with that removed by water.

Most Navajo Section soils formed in alluvium in washes and wind-reworked material derived from sandstone and sandy shale in sedimentary formations. These soils generally lack extensive soil profile development. Rock surfaces lacking soil material or having shallow soils also are common. Mapping units MA3, Sheppard-Fruitland-Rock Outcrop Association, and MA8, Fruitland-Camborthids-Torrifluvents Association, contain these soils.

The Painted Desert is in the west and south of the Navajo Section. The lithology is chiefly varicolored Chinle and Moenkopi formations of Triassic age. The sediments are relatively soft and nonresistant to erosion. In addition, the red sandstones of the lower Glen Canyon group are present. Because of little precipitation and sparse vegetation, the area is composed mostly of badlands and barren bedrock surfaces. Erosion is extremely rapid on the soft sediments. Colbert (1956) measured erosion in the Chinle Formation north of Cameron and in the Petrified Forest from July 1951 to June 1955 and reported that as much as 5 cm (2 in) of material was removed. Creep was the dominant erosive process.

The primary mapping unit in the Painted Desert is MA1, Badlands-Torriorthents-Torrifluvents Association. It consists of bare shale exposures of the Chinle Formation and a few outcrops of the Moenkopi and other formations. The soils in MA1 are shallow to deep, lack horizon development and form in material eroded from rock formations.

The Defiance Plateau and the Carrizo-Chuska Mountains are in the Navajo Section along the New Mexico state line. The Defiance Plateau, also called the Defiance Uplift, is a broad, elongate anticline about 160 km (100 mi) in length and 65 to 95 km (40 to 60 mi) wide (Refer to Figure 24). De Chelly Sandstone and sandstone beds of the Chinle Formation are exposed at elevations ranging from 2,125 to about 2,730 m (7,000 to 9,000 ft). The principal soil mapping unit on the Defiance Plateau is FH7, Cryorthents-Eutroboralfs Association. The soils are shallow to moderately deep and formed on the De Chelly Sandstone and related formations. They either lack horizon development or have moderately developed argillic horizons.

The Carrizo-Chuska Mountains extend as a bold range up to 2,730 m (9,000 ft) in elevation. The Carrizo Mountains on the north end of the range consist of a dioritic central mass and projecting laccolithic sills that were intruded into surrounding sedimentary rocks. These rocks have eroded into narrow ridges, sharp V-shaped canyons, hogbacks and buttressed and recessed cliffs. The Chuska Mountains consist of a long narrow mesa composed of thick, horizontally bedded Chuska Sandstone of Tertiary age that rests unconformably on the folded and beveled Mesozoic rocks of the Defiance

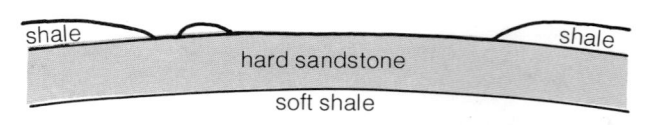

Stage 1: area gently folded into low arch.

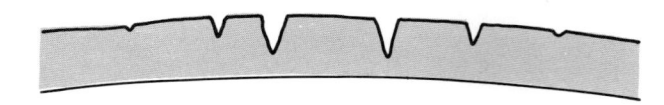

Stage 2: erosion begins to cut into higher part of arch.

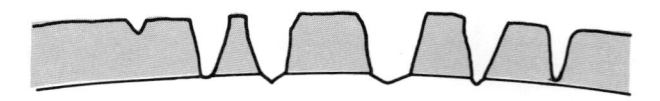

Stage 3: erosion cuts through sandstone layer and monuments form.

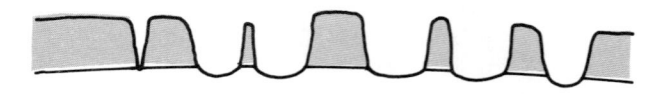

Stage 4: erosion continues widening and deepening valleys cut into the shale beneath the sandstone.

Stage 5: present condition.

FIGURE 26. *Schematic Representation of the Evolution of Typical Landforms in the Navajo Section of the Colorado Plateau (after W. L. Stokes, 1973)*

monocline. The Lukachukai Mountains are a narrow, curving bridge between the Carrizo-Chuska Mountains. They are dominated by rocks of the Chuska Sandstone and a capping volcanic flow underlain by thick sandstone and shale formations. The primary mapping unit in the Carrizo-Chuska Mountains is FH5, Mirabal-Baldy-Rock Outcrop Association. The soils of this unit are mostly residual, being generally moderately deep to deep.

Grand Canyon Section Plateaus. The Grand Canyon Section of the Colorado Plateau is subdivided by north-south trending faults and monoclines into several major plateaus north of the Grand Canyon. Plateau units marked off by these fault scarps differ in altitude, topography and climate. In general they rise step-like from west to east, each step being roughly 300 m (1,000 ft). Except for the volcanics on several of the plateaus, the dominant surficial rocks in the Grand Canyon Section include the Permian Kaibab Limestone with lesser amounts of Triassic Moenkopi and other formations.

The Kaibab Plateau is the highest of the plateaus. It ranges in elevation from 2,275 to 2,820 m (7,500 to 9,300 ft). The plateau surface is dissected by rounded valleys of gentle slope not more than 90 to 120 m (300 to 400 ft) deep. Despite fairly high amounts of precipitation, the valleys on top of the plateau are streamless (Huntoon, 1974b) but near the plateau edges the valleys deepen into ravines and carry streams.

Another feature of the Kaibab Plateau is north-south trending faults through the central highlands and along the western margins. Associated with these fault zones are narrow parks and meadows that may extend for kilometers (Strahler, 1944, 1948). The soils on ridges and gentle slopes reflect extensive soil formation. These soils are deep, clay-rich and carbonate-free, and evolved from the residue of Kaibab Limestone following the dissolution and removal of carbonate minerals. The soils also may be influenced by deposition of aeolian materials. The fairly advanced soil formation indicates that the geomorphic surface has been stable for a long time under climatic conditions as humid or more humid than present. On the steeper side slopes near the edges of the plateau, however, are calcareous soils. Deep soils high in organic matter are in parks and stream valleys. Kaibab Plateau soils are included in mapping unit FH4, Soldier-Lithic Cryoborolls Association. The Soldier and similar soils are on stable surfaces, while the Lithic Cryoborolls and related soils are on side slopes. Park and valley soils are included in this unit.

Contrasting with the Kaibab Plateau, the Shivwits, Uinkaret, Kanab and Coconino plateaus (except in the San Francisco volcanic field) have semiarid to arid climates. The Shivwits and Uinkaret plateaus are nearly level and dissected by shallow, open valleys in the Kaibab Limestone above which are mesas and tablelands capped by remnants of basalt flows. The Kanab Plateau is a smooth, sage-covered plain. Soil mapping units of the nonvolcanic, semiarid areas of the plateaus include MS2, Winona-Boysag-Rock Outcrop Association, and MS5, Roundtop-Boysag Association. These soils are derived from Kaibab Limestone and calcareous sandstones. The Roundtop and Boysag soils are noncalcareous in the upper horizons and have well-developed argillic horizons rich in clay in spite of the highly calcareous parent material. This indicates that these soils formed on stable geomorphic surfaces that provided a long time for the removal of most of the carbonate material. Aeolian materials also may have contributed significantly to these soils, in which case leaching has been more than adequate to remove any introduced carbonates. More humid climatic conditions of the past probably were important in the development of the soil properties. The Winona soils, on the other hand, are highly calcareous throughout the profile, contain less clay and lack the argillic horizon development of the Roundtop and Boysag soils, reflecting a less stable geomorphic surface.

Soils Formed from Volcanic Rocks. Volcanic rocks that are mostly basaltic in Arizona's Colorado Plateau form soils somewhat different than those derived from other parent materials under comparable climatic conditions. Basaltic rocks are subdivided into two groups based on soil-forming capabilities: pyroclastics (cinders and ash) and flow basalts.

Pyroclastics are extremely permeable and absorb nearly all the precipitation. Runoff and erosion are negligible. Laboratory tests prove that these soils have higher available water capacities than most other similar soils in the western United States (Berdanier et al, 1979). As a result, the pyroclastic soils supported higher plant densities and accumulated more organic matter than other soils of similar texture under comparable climatic conditions. As time passes, however, the cinders and ash weather, clay is produced and argillic horizons develop, reducing permeability. The resultant runoff will initiate gully erosion on the cinder cones. Colton (1967) noted that the amount of erosion of the surface of a cone in the San Francisco volcanic field is an indication of its age, within certain limits, and is one of the criteria he used to classify the relative ages of cinder cones.

Flow basalts have extremely limited available water capacities and initiation of soil formation takes much longer. For example, the SP Crater flow is believed to be 70,000 years old (Baksi, 1974) and is essentially bare rock with little or no soil. The arid climatic conditions do not favor chemical weathering and soil formation on this flow, but under more humid conditions soil formation might be expected to be more rapid. Soils have formed, however, on pyroclastic materials of comparable and younger ages near SP Crater.

Basalt eventually breaks down by physical weathering into smaller particles that yield more surface area for chemical weathering. Clay is produced by chemical weathering, increasing the water-holding capacity of the material. Under suitable climatic conditions, fine-textured soils evolve from flow basalts and clay may be redistributed in profiles to form argillic horizons. Aeolian materials may have been important in contributing to soils formed on flow basalts. Of particular significance is the possibility that a given basalt flow might have received volcanic ash from later volcanic events. Cheevers and Lund (1981) found evidence of the introduction of large volumes of aeolian material to soils formed on the Tappan Flow basalts (See Table 1). They suggested that the aeolian materials contained quartz derived from sedimentary rocks of the area and from the more siliceous volcanic rocks.

Semiarid and subhumid climatic conditions promote formation of montmorillonite in soils derived from both pyroclastic and flow-basaltic parent materials. The presence of expandable montmorillonite clay causes high shrink-swell capacities in soils. These soils swell, or increase in volume, when wet and shrink, or decrease in volume, when dry. Sometimes the magnitude of shrinking and swelling causes soils to develop the unique "self-swallowing" characteristic of Vertisols (Refer to Figure 17). Vertisols form deep, wide shrinkage cracks during the dry season. Soil material from the surface horizons may fall into the cracks. When the wet season comes, the cracks close because the soil expands. The extra material that fell into the bottom of the closed cracks creates stresses within the soil mass. To relieve stress, the soil mass tends to move toward the surface. The cycle is repeated over time and soil material from the bottom moves to the surface. This pedoturbation tends to inhibit formation of certain soil horizons such as argillic horizons. Vertisols in Arizona are derived primarily from basaltic parent materials.

Mapping units MS4, Rudd-Bandera-Cabezon Association, MS7, Cabezon-Thunderbird-Springerville Association, FH2, Sponseller-Ess-Gordo Association, and FH8, Gordo-Tatiyee Association contain soils evolved from volcanic parent materials. MS4 includes soils at the lower elevations of the San Francisco and White Mountain volcanic fields and the Hopi Buttes field. MS7 includes soils in the volcanic fields of the Uinkaret and Shivwits plateaus, the Mount Floyd field and portions of the Mormon Mountain field. FH2 soils are in the forested, higher elevations on the San Francisco, White Mountain and Mormon Mountain volcanic fields. FH8 soils are on the high-elevation meadows of the White Mountain field.

TABLE 2.
Erosional Cycles along the Little Colorado River in Northeastern Arizona

Age			Cycle	Erosional and Depositional Events	Height of Terraces Above River Level, meters (feet), Little Colorado River in the Cameron-Winslow Area	Approximate Age of Cutting of Principal Canyons				
Quaternary	Pleistocene	Middle and Late	Wupatki	Downcutting and Terracing	Five Terraces at 9(30), 15(50), 23-30 (75-100), 46-60 (150-200), 60-90 (200-300)	Eastern Grand Canyon	Marble Canyon	Glen Canyon	San Juan Canyon	Canyon de Chelly
		Early	Black Point	Downcutting and Terracing	Two Prominent Terraces at 120-150 (400-500), 220-243 (600-800)					
Tertiary	Pliocene	Late	Hopi Buttes Zuni	Formation of Zuni Surface and Deposition of the Upper Member of the Bidahochi Formation	303-455 (1,000-1,500)					
		Middle								
		Early		Formation of Hopi Buttes Surface and Deposition of the Lower Member of the Bidahochi Formation						
	Miocene		Valencia	Few Deposits	606-760 (2,000-2,500)					

3

Climate of Arizona

Introduction

Ranges of temperatures and precipitation in Arizona are extreme. Average annual temperatures vary from the middle 20s C (70s F) in the low desert areas along the Gila and Colorado rivers to less than 5 C (middle 40s F) in the pine country of central and east-central Arizona. And average annual precipitation ranges from about 76 mm (3 in) in the southwest to more than 760 mm (30 in) in the central and eastern mountains. A general schematic of atmosphere-to-soil precipitation distribution is shown in Figure 27.

Record high and low temperatures span nearly 71 C (167 F). An all-time low of -40 C (-40 F) was recorded January 7, 1971, at Hawley Lake in the White Mountains near McNary. The record high of 52.8 C (127 F) was set twice, once at Fort Mohave on June 15, 1896, and again at Parker on July 7, 1905. Both towns are along the lower Colorado River. In fact, summer temperatures above 49 C (120 F) have been reported at all towns along the Colorado River south of Hoover Dam and along the Gila River west of its confluence with the Salt River.

The state's warmest and coldest towns are, respectively, Mohawk, about 80 km (50 mi) east of Yuma, and Maverick in the southwest corner of Apache County on the Fort Apache Indian Reservation. Average July temperatures at these towns differ by almost 20 C (35 F), the average at Mohawk being 34.6 C (94.3 F) and at Maverick 15.5 C (59.9 F). Arizona's range of thermal climates, then, varies from the long, hot summers and short, mild winters of the warm deserts, to the short, cool summers and long, icy winters of the cold highlands.

This wide range of climatic conditions in Arizona is due to four important factors: 1) altitude, 2) latitude, 3) moisture source, and 4) orientation of mountain ranges (Kangieser, 1959).

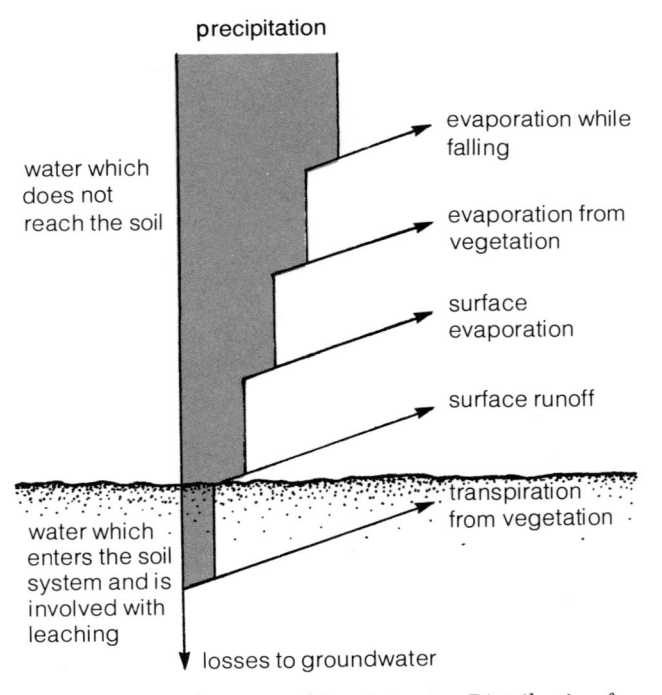

FIGURE 27. Schematic of Precipitation Distribution from Atmosphere to Soil (after E. M. Bridges, 1978)

Temperature

The great range of mean annual air temperatures in Arizona is mostly due to elevation. This was illustrated by Smith (1956) who plotted the mean annual air temperatures of selected Arizona weather stations as a function of elevation. The mean January and mean July air temperatures are strongly influenced by elevation also (Plates 4, 7 and 8). The coldest mean January air temperature of -7 to -4 C (20 to 25 F) are on the Kaibab Plateau, in the White Mountains and in the San Francisco Peaks. The warmest mean January air temperatures are greater than 13 C (55 F) along the lower Gila River.

Latitude also appears to affect temperatures since the weather stations in northeastern Arizona have mean January air temperatures about 5.5 C (10 F) colder than those at comparable altitudes in the southeast. Mean July air temperatures, on the other hand, drop rather uniformly with increased elevation. Temperatures decrease about .55 C (1 F) per 71 m (235 ft) increase in elevation (Sellers and Hill, 1974).

Precipitation

Roughly half of Arizona receives less than 250 mm (10 in) average annual precipitation. Areas in this regime are the southwestern, extreme western and a large part of the northeastern portions of Arizona as well as most of the San Simon Valley in the southeast. The wetter parts of Arizona receive up to 635 mm (25 in) or more average annual precipitation and are at the higher elevations along the Mogollon Rim, the southern part of the Kaibab Plateau, the White Mountains, the San Francisco Peaks and several mountains in the Basin and Range Province (Plate 5).

A unique feature of Arizona climate is the two periods of precipitation; one season is from December through March and the other during July, August and September. The proportion of winter precipitation decreases from west to east and the summer precipitation decreases from south to north (Plate 6). The biseasonal precipitation pattern appears to reach equality in the Flagstaff-Prescott area with half the average annual moisture falling during each period (Jameson, 1969).

Winter precipitation is the "Mediterranean" type, the same as that which provides the strong winter maximum moisture in California. The storms are associated with large-scale, mid-latitude cyclonic disturbances. Arizona generally receives only the fringes of these storms. Occasionally the semipermanent ridge of high pressure in the Pacific Ocean that is normally off the West Coast is displaced westward and a low pressure trough from the east is drawn over the western United States. When this happens, storms, rather than moving inland farther north, enter central and southern California and move inland across Arizona (Jurwitz, 1953). This weather pattern produced the wettest winters recorded in Arizona (Sellers and Hill, 1974). The maximum moisture from these storms falls on the central portion of the Mogollon Rim. Downwind from the Rim is a distinct rainshadow that is especially pronounced in the Little Colorado River Basin. The result is that the Mogollon Rim has a winter wet climate on the windward, or west, side and a winter dry climate on the lee, or east, side.

Summer rainfall has a tropical origin. The influx of warm, moist air masses from the Gulf of Mexico begin about July 1 and last until the end of August. When the Pacific high pressure cell off the West Coast moves rapidly northeast in late June, the Southwest receives a deep, gentle air flow from the Gulf of Mexico on the southwest side of a high pressure cell that protrudes from the Atlantic Ocean into the central part of the United States (Bryson and Lowry, 1955; Bryson, 1957). The unstable air advances into Arizona from the southeast over strongly heated land surfaces and yields moderate to heavy afternoon or evening thundershowers. This convective activity is most pronounced near mountains (Sellers and Hill, 1974). This phenomenon has been referred to as a monsoon (Huntington, 1914).

Summer rains are brief, sometimes intense, particularly when the storm begins, and scattered, seldom affecting more than several square kilometers (a few square miles) (Humphrey, 1933; Osborn, 1983). These storms tend to occur in a region for several days, then several days of dry weather follow. Monsoon air masses do not show distinct frontal characteristics and no rainshadow effect develops, as with winter storms. Areas where topography changes abruptly receive more rain. As it is in winter, summer precipitation is heaviest on the central Mogollon Rim (Jameson, 1969).

Hurricane influences also appear in Arizona and may contribute to the late summer rainfall. These hurricanes are not from the Caribbean, but from off the west coast of Mexico and occasionally drift up into northern Mexico and southern Arizona. Although they lose surface circulations characteristic of true hurricanes as they move inland, they do bring deep, moist air masses that have given the Southwest and northern Mexico some of the heaviest rainfalls on record (Thorud and Ffolliot, 1973; Sellers and Hill, 1974). These storms, when they occur, are in late August and September.

Records at Arizona climate stations show that precipitation is quite variable from year to year, but arrival time and expected amount of summer precipitation are less variable than those of winter storms at most stations. McDonald (1956) computed coefficients of variation of winter, summer and yearly amounts of precipitation for Yuma, Phoenix, Tucson and Flagstaff (Table 3). The coefficients of variation

TABLE 3.

Computed Coefficients of Variation of Winter, Summer and Yearly Amounts of Precipitation for Four Arizona Climate Stations.
(McDonald, 1956)

Station	C_w	C_s	C_y
Yuma	0.75 ± .06	0.94 ± .07	0.62 ± .05
Phoenix	0.57 ± .05	0.56 ± .05	0.40 ± .03
Tucson	0.54 ± .04	0.40 ± .03	0.30 ± .02
Flagstaff	0.37 ± .04	0.28 ± .04	0.23 ± .02

C_w—Coefficient of variability of winter precipitation
C_s—Coefficient of variability of summer precipitation
C_y—Coefficient of variability of yearly precipitation

represent the ratios of the standard deviation amounts to the mean amounts. The unusually high coefficient of variation of summer precipitation for Yuma is due to the extremely low amounts of summer precipitation. At Tucson and Flagstaff the coefficients of variation of summer precipitation clearly are lower than the variations of winter precipitation.

Arizona Climate Zones

Arizona has three basic climatic regimes based on average annual precipitation and mean temperature. Desert regimes cover about 30 percent of Arizona, the steppes about 53 percent and the highlands about 17 percent. These three regimes are each subdivided further into warm and cold, depending on whether the average temperature of the coldest month is above or below 0 C (32 F) and are illustrated in Figure 28. (The reader is invited to compare these regime boundaries with those of other plates and figures in this book.)

Deserts and steppes are characterized by limited amounts of precipitation; the former more so than the latter. Practically all of the moisture that falls in these regions evaporates; appreciable runoff and subsurface storage seldom occur (See Figure 29 and Plates 5 and 13). As result, the vegetation cover consists mostly of creosotebush, cacti and sagebrush on the deserts and mesquite, pinyon-juniper and various grasses on the steppes (see Plates 11 and 12). Irrigation is a must for successful farming in these dry regions. On the other hand, the highlands, particularly the cold highlands, normally receive sufficient precipitation during the year to support moderately dense vegetation and to provide substantial runoff to the surrounding drier areas (See Figure 30 and Plates 5 and 13). Some of the finest and most extensive pine forests in the world are in Arizona's cold highlands where precipitation is reasonably dependable from year to year. Precipitation is less dependable in the warm highlands, varying greatly in amount and intensity from year to year.

Most precipitation in Arizona falls during summer or winter. Only rare late or early storms occur during fall or spring and the late spring is especially dry.

Summer rains are heaviest and most dependable in the highlands. A dry afternoon between the second week of July and the first week of September is rare. Occasional cloudbursts send water raging down into the surrounding valleys, filling washes and arroyos to overflowing and doing considerable damage to roads and poorly sited buildings. These storms also occur over desert towns, where they can be especially destructive. Sewer systems in most of these communities were not built to handle more than moderate amounts of runoff and are inadequate to channel the large volumes of runoff produced by the intense summer storms.

The driest regions in winter generally are the cold steppes, particularly those north of the Little Colorado River. These areas are "protected" from major sources of winter moisture by high mountain ranges and plateaus including the White Mountains, San Francisco Peaks and the Mogollon Plateau to the south and southwest, the Wasatch Mountains of Utah and the Rocky Mountains to the northeast. These barriers, however, do not block unusually strong winds, particularly during winter and spring. When a storm system passes over the area, winds exceeding 50 kmh (30 mph) may last for several days, first warm from the south, then bitterly cold from the north.

At least a trace of snow has been recorded in all parts of Arizona, but usually only in the mountain sections are snow depths measurable. Seasonal totals vary from 300 to 1,500 mm (1 to 5 ft). Slopes of the higher mountains provide excellent opportunities for winter sports, even those rising from the desert regions. Early season snowstorms are very unpredictable; therefore, it is a good policy for hunters, hikers and campers to be prepared for snow when they go into the mountains in the late fall months. Motorists also should be prepared to encounter unexpected snowfall while driving in the mountains during this season.

The southern Arizona desert has mild, dry winters that are very important in its rapid population growth. Temperatures in the coldest month usually range in the early morning from 0 to 2 C (middle 30s F) to 18 to 20 C (high 60s F) in the midafternoon. Below-freezing temperatures are rare. When they do occur they are limited for the most part to the time just before and during sunrise, and to low areas that receive drainage of cold, dense air from higher surrounding terrain during the night.

Many persons find Arizona desert summers less uncomfortable than those of the eastern United States despite the high temperatures. The extreme dryness of the air, with relative humidities often falling below 10 percent in May and June, relieves intensity of the heat by increasing evaporation, a mechanism that helps cool the body. And virtually all buildings are cooled either by refrigeration or by evaporative coolers. When humidity is low even the least expensive evaporative coolers are quite effective. But when humidities are high, particularly during the July-August rainy season, evaporative coolers are rather ineffective.

Arizona's warm steppes are slightly more cool, wet and humid throughout the year than the warm deserts but they still generally are characterized as being hot and arid. These regions are close to the highlands that provide refreshingly mild days and cool nights during the hot summer months. The warm steppes, particularly those in southern Arizona, are popular centers of winter recreational activities. The temperature variation between day and night in the cooler months often exceeds 22 C (40 F). The early morning minimum is between -2.8 and -1.7 C (upper 20s F) and the afternoon maximum between 21.5 and 23 C (low 70s F).

Perhaps the most rugged climate is in the Arizona cold steppe regions that are almost entirely in the northeast. Winters are cold, dry and windy; the temperature normally falls below -17.8 C (0 F) on four to six days between November and March. Summers are quite warm, especially in the northern regions, where summer afternoon temperatures above 32 C (90 F) may be expected 50 to 65 days. But summer nights are cool and temperatures usually fall to the 10s C (50s F) by sunrise.

ARIZONA SOILS

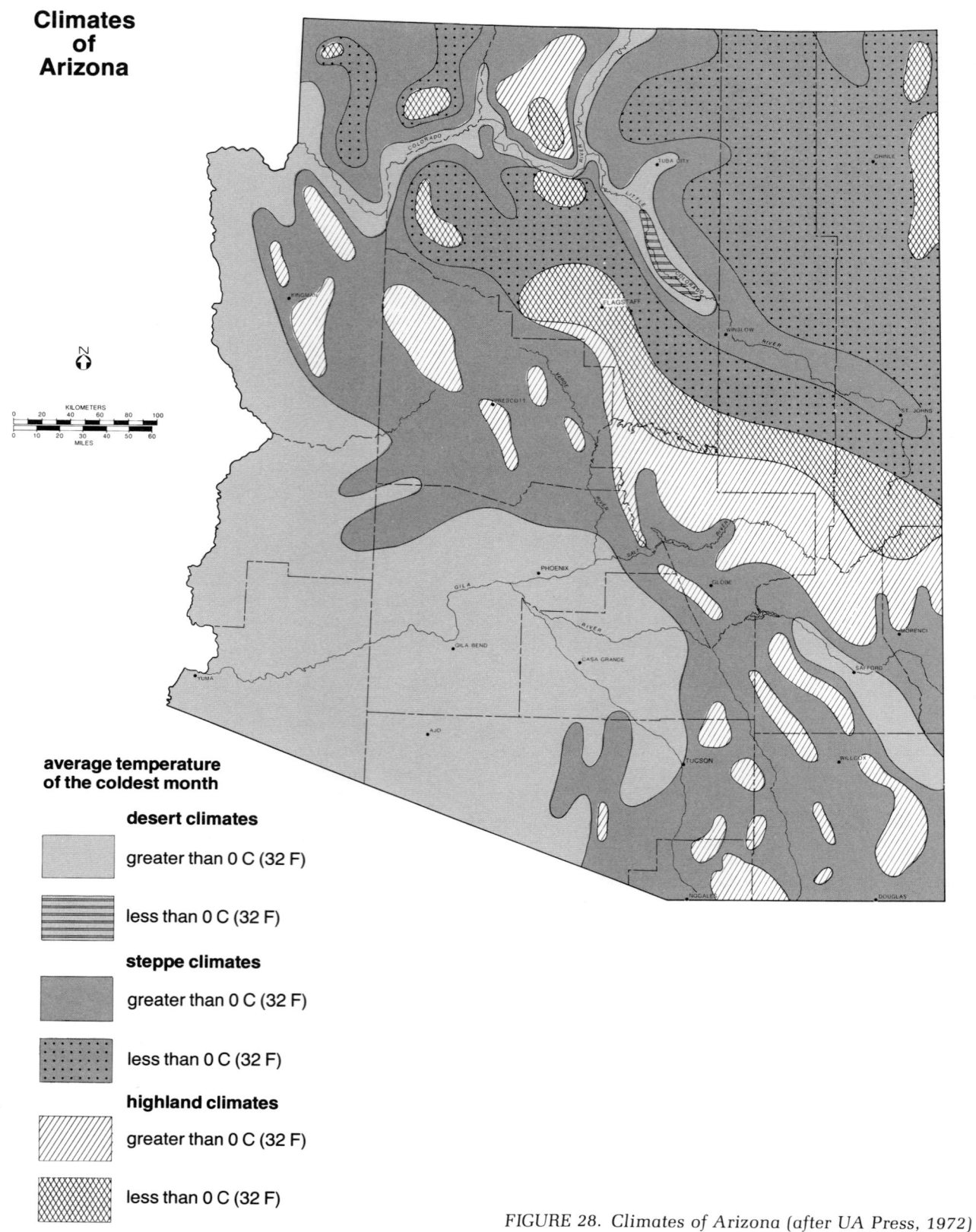

Climates of Arizona

average temperature of the coldest month

desert climates

greater than 0 C (32 F)

less than 0 C (32 F)

steppe climates

greater than 0 C (32 F)

less than 0 C (32 F)

highland climates

greater than 0 C (32 F)

less than 0 C (32 F)

FIGURE 28. *Climates of Arizona (after UA Press, 1972)*

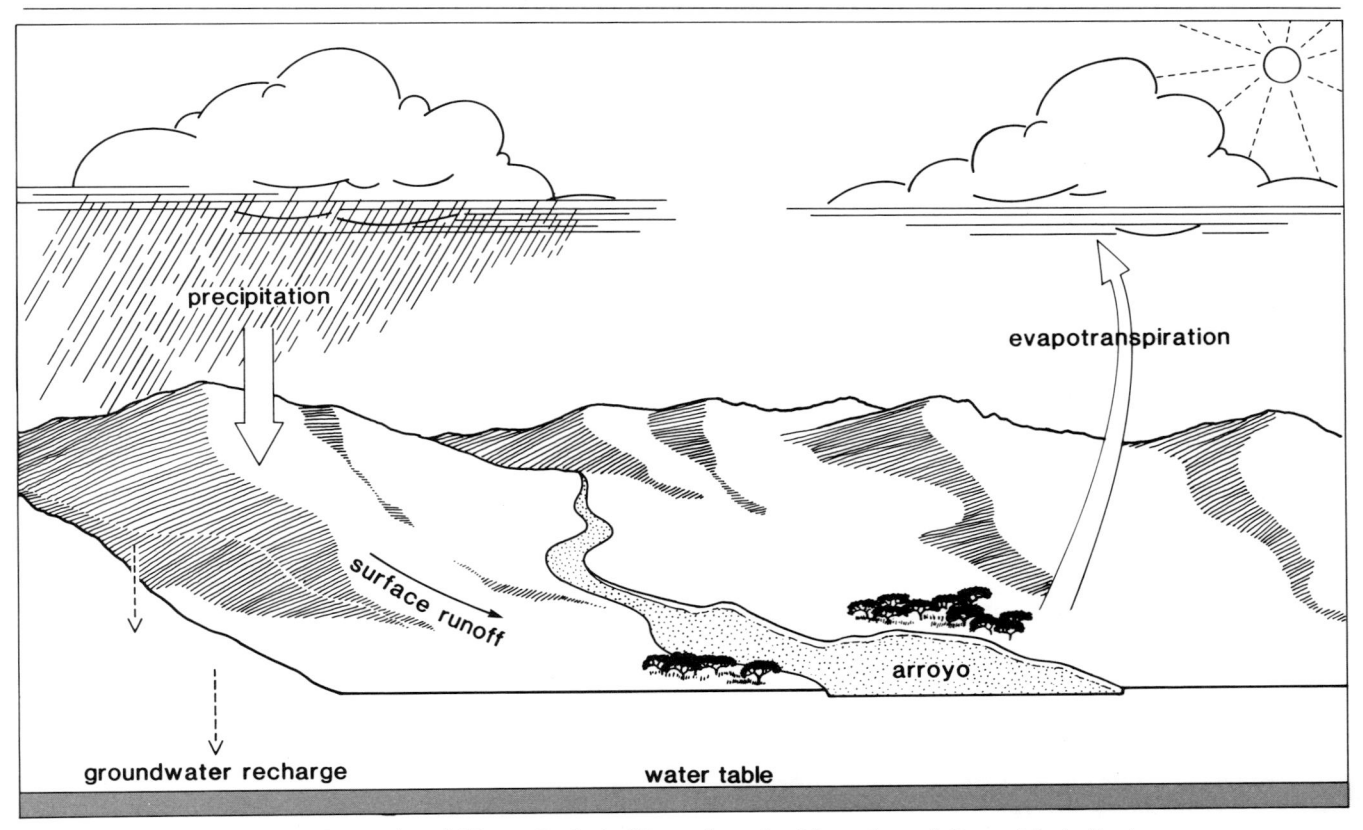

FIGURE 29. Generalized Water Cycle in Hyperthermic, Thermic and Some Mesic Regions

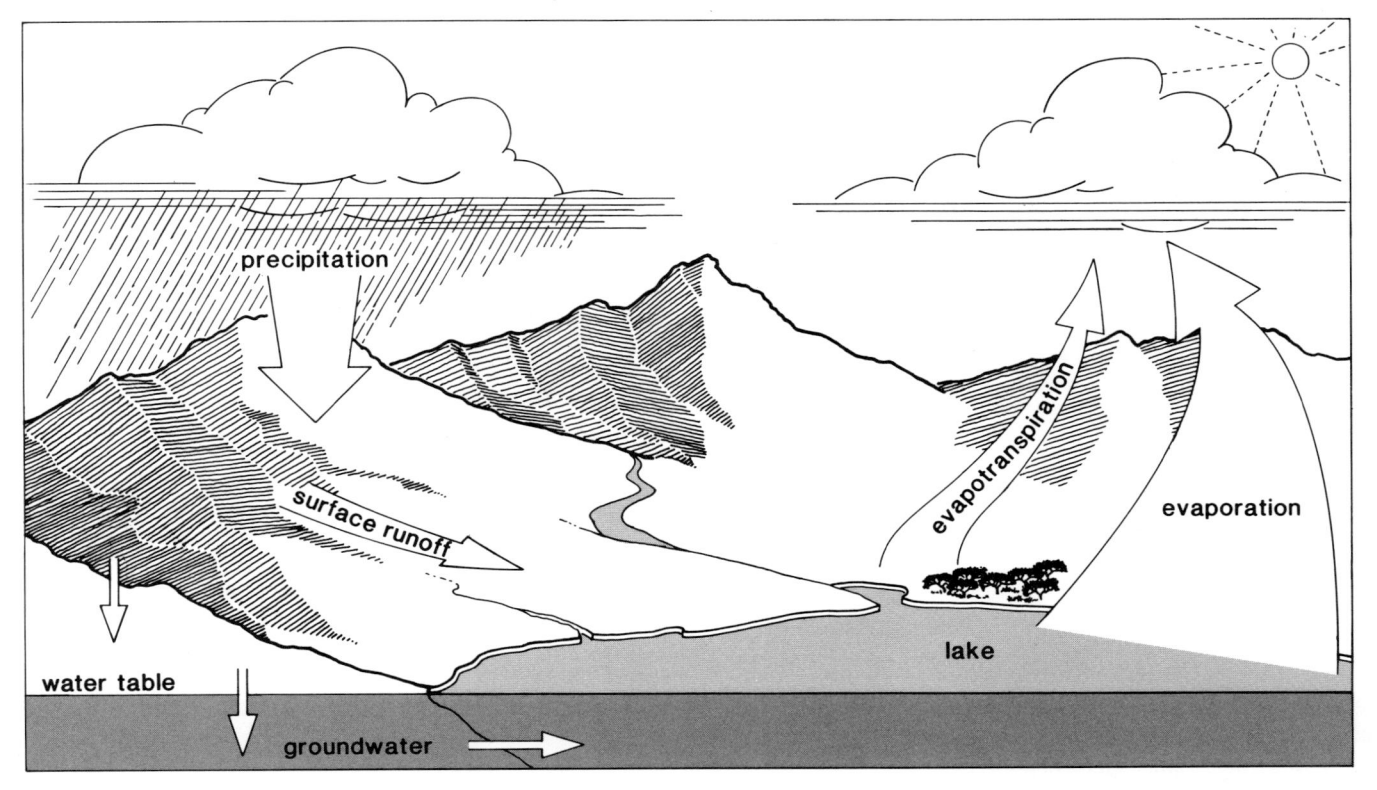

FIGURE 30. Generalized Water Cycle in a Frigid Subhumid Region

Climatic Zones of the State Soil map

The Arizona General Soil Map mapping units are grouped into seven climatic zones. These zones are a combination of the four soil temperature regimes and three precipitation zones, arid, semiarid and subhumid. The seven climatic zones are defined in Table 4.

Soil Temperature Regimes

It has been shown that mean annual soil temperature is rather closely related to mean annual air temperature (Smith et al, 1964). Factors that may affect this relationship include amount and distribution of rain and snow, shade and litter layers in forests, percent and direction of slope and irrigation. Methods for calculating soil temperature with multiple regression equations by using site factors such as mean annual air temperature, elevation, latitude, slope, aspect and certain soil characteristics have been proposed (Arkley, 1971; Ouellet, 1973; Konstantinov and Popovich, 1980; Meikle and Treadway, 1981).

Soil temperature is an important property that controls or has a strong influence on plant growth and soil formation. Every pedon has a mean annual temperature that is essentially the same at all depths, but at any given moment the temperature of any two horizons is rarely the same because of daily, short-term and seasonal fluctuations (Smith et al, 1964). Soil temperature data for Arizona are limited. U.S. Forest Service soil scientists, however, recently have started taking soil temperature measurements under a variety of environmental conditions. In some cases several years of data are available (Robbie, Brewer and Jorgensen, 1982).

Each pedon has a characteristic temperature regime that can be measured and described. For most practical purposes, the regime can be described by the mean annual soil temperature, the average seasonal fluctuations from that mean, and the mean warm or cold seasonal soil temperature gradient at depths from 5 to 100 cm (2 to 40 in), the main root zone (Soil Survey Staff, 1975).

Four soil temperature regimes are recognized in Arizona according to *Soil Taxonomy* (Soil Survey Staff, 1975) definitions. They are hyperthermic, thermic, mesic and frigid.

Hyperthermic soils have mean annual soil temperatures of 22 C (72 F) or more and a greater than 5 C (9 F) difference between mean summer (mean temperature of June, July and August) and mean winter (mean temperature of December, January and February) temperatures at a depth of 50 cm (20 in) or at the soil-rock interface in shallow soils, soils less than 50 cm (20 in) deep.

Thermic soils have mean annual soil temperatures of 15 C (59 F) or more, but less than 22 C (72 F), and the difference between mean summer and mean winter temperatures is greater than 5 C (9 F) at a depth of 50 cm (20 in) or at the soil-rock interface in shallower soils.

Mesic soils have mean annual soil temperatures of 8 C (47 F) or more, but less than 15 C (59 F), and a difference between mean summer and mean winter temperatures greater than 5 C (9 F) at a depth of 50 cm (20 in) or at the soil-rock interface in shallower soils.

Frigid soils have mean annual soil temperatures of more than 0 C (32 F), but less than 8 C (47 F) and a difference between mean summer and mean winter temperatures greater than 5 C (9 F) at a depth of 50 cm (20 in) or at the soil-rock interface in shallower soils.

Arizona also has cryic soils. These soils, like the frigid soils, have mean annual temperatures between 0 C (32 F) and 8 C (47 F), but they are considered to be colder than the frigid soils since they are colder in summer. Pergelic soils have mean annual soil temperatures less than 0 C (32 F) and may be in Arizona at the highest elevations of the San Francisco Peaks.

The distribution of the four soil temperature regimes throughout Arizona is shown in Plate 14. The cryic and Pergelic soils, if any, are included in the Frigid Soil Zone.

Another aspect of the influence of temperature on soils is the number of diurnal freeze-thaw cycles per year. This aspect may be significant in soil formation because of frost weathering (Potts, 1970). The White Mountains have the most days in which the temperature fluctuates across the 0 C (32 F) mark (Figure 31).

TABLE 4.

Temperature-Precipitation Zones of the Arizona General Soil Map

Symbol	Unit Name	Soil Temperature*	Precipitation
HA	Hyperthermic Arid	22 C (72 F) or greater	250mm (10 in) or less
TA	Thermic Arid	15 to 22 C (59 to 72 F)	130 to 250 mm (5 to 10 in)
TS	Thermic Semiarid	15 to 22 C (59 to 72 F)	250 to 460 mm (10 to 18 in)
MA	Mesic Arid	8 to 15 C (47 to 59 F)	150 to 250 mm 6 to 10 in)
MS	Mesic Semiarid	8 to 15 C (47 to 59 F)	250 to 410 mm (10 to 16 in)
MH	Mesic Subhumid	8 to 15 C (47 to 59 F)	410 mm (16 in) or greater
FH	Frigid Subhumid	8 C (47 F) or less	410 mm (16 in) or greater

*The soil temperature refers to the mean annual soil temperature at a depth of 50 cm (20 in) or at the soil-rock interface in shallow soils.

CLIMATE OF ARIZONA

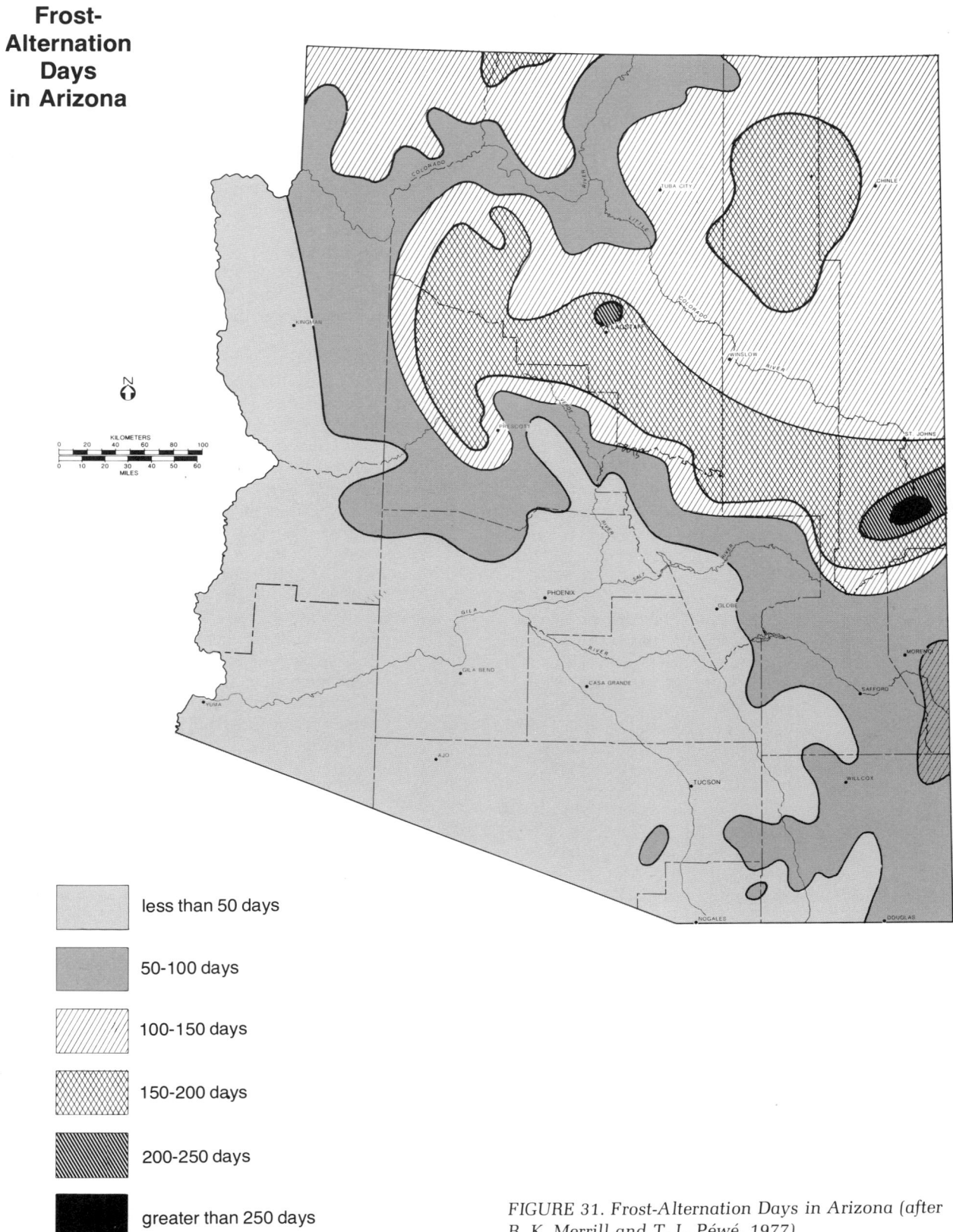

Frost-Alternation Days in Arizona

- less than 50 days
- 50-100 days
- 100-150 days
- 150-200 days
- 200-250 days
- greater than 250 days

FIGURE 31. Frost-Alternation Days in Arizona (after R. K. Merrill and T. L. Péwé, 1977)

39

Soil Moisture Regimes

Applying the soil moisture regime to classify and map Arizona soils is difficult because of the lack of data. And obtaining data about how moisture in the moisture control section changes with the seasons is difficult also. Consequently less direct methods have to be used to approximate soil moisture regimes. Soil moisture balance calculations based on climatic data as described below commonly are used (Newhall, 1980). Soil moisture and soil temperature regimes also are sometimes inferred from vegetation. Plant communities, or alternatively a few index species, are associated with particular soil moisture and temperature regimes. Since the moisture balance methods have shortcomings and since few studies of natural vegetation and its soil moisture and temperature requirements have been attempted, there is a general lack of agreement on identifying soil moisture regimes, especially in the western United States (Daugherty, 1982). As Daugherty (1982) also pointed out, the descriptions of soil moisture regimes in *Soil Taxonomy* (Soil Survey Staff, 1975) were based mostly on cultivated areas outside the western states and do not seem to apply as well in the West. Finally, the authors of *Soil Taxonomy* (Soil Survey Staff, 1975) recognized that the definitions of the soil moisture regimes were far from perfect and that revisions would be expected.

Nonetheless, soil moisture regimes can be determined in part by precipitation and evapotranspiration and may be influenced by topography and its effect on runoff, consequently on runon (Plate 13). Soil properties such as texture, organic matter content, type of clay, structure and soil depth as they affect infiltration and retention of moisture also influence soil moisture regimes. Three categories, dry, moist and saturated, are used to define soil moisture regimes (Soil Survey Staff, 1975). Soil is considered dry when moisture tension is greater than 15 bars. Available water in soils that are moist is held between 0 and 15 bars. Soils are saturated when there is no tension on the water and essentially all the void space is filled with water. The depths in the soil at which soil moisture criteria are applied are established by the concept of the moisture control section. The upper boundary of the moisture control section is the depth to which a dry soil will become moist 24 hours after a 25 mm (1 in) rain. The lower boundary is either the depth to which 75 mm (3 in) of water, introduced through the soil surface, will moisten a dry soil in 48 hours, or the depth to a pan, bedrock or other root-restricting layer, whichever is shallower.

Soil moisture regimes described in *Soil Taxonomy* (Soil Survey Staff, 1975) are based on the time and season that the soil moisture control section is saturated, moist or dry. The definitions of these soil moisture regimes were summarized by Buol, Hole and McCracken (1980) and are repeated below.

Aquic Moisture Regime. These soils are water-saturated for at least enough time, usually several days, so that anaerobic conditions exist. Soil mottling normally is indicative of this condition.

Aridic or Torric Moisture Regime. These soils are both dry more than half the time when not frozen and never moist more than 90 consecutive days when soil temperatures are above 8 C (47 F) at 50 cm (20 in) depth.

Perudic Moisture Regime. In most years precipitation on these soils exceeds evapotranspiration every month of the year.

Ustic Moisture Regime. In most years, these soils are dry for more than 90 cumulative days but less than 180 days. In temperate climates, they are usually moist at least 45 consecutive days in the four months after the winter solstice and not dry 45 consecutive days in the four months after the summer solstice.

Xeric Moisture Regime. These soils are only in the temperate areas where summers are dry and winters moist. These soils usually are dry more than 45 consecutive days in the summer and moist more than 45 consecutive days in the winter.

Soils with aridic (torric) and ustic moisture regimes are the most widespread in Arizona. Aridic soils constitute essentially all Hyperthermic Arid (HA), Thermic Arid (TA) and Mesic Arid (MA) soils, and most of Thermic Semiarid (TS) and Mesic Semiarid Soils (MS) shown on the Arizona General Soil Map (Plate 1). Some ustic soils are included in Thermic Semiarid (TS) and Mesic Semiarid (MS) soils. Ustic and Udic soils constitute the Mesic Subhumid (MH) and the Frigid Subhumid (FH) soils with udic soils in the higher rainfall areas. Small areas of aquic soils are in topographically low areas that have poor drainage and limited runoff. Soils with xeric regimes are not recognized in Arizona. Areas in northern Arizona near the Utah border contain soils that might approach having a xeric moisture regime but are considered ustic because they usually receive enough summer rain to prevent the soil moisture control section from becoming dry for a sufficient length of time. Perudic moisture regimes are not present in Arizona soils.

Water Balance and Thornthwaite's Climatic Classification

Temperature and precipitation alone are poor descriptors of climate. The amount of precipitation does not indicate whether a climate is moist or dry unless the water need of the site can be compared with it. And temperature does not really reveal the energy that is available for plant growth and development unless the moisture condition of the soil is known. Thus, one of the major objectives of the water balance approach in characterizing climate is to arrive at a better way of determining whether a climate is moist or arid by comparing the climatic moisture supply with the moisture needs. Water balance estimations also have been quite useful for correlating some climatic characteristics of a region that are important in soil formation (Hurst, 1951; Arkley, 1963, 1967), in soil classification (Cox, 1968), in plant growth (Arkley and Ulrich, 1962), in distribution of vegetation (Mather and Yoshioka, 1968) and in determination of soil moisture regimes (Newhall, 1980).

Many scientists have studied the problem of how to express the daily or seasonal water budget of a site or region. The ability of a scientist to describe water budgets has been advanced significantly within the last 30 years or so by the work of Penman (1956) in England, Budyko (1958) in the Soviet Union and Thornthwaite (1948) in the United States. Each approach is quite different from the other and each has certain limitations. The formula developed by Thornthwaite (1948), as modified by Thornthwaite and Mather (1955), is used here, not necessarily because it is superior, but because it has been widely used in the U.S. West.

Thornthwaite's (1948) procedure makes it possible to estimate soil moisture conditions of a site from the gains and losses of soil moisture over a certain interval of time, day, week or month. Gains are from precipitation, the data of which are obtained from weather stations. Losses are evapotranspiration, the combined loss from the ground and plant surfaces (evaporation) and the loss of water from living plants (transpiration). Potential evapotranspiration is the amount of moisture lost by soil when it is covered by vegetation and amply supplied with water. Thornthwaite (1948) studied water-use rates in irrigation projects and catchment runoff records. He discovered a close relationship between mean monthly temperature and potential evapotranspiration if adjustments are made for variations in day length.

From the findings of these studies Thornthwaite (1948) developed a rather complicated formula for computing potential evapotranspiration of a site if the latitude is known and temperature records are available. The computations involved are simplified by the use of tables (Thornthwaite and Mather, 1957) and/or graphs (Palmer and Havens, 1958). Potential evapotranspiration differs from the actual evapotranspiration (see Plates 9 and 10). The actual water loss to the atmosphere equals potential evapotranspiration only during those periods when precipitation is greater than the potential evapotranspiration. When precipitation is less than potential evapotranspiration, actual evapotranspiration is equal to the sum of precipitation and water lost by soil to the atmosphere.

When precipitation is equal to water need or potential evaporation, the soil is at field capacity and no leaching takes place. When the soil water supply is greater than the need, there is a surplus of moisture for leaching the soil and subsequent lowering of the soil pH over time. When it is less, the soil moisture in the root zone is drawn upon by vegetation until the wilting point is reached and the moisture budget becomes deficient. After the dry season, the soil-moisture reservoir must be replenished to field capacity before there can be a surplus again.

Thornthwaite (1948) originally assumed that except in shallow soils, soils at field capacity have 10 cm (4 in) of water available for the atmosphere and vegetation to draw on if precipitation does not fall. Arkley (1967) and Steila (1972) suggested 15 cm (6 in) as a better value of soil-moisture storage capacity. Thornthwaite and Hare (1955), however, suggested that at least 30 cm (12 in) of water are available to deep-rooted, mature plants in most soils. Thornthwaite (1948) also originally assumed that all stored moisture, 10 cm (4 in), was equally available. However, it is well known that as soil dries, it becomes increasingly difficult for water to be lost to evaporation and transpiration. Thus, as the soil-moisture content decreases, so too does the rate of evapotranspiration. Thornwaite and Mather (1955), Baier and Robertson (1966) and Baier (1969) suggested different relationships between actual evapotranspiration and soil-moisture content. Thornthwaite and Mather (1955) assumed a linear relationship between water loss and soil-moisture content, meaning that when the soil moisture is reduced to 50 percent of capacity, the actual evapotranspiration rate is only 50 percent of the potential rate.

Baier (1969) measured moisture of soils used for dryland wheat cultivation in Canada for 10 years. He found essentially no reduction in soil moisture depletion rates from 100 percent to 70 percent available water. With further soil drying, an approximate linear relationship was observed between moisture loss and moisture content from 70 percent to 0 percent available water.

It is now recognized that the Thornthwaite calculation method underestimates potential evapotranspiration in arid regions. This was illustrated by Gay (1981) who compared potential evapotranspiration estimates for Tucson by several methods. The Thornthwaite method produced the lowest values when compared with other calculation methods and with pan and adjusted pan evaporation rates.

Average annual water budgets of 14 weather stations representative of the seven temperature-precipitation zones outlined in Table 4 are illustrated in Figure 32. The potential evapotranspiration values for each month were calculated by the Thornthwaite (1948) method using Palmer-Havens (Palmer and Havens, 1958) graphics techniques. A soil-moisture storage capacity of 10 cm (4 in) and a linear relationship between water loss and soil moisture content were assumed (Thornthwaite and Mather, 1955). Climatic data used in the calculations are from Sellers and Hill (1974). The annual water budgets depicted in Figure 32 should be considered to be qualitative representations of changes in soil-moisture conditions since the estimates of potential evapotranspiration may be too low, especially for the more arid stations. Potential evapotranspiration values and water budget tabulations for other Arizona stations were compiled by Buol (1964).

Yuma, representative of the Hyperthermic Arid region, is one of the hottest and driest places in Arizona as illustrated in Figure 32. The potential evapotranspiration is greater than precipitation every month and the soil-moisture budget is deficient throughout the year. Phoenix, also in the Hyperthermic Arid region, receives slightly more moisture than Yuma that provides some soil-moisture recharge during the winter and some soil-moisture use in the early spring. All other stations, except Fort Valley and Maverick in the Frigid Subhumid region, have soil-moisture recharge during winter and soil-moisture use and deficit during spring and summer. The amount of moisture available during spring is higher at stations that have more winter precipitation and lower temperatures. Thus, the soil-moisture recharge during winter is greater at Kingman than at Lees Ferry, at Bagdad than at Douglas, at Tuweep than at Springerville and at Whiteriver than at Canelo. The differences between soil-moisture budgets at stations in similar climatic regions are due to

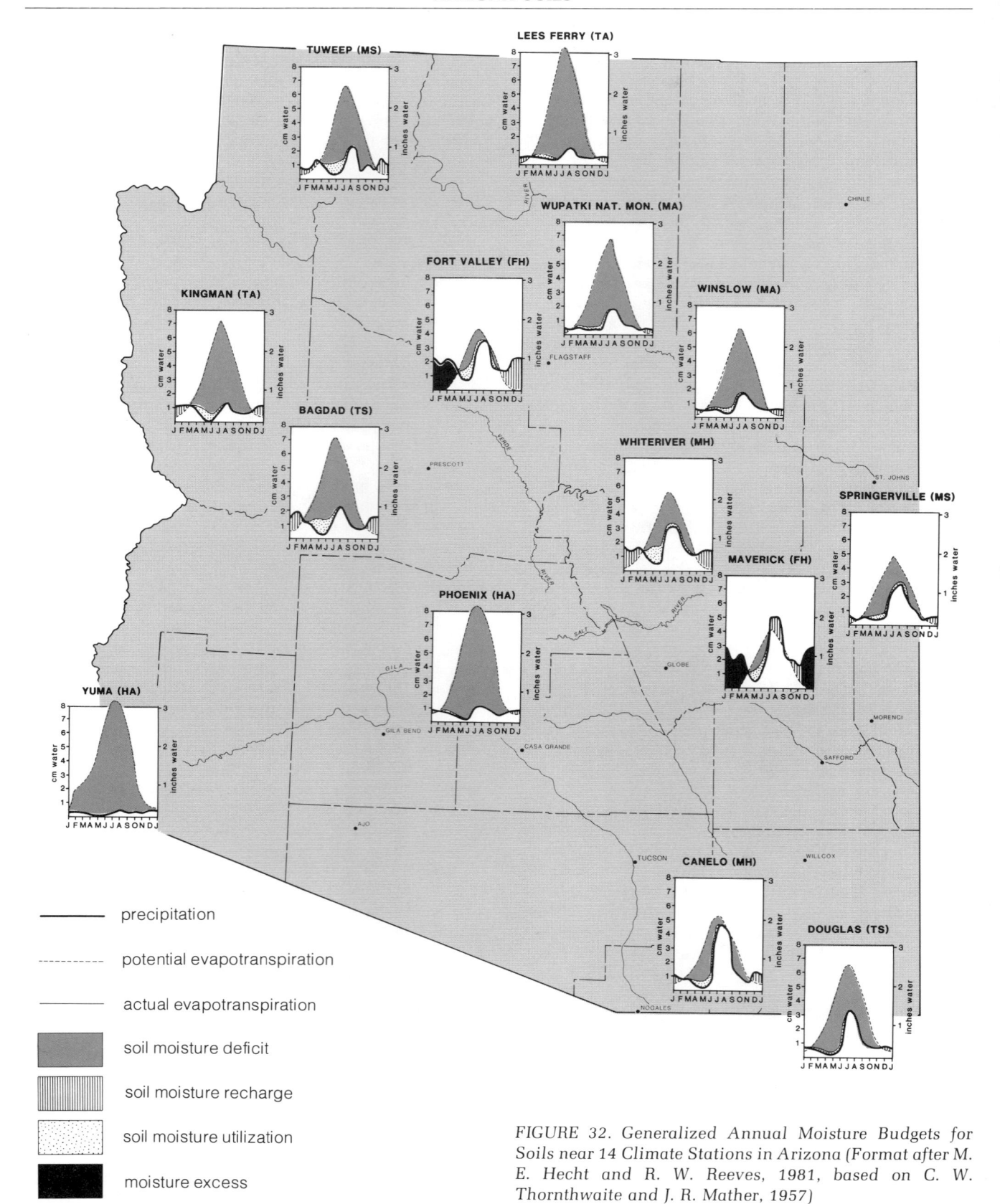

FIGURE 32. Generalized Annual Moisture Budgets for Soils near 14 Climate Stations in Arizona (Format after M. E. Hecht and R. W. Reeves, 1981, based on C. W. Thornthwaite and J. R. Mather, 1957)

different seasonal precipitation patterns. Fort Valley and Maverick have surplus water for leaching soils during the winter months. There follows a period of soil-moisture use and then a moderate deficit. Maverick, with its usually abundant July and August precipitation, has summer recharge.

Many stations, particularly those in southeastern Arizona, receive more than 50 percent of the average annual precipitation during July, August and early September when potential evapotranspiration is high. In spite of moderate to high amounts of precipitation, therefore, the water budget method may forecast fairly large soil-moisture deficits, indicating that precipitation and water stored in the soil are inadequate to meet water needs.

Actual evapotranspiration may be fairly high during the summer rainy season, but lower than potential evapotranspiration because of vegetative growth during July, August and September in the subhumid, semiarid and to a lesser extent in the arid regions of Arizona. This observation supports the contention that the soil-moisture deficit during the summer rainy season in Arizona is misleading as calculated by Thornthwaite's (1948) water budget technique.

It is emphasized that water budgets shown in Figure 32 are based on averages. Since precipitation at the climatic stations varies considerably from year to year, water budgets also vary from year to year. Extremely wet seasons, such as the winters of 1973 and 1980 in Arizona, produce greater excess soil moisture for leaching and/or recharge of soil moisture for use in the spring. Dry years, on the other hand, have less intense leaching and less moisture for recharge.

Paleoclimate

Climatic conditions in the past, however, may have been more influential than those of more recent times in determining many properties of Arizona soils. Knowledge of past climatic conditions, therefore, is highly desirable in interpreting the influence of climate on soil properties and gaining a greater understanding of the relationships between soils.

A wealth of evidence is accumulating from which generally consistent inferences about the climate of Arizona during the past 50,000 years or so can be made. This evidence comes chiefly from paleobotanical, palynological, dendrochronological, paleozoological, glacial and pluvial studies in the Southwest.

Pleistocene alpine glaciation in Arizona occurred in the San Francisco Peaks near Flagstaff (Sharp, 1942; Péwé and Updike, 1970; Updike and Péwé, 1974; Duncklee, 1978) and in the White Mountains near Mount Baldy (Melton, 1961; Merrill and Péwé, 1972, 1977). The glaciations are evidence of colder climatic conditions, but only very small areas in mapping unit FH2, Sponseller-Ess-Gordo Association, were directly affected.

Periglacial phenomena occurred in nonglaciated high elevations during glaciation on Kendrick Peak near Flagstaff (Barsch and Updike, 1971) and in the Chuska Mountains (Blagbrough, 1971), and were suggested by Melton (1965) in the Pinaleno and other high mountains in southern Arizona. Galloway (1970) and Melton (1965) indicated that physical weathering associated with periglacial environments in the mountains of the Southwest was important in providing rock and other coarse fragments common in many colluvial and alluvial fan deposits and soils.

During periods of the Pleistocene many extensive and relatively deep lakes filled basins in the Southwest that now are dry or contain shallow, saline lakes (Feth, 1961). Pluvial has been used to describe the times when the lakes existed and implies that precipitation was somewhat greater than now. Most authorities believe that the more moist climatic conditions during glacial-pluvial times were caused by the combined effects of increased precipitation and decreased potential evapotranspiration due to lower temperatures.

Biotic communities in Arizona were displaced about 300 m (1,000 ft) or more to lower elevations and hundreds of kilometers (miles) south during glacial-pluvial times. Evidence of this displacement consists mostly of pollen stratigraphy and plant materials preserved in fossil packrat middens.

A number of pollen-stratigraphic studies made in the Southwest were reviewed by Hevly and Karlstrom (1974). The results of these studies showed that the Southwest paleoclimate was in phase with those of the Pacific Coast and mid-continental North America. When continental glaciers were expanding, Southwest biotic communities were displaced to lower elevations and more southerly latitudes. Pluvial lakes expanded and alluvial deposition was augmented by the actions of more moist and cooler climates. As continental glaciers waned, the reverse biotic, geologic and climatic phenomena occurred.

Widespread occurrences and discoveries in the Southwest of extremely old packrat (genus *Neotoma*) middens containing abundant, well-preserved plant macrofossils recently have provided good information about the nature of paleoplant communities and inferred climatic conditions (Lanner and Van Devender, 1974; Wells, 1976). Analyses of fossil plant remains yield more reliable information than pollen analyses. Packrats thoroughly sampled vegetation within 100 m (330 ft) of the midden sites. But with pollen stratigraphy there is always the possibility that some pollen was introduced to a site from some distance away. Analyses of plant remains also often allow for more specific identification in certain genera. Radiocarbon ages of middens from a number of sites in the Mohave, Sonoran and Chihuahuan deserts have been reported to range in age from less than 4,000 to 40,000 years. The results of all of these studies show vegetation changes indicative of cooler and/or more moist climates in the three deserts during the late Pleistocene.

Based on interpretations of packrat midden data and to a lesser extent on pollen stratigraphy and geologic evidence of snowline lowering, three principal models were developed of climatic conditions during the glacial maximum of the late Wisconsin in the Southwest.

The first model is that lower temperatures alone accounted for the more moist climate. Brakenridge (1978) concluded that the mean annual precipitation was very much like today, while Galloway (1970, 1983) maintained that it was somewhat less. Brakenridge (1978) and Galloway (1970) suggested a 7 to 8 C (13 to 14 F) and 11 C (20 F) annual cooling, respectively, during maximum Wisconsin glaciation. On a global scale Budel (1959) agreed with Galloway by suggesting that during periods of glaciation regions outside the tropics were particularly colder and drier, while tropical regions were more humid and only moderately colder than today.

In the second climatic model, Wells (1979) disagreed with the Brakenridge (1978) and Galloway (1970, 1983) interpretations. He believed that the glaciopluvial climates of the Southwest desert lowlands had more equable temperatures. Wells (1979) also suggested considerably more summer rainfall than now, but with a strong southeast to northwest gradient of decreasing rain with increasing distance from the Gulf of Mexico, similar to the present gradient. Thus, the Mohave Desert to the northwest was relatively cool and dry and the Chihuahuan Desert to the southeast was relatively warm and moist. Wells (1979) also did not believe that winter precipitation increased although its effectiveness would have been enhanced by cooler temperatures.

The third climatic model proposed by Van Devender (1977) and Van Devender and Spaulding (1979), disagreed with Wells (1979). They believed that the late Wisconsin climate of the Southwest was characterized by more winter precipitation than now, probably because of more numerous frontal storms south of the crest of the Sierra Nevada (36°N) that extended as far east as the Trans-Pecos in Texas. These frontal storms also may have begun earlier in the fall than now and lasted later in the spring. Van Devender and Spaulding (1979), also suggested that the late summer-early fall hurricanes were moved farther south by colder sea surface temperatures so that that source of precipitation was eliminated from the Southwest during the late Wisconsin. Moreover, the winter temperatures, according to Van Devender and Spaulding (1979), were rather mild, perhaps comparable with the present, while the summers were cool. The mild winters could have been due to the thick continental ice sheets that altered atmospheric air currents, drawing warmer currents from the Pacific and preventing cold Arctic air masses from entering the mid-continent and the Great Basin areas. Cole's (1982) packrat midden data also indicated that precipitation in the eastern Grand Canyon fell predominantly in the winter but that the Wisconsin glacial climate was colder in all seasons than it is today. Cole's (1982) data suggested a wider range between summer and winter mean temperatures.

The elevational lowering of vegetation zones implied by Hevly and Karlstrom (1974) is not as straightforward as it might seem. Brakenridge (1978) suggested the telescoping effect. The upper elevational limit of a given vegetation zone was lowered more than the lower boundary. He indicated that this effect was most marked on ponderosa pine and severely restricted its vertical range. Cooler summers depressed the upper boundary, while expansion to lower elevations was limited by soil moisture deficits. Cole (1982) found evidence that glacial vegetation of the Grand Canyon was similar to modern vegetation in northern Utah and that the composition and elevational zones of the plant associations changed. Thus, Cole (1982) concluded that Pleistocene vegetational zones were not simply depressed versions of modern zones but rather reflected a latitudinal shift of climate in the Grand Canyon region.

One difficulty in evaluating climatic conditions during glaciation is that lesser glaciations with relatively higher temperatures occur during a major glacial period. Such periods are called interstadials, but as yet no universally acceptable definition distinguishes an interstadial from an interglacial (Goudie, 1977).

Earlier glacial stages in Arizona probably produced climatic conditions similar to those of the last glaciation, the Wisconsin, described above. Gray (1961) provided paleobotanical and sedimentary evidence in the Safford Valley in southeastern Arizona to support this theory. Her evidence indicated a cooler and/or wetter climate during the early Pleistocene that she tentatively correlated with the Nebraskan glacial stage. She further suggested that winter precipitation played a more significant role in southeastern Arizona then than now. Segota (1967) believed that temperatures of the glacial ages were not the same, that each younger glacial age was a little colder than the preceding one. His pollen-stratigraphy data are for Europe but probably are applicable worldwide.

An interglacial was defined by Suggate (1965) as being a "warm period between two glaciations during which the temperature rose to that of the present day." The term interglacial is used also to describe a warm episode between two cold ones in nonglaciated regions when the cold episodes correspond to those truly glacial effects that occurred at higher latitudes or altitudes (Suggate, 1974). Turner and West (1968) and Wright (1972) pointed out, however, that the climate of an interglacial was not uniform, that each interglacial consisted of distinct subperiods of vegetational development and presumed climatic conditions as evidenced by the pollen record in Europe. The last interglacial is believed to have lasted from about 128,000 to between 90,000 and 73,000 years ago (Suggate, 1974).

Some Arizona soils undoubtedly have passed through one or more interglacial. The high terraces described by Morrison (1965), McFadden (1981) and Menges and McFadden (1981) are thought to date back to the middle to early Pleistocene. Soils on these and similar surfaces in Arizona, therefore, have been under the climatic influences of at least two glacial and two interglacial periods.

Although moisture is not included in the definition of an interglacial, Melton (1965) suggested that the Yarmouth in southern Arizona was semiarid, and that the Sangamon was considerably more humid than now as well as being warm. Melton's (1965) Sangamon climate was postulated to explain formation of well-developed and deeply weathered red soils on old alluvial fan surfaces.

No agreed upon time that is applicable worldwide has been established for the beginning of the Holocene, time when the last glacial age ended (Mercer, 1972). Packrat midden paleobotanical evidence indicates that a rather sharp change in climate occurred about 11,000 years ago that marked the beginning of the Holocene in Arizona (Van Devender, 1973, 1977; Van Devender and Spaulding, 1978; Cole, 1982). Several investigators including Morner (1972), Richmond (1972) and Wright (1972) considered the Holocene to be an interglacial age, Flandrian or Present Interglacial, and that another period of glaciation will occur.

The worldwide climate of the Holocene has changed as documented by the advances and retreats of glaciers (Denton and Karlen, 1973; Grove, 1979; Beget, 1983). In North America, the Holocene was subdivided into three phases (Bryan and Gruhn, 1964): the Anathermal, from 10,000 to 7,500 years ago; Altithermal, from 7,500 to 4,000 years ago; and Medithermal, from 4,000 years ago to the present. There are several views of the climate in Arizona during these three phases of the Holocene. Bryan (1941) and Antevs (1955, 1962) believed that the Anathermal climate had temperatures initially like those of the present, but that temperatures increased over time and the moisture levels became subhumid to humid. Martin (1963) believed that the climate of 10,500 to 8,000 years ago, the Anathermal, was dry as now. Van Devender and Spaulding (1979) indicated that the late Wisconsin winter precipitation and colder winter temperatures continued from 11,000 to 8,000 years ago. The colder temperatures were caused possibly by a decrease in the continental glaciers so that they no longer prevented cold polar air masses from entering the mid-continent. Cole (1982), on the other hand, suggested an early Holocene, about 9,000 years ago, increase in the summer monsoon precipitation in the eastern Grand Canyon.

Bryan (1941) and Antevs (1955, 1962) believed that the Altithermal was a long warm, dry period. Martin (1963) challenged the Bryan-Antevs interpretations, especially in southern Arizona where the climate of roughly 8,000 to 4,000 years ago was subpluvial, characterized by heavy summer rains. The differences between the two views lie largely in the interpretations of arroyo cutting and filling in response to climatic change and in the interpretation of pollen data. Mehringer (1967a) attempted to reconcile Martin's (1963) and Antevs' (1955, 1962) views by suggesting that the Altithermal ended 500 to 1,000 years earlier than Antevs proposed. Thus, according to Mehringer (1967a), the early Medithermal that Antevs reported as being cool-moist was actually late Altithermal. Furthermore, the pollen record in southern Arizona between Altithermal years 7,000 to 5,000 is sketchy due to limited samples (Mehringer 1967b; Haynes, 1968). Haynes (1968) suggested two climatic episodes within the Altithermal that may have contributed to the different interpretations of pollen data. The first part of the period was relatively dry while the latter part was characterized by more effective moisture, according to Haynes (1968). More recent fossil plant data from packrat middens led Van Devender (1977) and Van Devender and Spaulding (1979) to conclude that the present climatic pattern was established after about 8,000 years ago. At that time the amount of winter precipitation diminished and the summer monsoon expanded, giving Arizona its characteristic biseasonal moisture pattern. Van Devender and Spaulding (1979) also tended to support Martin's (1963) interpretation of the middle Holocene climate by suggesting that more monsoon rain fell than now.

Bryan (1941) and Antevs (1955, 1962) believed that the Medithermal in the Southwest was moderately warm with periods more moist and others more dry than now. Martin (1963) considered the climate of this period to resemble closely the current conditions.

Regardless of the finer points of interpreting Holocene climatic conditions in Arizona, the climate did fluctuate both in effective moisture and in temperature. And these differing climatic conditions had an effect on Arizona soils.

The mouth of Madera Canyon, Santa Rita Mountains on the Santa Rita Experimental Range. In the foreground is Sonoran desert scrub; on the alluvial fan, desert grassland; and on mountains in the background, oak-pine woodland. (Photo by Michael Parton)

4

Natural Vegetation of Arizona

Introduction

Arizona has 3,666 species of native and naturalized plants in 1,003 genera and 145 families (Lehr and Pinkava, 1980). This rich variety of flora is equalled by few other regions of the United States. The composites, the grasses, and the legumes constitute the three largest families. In 1969, Texas and California were the only states with more than the 381 grass species recorded for Arizona. The cactus family, especially well-represented in Arizona, has 74 native species. Arizona also is rich in members of the fern family and fern allies. Nearly 100 species are known.

At Yuma, in the southwest corner of Arizona, the elevation is about 30 m (100 ft) above sea level, while slightly more than 400 km (250 mi) northeast, the San Francisco Peaks rise to more than 3,820 m (12,600 ft). The ecologist recognizes a rule-of-thumb that states that every 300 m (1,000 ft) in rise in altitude is equivalent to moving about 480 km (300 mi) distance north. Thus, while Yuma and the San Francisco Peaks are only some 400 km (250 mi) apart geographically, they are some 6,000 km (3,750 mi) apart ecologically. Yuma has a subtropical climate, but on top of the San Francisco Peaks the climate is similar to that of northern Canada and Alaska.

Distribution of plant life is dependent on a number of interacting environmental factors that dictate the type of vegetation established. Temperature, precipitation, slope, soil, animals and man all are part of the complex of interacting factors. Some of these factors in the Southwest are extreme and often are decisive in determining the type of vegetation that grows in a given area.

Forests generally grow where the soil is moist year-round, but especially when it is moist during the growing season. Grasslands, woodlands and chaparral occupy areas where an extreme dry period occurs during the year, or where the soil does not remain moist because of limited precipitation. Deserts, regions of sparse vegetation, are in areas with warm temperatures and precipitation so limited that the soil usually is not moistened beyond a depth of several centimeters (a few inches).

Quite often deserts are thought of as areas without vegetation. If this were true, deserts would be rare because even under the most extreme climatic conditions some plant life can exist. Shifting sand dunes, the popular conception of a desert, are deserts of extreme barrenness because of the shifting sands rather than lack of moisture. The Southwest "deserts" are better thought of as semideserts or semiarid areas. Vegetation on Arizona deserts is sparse compared with vegetation in the eastern United States, but Arizona deserts rarely are devoid of all vegetation.

Some of Arizona's flora comes from other regions to the north, east, south and west. Regions to the north contribute species to two main vegetation associations, the coniferous forests and the cold desert. The coniferous forest region in Arizona culminates along the high Mogollon Rim, the escarpment forming the southern limit of the Colorado Plateau. This great belt of coniferous forest stretches along the rim from the New Mexico boundary to the vicinity of Williams. Less extensive forests grow on the higher regions north and south of the Mogollon Rim country.

In addition to the southern extensions of the northern coniferous forest flora there is also an extension of the Great Basin Desert flora, mostly from Utah. This cold, winter desert follows the low-lying land of the eroded river valleys in northern Arizona. The Great Basin Desert landscape is outstanding along the northern reach of the Colorado River and the Little Colorado River, its main tributary in Arizona.

Grasslands in northern Arizona are closely related to the Great Plains. This vegetation is usually thought of as being the western extension of the short-grass plains that occupy a region just east of the Rocky Mountains.

Both desert and grassland vegetation extend into southeastern Arizona from Chihuahua. Some authorities believe that the small area of southeastern Arizona labeled on some maps as the northwestern-most extension of the great Chihuahuan Desert actually is a transition zone between the Chihuahuan and Sonoran deserts. Others believe that it is true Chihuahuan Desert. Nonetheless, whether it is Chihuahuan or transition, the desert scrub in that part of Arizona is similar to that in northern Mexico. The semidesert grasslands of northeastern Sonora and western Chihuahua along the east side of the Sierra Madre Occidental extend into southeastern Arizona.

Desert vegetation in Arizona is floristically related to areas south and west. Sonoran Desert vegetation occupies the southwestern quarter of the state and extends southward into Sonora and Baja California. The Mohave Desert of California extends from the west into northwestern Arizona up to the western base of the Hualapai Mountains in Mohave County and into the lower Grand Canyon. The interior chaparral, which occurs in central Arizona below the forests and above the deserts, has floristic affinities with the California chaparral to the west.

Arizona is well-endowed with national parks and monuments (See Figure 18). Two monuments were established exclusively because of the vegetation they contain. Saguaro National Monument near Tucson has many species of desert vegetation including a large stand of the giant saguaro cactus. Organ Pipe Cactus National Monument in southwestern Pima County along the Mexican border also has a large and varied array of desert vegetation including the only abundant stand of organ pipe and senita cacti in the United States.

Grand Canyon National Park and Chiricahua National Monument (in Cochise County) were established because of topography, but each has vegetation of considerable interest. Several national forests cover extensive areas of Arizona.

Cities, counties, the state and other organizations also have established parks and recreation areas that exhibit natural vegetation. Tucson Mountain Park has abundant and diverse desert vegetation. The Arizona-Sonora Desert Museum near Tucson has assembled a large collection of plants labeled in detail, while the Desert Botanical Garden at Papago Park near Phoenix has a large collection of labeled cacti. Native desert plants are the focus at the Tucson Botanical Garden. The Boyce Thompson Southwestern Arboretum, a research and teaching arboretum that has many public events and tours, is about 96 km (60 mi) east of Phoenix near Superior. The arboretum exhibits about 1,500 plant species belonging to 600 genera in 125 families. These plants are not only from the arid and semiarid regions of the U.S. Southwest, but from similar regions around the world.

Bioclimatic Classification of Vegetation

Because of its vegetal diversity, Arizona is and has been of particular interest to plant ecologists, as evidenced by some of the earliest research on bioclimatic classification of vegetation (Merriam, 1890) in the United States. Merriam (1890) studied the San Francisco Peaks and surrounding north-central Arizona to develop his classical concept of vertical life zones. Figure 33 is a diagram of the vertical zonation of these life zones in the San Francisco Peaks. Figure 34 is a similar representation of the Pinaleno and Santa Catalina mountains in southeastern Arizona. Although the Merriam life zone concept has shortcomings, as reviewed and discussed by Lowe (1972), it was expanded upon by other early investigators and has been used considerably. It is particularly useful in the western United States by providing a direct method for describing vegetation distribution.

More recently, Shreve (1942) named and defined nine principal types of vegetation based on altitude. Nichol (1937, 1952) described and mapped the natural vegetation of Arizona while Kuchler (1964) prepared a map and accompanying manual of the potential natural vegetation of the conterminous United States that is well known and often referenced. A 1:500,000-scale vegetation map of Arizona was prepared by Brown (1973) that was accompanied by a manual published by Lowe and Brown (1973) that described and illustrated Arizona's natural vegetative communities (Plate 11). Brown, Lowe and Pase (1977) published a map of biotic communities of the Southwest. This map has a scale of 1:1,000,000, the same scale as the Arizona General Soil Map. It includes all of Arizona and New Mexico, and parts of Southern California, southern Nevada, southern Utah, southern Colorado, West Texas and northern Mexico. A recent book (Brown, 1982) describes in detail the vegetation types shown on this map.

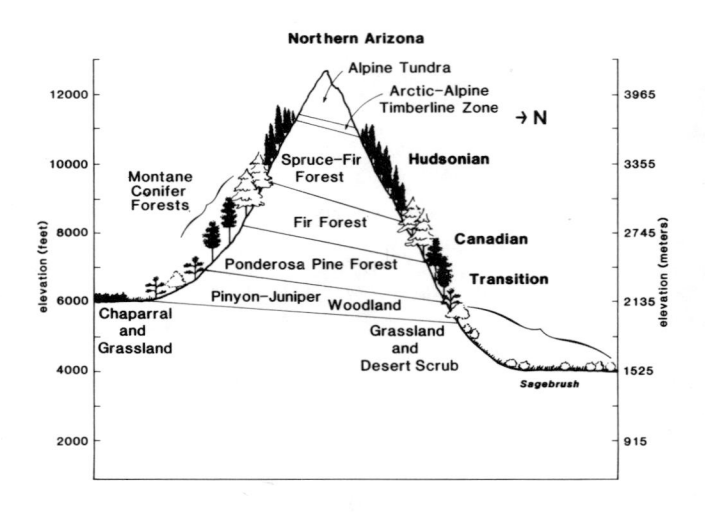

FIGURE 33. Altitudinal Distribution of the Major Vegetation Zones of the San Francisco Peaks in Northern Arizona (after C. H. Merriam, 1890)

The following descriptions of the Arizona natural vegetation zones are based mostly on those of Lowe and Brown (1973), although supplemented by additional references.

Alpine Tundra

The alpine tundra zone in Arizona is restricted to about 650 ha (1,600 ac) on the San Francisco Peaks above the timberline at elevations exceeding 3,340 m (11,000 ft) on Humphreys, Agassiz and Fremont peaks. Two plant associations were described, the alpine rock field and the alpine meadow (Little, 1941), but the zone has no forest overstory species. The alpine rock field association is the more extensive areally, covering most of the peaks above the timberline, but has much less dense cover and occupies a more unstable and less weathered substrate than the alpine meadow association. It is characterized by lichens and mosses on rock outcrops and in crevices, and by vascular plants scattered among boulders if soil is sufficient. Normally, the alpine rock field association merges into and is succeeded by the alpine meadow association, the species of which start as pioneers on rocky slopes and at the base of rock slides, then spread vegetatively to form mats.

These vegetative communities represent a relict alpine tundra flora definitely related to the high peaks tundra of the Rocky Mountains north and northeast (Moore, 1965). Twenty of about 50 species of the alpine tundra flora on San Francisco Peaks are arctic-alpine disjuncts that also live in arctic tundra zones. Fifteen of the 20 are circumpolar, growing in Arctic Eurasia as well as Arctic North America. Moore (1965) believed that at least 90 percent of the alpine tundra vascular species on San Francisco Peaks migrated from the north during Pleistocene time, possibly as recently as 65,000 to 75,000 years ago, coinciding with the last period of glaciation described by Sharp (1942). Updike and Péwé (1976) provided evidence of more recent glaciation. However, this last glaciation in the San Francisco Peaks was fairly limited in areal extent. Moore (1965) suggested further that relict alpine tundra on San Francisco Peaks has been losing its true alpine tundra character for at least 10,000 years.

The climate is characterized by relatively high mean annual precipitation that ranges from 890 to 1,020 mm (35 to 40 in), most of which is snow, and cold temperatures. This zone probably has a pergellic soil temperature regime, a mean annual soil temperature of less than 0 C (32 F) at a depth of 50 cm (20 in) or the soil-bedrock interface. The soils are shallow and rocky. The area is included in mapping unit FH2, Sponseller-Ess-Gordo Association.

Spruce-Alpine Fir Forest

Spruce-alpine fir forests cover about 97,130 ha (240,000 ac) on and around the summits of the highest mountains, including San Francisco Peaks and the Chuska, White, Pinaleno and Chiricahua mountains, and on the large summit area of the Kaibab Plateau. These Rocky Mountain forests reach their southernmost extension in Arizona and New Mexico (Dye and Moir, 1977). Spruce-alpine fir forests generally lie between 2,430 to 2,730 m (8,000 to 9,000 ft) and extend to the mountain summits, except for San Francisco Peaks where the upper limit is approximately 3,490 m (11,500 ft). The mean annual precipitation ranges from 760 to 1,140 mm (30 to 45 in), much of it as snow, and exceeds mean annual potential evapotranspiration (Beschta, 1976).

Seven coniferous and one deciduous species variously mixed characterize these forests. The principal boreal conifers are Engelmann spruce, blue spruce, corkbark fir, white fir, Douglas fir, bristlecone pine and limber pine. Quaking aspen is the dominant deciduous species, both intermixed with various coniferous species and in pure stands. Dense overstories common to these forests severely limit or prevent growth of herbaceous vegetation. Quaking aspen is considered to be a seral species that invades an area following a disturbance such as fire.

Moir and Ludwig (1979) have classified the Lowe and Brown (1973) spruce-alpine fir forests into eight spruce-fir and 11 mixed conifer habitat types based on the concept of Daubenmire and Daubenmire (1968). The dominant climax species within the spruce-fir habitats are either Engelmann spruce or corkbark fir. Climax dominants or codominants in the mixed conifer habitats include white fir, blue spruce and Douglas fir. Kuchler's (1964) southwestern spruce-fir forest and spruce-fir-Douglas fir forest zones are included within the spruce-alpine fir forest zone on the vegetation map (Plate 11).

The spruce-alpine fir forests are primarily in Plate 1 mapping units FH2, Sponsellor-Ess-Gordo Association, and FH4, Soldier-Lithic Cryoborolls Association.

Montane Conifer Forest

Most of the montane conifer zone vegetation grows along the southern rim of the Colorado Plateau in central Arizona for nearly 360 km (225 mi) as an unbroken ponderosa pine forest. In southern Arizona, the montane conifer forest grows

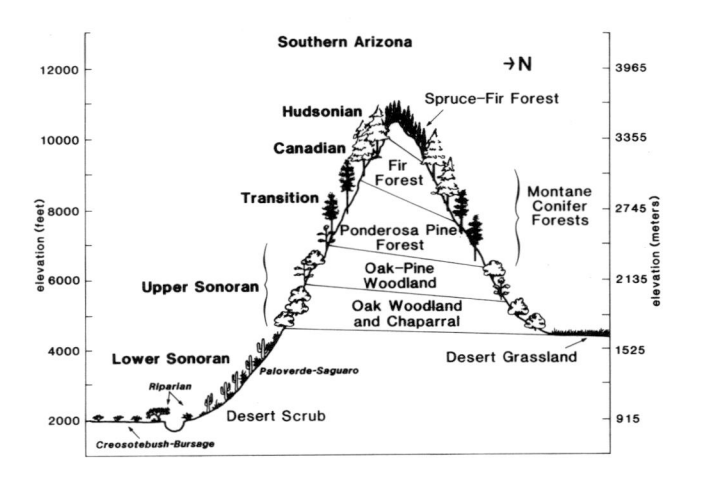

FIGURE 34. *Altitudinal Distribution of the Major Vegetation Zones of the Santa Catalina Mountains in Southern Arizona (after C. H. Lowe, 1972)*

primarily on the larger mountains as "islands." Characteristically this conifer association grows at elevations between 1,820 to 2,430 m (6,000 to 8,000 ft), but may be as low as 1,670 m (5,500 ft) on north-facing slopes or as high as 2,730 m (9,000 ft) on south-facing slopes and in the southern mountains. This association covers about 2,023,500 ha (5 million ac) in Arizona.

The montane conifer forest grows in a climate where moisture is relatively limited. The average annual precipitation is 510 to 760 mm (20 to 30 in) and is rarely lower than 460 m (18 in) or higher than 840 mm (33 in) (Beschta, 1976).

For the most part, ponderosa pine dominates the montane conifer forest, particularly in the warmer and drier areas. On north-facing slopes and at higher elevations ponderosa pine and Douglas fir and white fir grow in varying mixes, but the firs dominate north-facing slopes at the highest elevations. Other tree species include limber pine, southwestern white pine, Gambel oak, silver-leaf oak, madrone, locust, bigtooth maple and quaking aspen. Many stands of ponderosa pine are relatively open or park-like, in contrast to the closed-canopied spruce-alpine fir forest. This permits the growth of grasses, forbs, shrubs and broadleaf trees as understory. Sometimes the understory may be multilevel.

The montane conifer forests coincide approximately with Kuchler's (1964) Arizona pine forest.

The montane conifer forests grow in soils in parts of all FH, Frigid Subhumid, mapping units on Plate 1.

Riparian Deciduous Woodland

Riparian vegetation occurs in or near drainageways and floodplains and is characterized by plant species different from immediately surrounding nonriparian vegetative species (Lowe, 1972). Although riparian communities comprise a limited geographic area in total, they are significant because of their landscape importance and recreational use, and their value as wildlife habitats. Estimates of the size of this zone in Arizona range from 113,320 to 129,500 ha (280,000 to 320,000 ac).

Three elevational divisions of riparian vegetation in Arizona are recognized (Campbell, 1970). The communities are those below 1,060 m (3,500 ft), those between 1,060 and 2,120 m (3,500 and 7,000 ft) and those between 2,120 and 3,030 m (7,000 and 10,000 ft) above sea level. Below 1,060 m (3,500 ft) many ephemeral streams have broad alluvial floodplains that support high densities of deep-rooted trees and shrubs, including mesquite, acacia, saltcedar, paloverde, cottonwood, willow, sycamore and other species (Figure 35). Between 1,060 and 2,120 m (3,500 and 7,000 ft) the riparian vegetation consists of the greatest number of plant species and has the greatest canopy cover. Typical vegetation at these elevations is cottonwood, willow, sycamore, ash and walnut with three or four species often together. Above 2,120 m (7,000 ft) willow, chokecherry, boxelder, Rocky Mountain maple and various conifer tree species are dominant along stream channels. Climates of the riparian associations vary greatly because of large differences of elevation, latitude and distribution of mountains and highlands.

Vegetation native to many riparian areas in Arizona has changed markedly during the past 100 years, primarily the result of changes brought about by immigrants. Riparian woodlands have dwindled as water tables fall, and flood control and various other water management activities expand. And in some areas stands of mesquite, cottonwood and willow have been nearly replaced by saltcedar (Hasse, 1972).

Riparian deciduous woodland communities grow in the soils of some of the floodplain mapping units such as HA1 and TS2 (Plate 1), Torrifluvents associations, and along stream channels in many other mapping units.

Pinyon-Juniper Woodland

The pinyon-juniper woodland zone encompasses more than 5,665,800 ha (14 million ac) and is adjacent to and surrounds montane conifer and spruce-alpine fir forests in Arizona. Pinyon-juniper woodlands are mostly in the northern half of Arizona at elevations between 1,370 and 2,280 m (4,500 and 7,500 ft). In southern Arizona these woodlands merge with the chaparral zone.

Pinyon-juniper woodlands are in semiarid to dry, sub-humid climatic regions that have relatively broad fluctuations of precipitation, temperature, evapotranspiration and

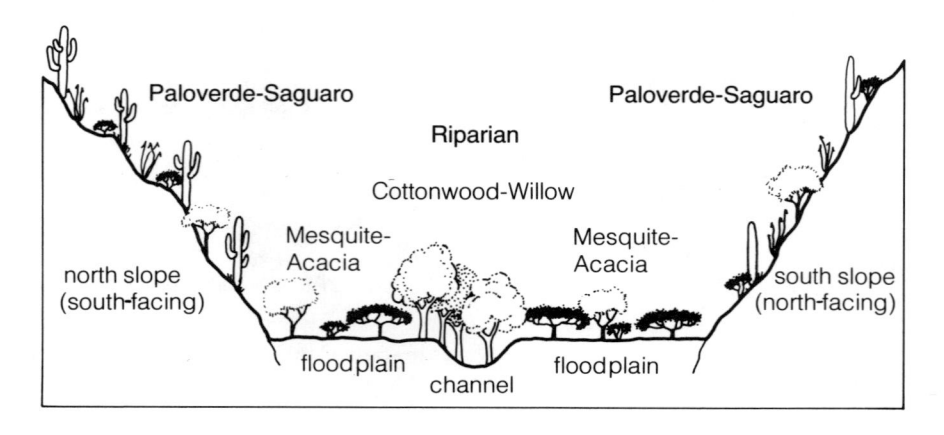

FIGURE 35. Low Elevation Riparian Woodland (after C. H. Lowe, 1972)

wind. Annual precipitation varies from 300 to about 530 mm (12 to about 21 in). Generally mean annual air temperatures are about 3.3 C (6 F) above the adjacent, higher montane conifer forests and about 2.8 C (5 F) below the lower vegetation zones (Beschta, 1976).

Three juniper and two pinyon species are the dominant trees in this zone. North of the Mogollon Rim, Utah and one-seed juniper are intermixed with pinyon and to the south alligator juniper grows. Common pinyon (Pinus edulis) is the characteristic species throughout nearly the entire zone. Singleleaf pinyon grows locally intermixed with Utah juniper, mostly in northwestern Arizona. Junipers commonly are dominant below 1,820 - 1,970 m (6,000 to 6,500 ft), while pinyon attain greatest size and density above 1,970 m (6,500 ft) (Lowe and Brown, 1973). Grassland, chaparral or desert scrub may form an understory beneath and between woodland trees, depending on the area.

Pinyon-juniper woodlands are mostly in the MS, Mesic Subhumid Soils, mapping units on Plate 1.

Encinal and Mexican Oak-Pine Woodland

These woodlands, for the most part, are in southeastern Arizona. Their greatest development is on foothills and lower slopes of the larger mountains at elevations between 1,210 and 2,120 m (4,000 and 7,000 ft). These communities in Arizona represent their northern extension from the Sierra Madre region in northwestern Mexico and cover about 870,270 ha (2,150,400 ac).

Encinal and Mexican oak-pine woodlands receive mean annual precipitation of between 300 and 610 mm (12 and 24 in). The woodland habitat has warmer winter temperatures than the equivalent precipitation zone farther north that supports pinyon-juniper communities. Also, more precipitation is associated with the monsoon season during July through September.

The encinal communities are composed of evergreen oaks or of mixtures of oak, juniper and Mexican pinyon. The dominant oak species are Emory, Arizona and Mexican blue oak. Mexican oak-pine woodland communities also contain evergreen species including Chihuahua and Apache pine, Mexican pinyon and Arizona madrone. Silverleaf oak is the dominant oak species and Emory and Arizona oak also are present. The Mexican oak-pine woodland is above the encinal on mountain gradients, although not as widely distributed as the encinal woodland.

Kuchler's (1964) oak-juniper woodland corresponds approximately with the encinal and Mexican oak-pine woodland vegetation class.

These vegetation communities are in the Plate 1 MH, Mesic Subhumid Soils, mapping units in southeastern Arizona.

Interior Chaparral

The chaparral vegetation zone is mostly on rough, discontinuous mountainous terrain south of the Mogollon Rim, generally extending in a discontinuous band across the state from Seligman in the northwest to Safford in the southeast. Chaparral grows at elevations ranging from 910 to 2,430 m (3,000 to 8,000 ft) below woodland or coniferous forest and above grassland or desert scrub. Estimates of the area covered by chaparral in Arizona vary from 1,214,000 ha (3 million ac) to 2,428,200 ha (6 million ac), although the former and more recent value is thought to be the more accurate (Carmichael et al, 1978). The broad variance of estimates is caused by lack of common classification criteria.

Mean annual precipitation in the chaparral communities ranges from 380 mm (15 in) to more than 640 mm (25 in). The climate is further characterized by a cool, wet period from November until March, followed by a warm, dry period until the summer rains begin in July.

Chaparral consists of deep-rooted evergreen shrubs and trees that have broad, sclerophyllous leaves. They develop best on deep soils or on deeply weathered rock mantles. Although 50 or more shrub species are in the chaparral vegetation zone in Arizona, generally fewer than 15 are important in terms of density (Charmichael et al, 1978). Shrub canopy cover may vary from less than 40 percent on dry sites to more than 80 percent on the wetter sites. Annual and perennial grasses and forbs may grow where the overstory canopy is only moderately dense or is open.

Chaparral grows mostly below the Mogollon Rim in central Arizona within the soils of Plate 1 MH, Mesic Subhumid Soils, mapping units.

Grasslands

About 30 years ago, grasslands covered nearly 25 percent of the state, an estimated 7,369,000 ha (18,210,000 ac). But invasion by plant species from other proximate vegetation associations may have reduced the area covered by grasslands. Arizona has three types of grassland: mountain meadow, plains and desert.

Mountain meadow grasslands are scattered throughout the spruce-fir and montane forests at elevations ranging from 2,280 to 3,030 m (7,500 to 10,000 ft). Unlike plains and desert grasslands, mountain meadow grasslands receive relatively high average annual precipitation, 510 to 890 mm (20 to 35 in). The most extensive areas are in the White Mountains and on the Kaibab Plateau. These grasslands also occur in some higher, isolated ranges, such as the Pinaleno and Chiricahua mountains in southeastern Arizona.

Many grass species grow in the meadows. Grasslands in drier, warmer montane coniferous forest zones contain species such as Arizona fescue, pine dropseed and blue grama. Boreal grasses, such as mountain timothy, meadow hairgrass and mountain bluegrass, tend to dominate the higher, wetter, colder spruce-fir forests. In some wetter areas, including those with high water tables, forbs, mountain clover, wild daisy, mountain dandelion, sedges and a number of other species may outnumber grasses by as much as 10 to 1.

Mountain meadow grasslands dominate the soils of Plate 1 mapping unit FH 8, Gordo-Tatiyee Association, and are in soils that are important inclusions of mapping unit FH 4, Soldier-Lithic Cryoborolls Association.

The plains grassland extends southwest into Arizona where it grows mostly in the northeast as nearly uninterrupted ground cover between 1,520 and 2,120 m (5,000 and 7,000 ft). The desert grassland is principally in southeastern Arizona between about 910 to 1,520 m (3,000 to 5,000 ft) as Chihuahuan Desert Grassland. It occurs also in the northwestern quarter of the state, where it extends to the eastern edge of the Mohave, southern edge of the Great Basin and northern edge of the Sonoran deserts.

The plains grassland zone receives about 430 mm (17 in) mean annual precipitation with extremes of 250 and 510 mm (10 and 20 in), while the desert grasslands are the most arid of all North American grasslands, receiving only 200 to 380 mm (8 to 15 in) precipitation.

Plains grasslands in Arizona consist primarily of short grama grass species, blue, black and sideoats, and shrubs are absent or nearly absent. Desert grasslands are essentially pure stands of grass in some places, in other places, an open savanna with grasses beneath oaks or mesquites is common, and in still other places, the grasses are interspersed with a variety of low-growing trees or shrubs. Threeawn and tobosa species together with grama grasses dominate desert grasslands.

Changes in desert grassland vegetation during the last 100 years were two principal kinds: invasion by woody species, and change in the mix and density of nonwoody species (Humphrey, 1958). The first of these changes is well-documented and includes photographic evidence (Hastings and Turner, 1965; Martin and Turner, 1977). One or more of five explanations for these changes were suggested by various researchers, as reviewed by Humphrey (1958), and include 1) grazing by domestic livestock, 2) interspecies competition or lack of it, 3) effects of rodents, 4) changes in climate, and 5) suppression of natural grassland fires. Although the most popular explanation is overgrazing by livestock, Humphrey (1958) suggested that suppression of natural fires may be the most important factor. Before immigrant settlement of the Southwest, fires were frequent and widespread and may have restricted shrub invasion.

Plains grasslands characteristically are associated with soils in the Plate 1 MS, Mesic Semiarid Soils, mapping units. Desert grasslands grow mostly on more humid soils in southeastern and northwestern Arizona within the TS, Thermic Semiarid Soils, mapping units of Plate 1.

Desert Scrub

Desert scrub vegetation occupies about 14,447,800 ha (35.7 million ac) in Arizona. Making a clear distinction between desert scrub and grassland vegetation zones, however, often is difficult because of the invasion of the grasslands by desert scrub vegetation. For the purposes of this discussion, desert scrub vegetation has been broken into four classifications: Great Basin, Mohave, Sonoran and Chihuahuan (Plate 12).

Great Basin desert scrub is limited mostly to elevations between 910 and 1,970 m (3,000 and 6,500 ft) north of the Colorado and Little Colorado rivers where average annual precipitation is between about 180 to 300 mm (7 and 12 in) and is more evenly distributed throughout the year than in the other Arizona desert regions. The Great Basin Desert is dominated mostly by shrubs of relatively low stature. Pure, unbroken stands of big sagebrush are commonly associated with the Great Basin Desert. Blackbrush and shadscale are characteristic and rabbitbrush, horsebrush, winterfat and Mormon-tea are important shrub species. Much of the Great Basin desert scrub in Utah and Nevada is salt tolerant since numerous salt-affected soils are associated with the several large, internal drainage basins in those states. This situation occurs on a much smaller scale in Great Basin areas in northern Arizona.

Great Basin desert scrub grows mostly in soils in the Plate 1 MA, Mesic Arid Soils, mapping unit, and to a lesser extent in the Grand Canyon, Plate 1 mapping unit TA1, Torriorthents-Camborthids-Rock Outcrop Association.

Mohave desert scrub extends into northwestern Arizona, mostly Mohave County, from southeast-central California and southern Nevada. Mohave desert scrub in Arizona grows usually between about 300 and 1,210 m (1,000 and 4,000 ft) elevation. Average annual precipitation is 130 to 280 mm (5 to 11 in), most of which falls in the winter. Mohave sage and woolly-fruited bursage are near-endemic shrubby species. Although the Mohave Desert is shrub-dominated and lacks giant cacti and many subtropical desert tree species, several large plants such as the Joshua tree and Mohave yucca are endemic, and catclaw and mesquite grow along washes. Creosotebush and/or white bursage often dominate extensive areas. Some species, such as blackbrush and winterfat, associated with the Great Basin Desert, grow in the northern Mohave Desert.

The Mohave desert scrub grows in soils in several Plate 1 HA, Hyperthermic Arid Soils, and TA, Thermic Arid Soils, mapping units in northwestern Arizona, mostly near the Colorado River.

The Sonoran Desert, a large region with more than two-thirds lying in northwestern Mexico (Sonora, Baja California), contains several subdivisions. The northernmost two are in Arizona: the Arizona Upland in southern Arizona and the Lower Colorado desert subdivision in southwestern Arizona (Shreve and Wiggins, 1964).

The Arizona Upland subdivision has the most structurally diverse vegetation in the United States. It includes one of the most famous species of succulents, the giant saguaro cactus (Lowe and Brown, 1973). Other important species are organ-pipe, ocotillo and cholla cacti, foothill and blue paloverde, ironwood, mesquite and creosotebush.

This Sonoran subdivision is mostly between 150 and 1,210 m (500 and 4,000 ft) where average annual precipitation ranges from about 130 to 150 mm (5 to 6 in) in Yuma County to about 280 to 330 mm (11 to 13 in) on some southeastern mountain ranges. Although winter and summer precipitation amounts are approximately equal, the proportion of summer rain does increase across the state in an easterly direction.

The Lower Colorado subdivision is the largest of the Sonoran Desert. In Arizona it encompasses the lower drainages of the Colorado and Gila rivers. Elevation ranges from about 30 m (100 ft) near the Colorado River to 910 m (3,000 ft) in the eastern valleys. It is one of the most arid regions in the

Sonoran Desert, with average annual precipitation that varies from 80 to 250 mm (3 to 10 in). Precipitation increases from west to east and with elevation.

Vegetation of the Lower Colorado subdivision is dominated by creosotebush and its major associate, white bursage. Low, open stands, of which these two shrub species constitute at least 90 percent, often cover thousands of hectares on the intermontane plains. Smaller areas that have low, undrained and salt-affected soils commonly are dominated by saltbush, desert-thorn and sometimes mesquite.

In addition to perennial vegetation, the Sonoran Desert also has annual species, sometimes referred to as "ephemerals" since they grow only after brief moist periods and are short-lived. The biseasonal precipitation distribution in the Sonoran Desert, especially in the Arizona Upland subdivision, produces two distinct floras of ephemerals (Lowe, 1959). One is derived from the ancient Madro-Tertiary flora to the south, being naturally attuned to summer moisture and hot temperatures. The other is derived from the Arcto-Tertiary flora of the north, being naturally attuned to winter and early spring moisture and cool temperatures. Shreve and Wiggins (1951) concluded that this pattern of summer and winter ephemerals, so clearly related to the seasonal distribution of precipitation, represents a geographic and climatic segregation that has endured since early Pliocene time.

The Arizona Upland subdivision desert scrub grows in the higher elevation Plate 1 HA, Hyperthermic Arid Soils, mapping units and the lower elevation TS, Thermic Semiarid Soils, mapping units in southern Arizona. Lower Colorado subdivision desert scrub grows mostly in valley areas of the Plate 1 HA, Hyperthermic Arid Soils, mapping units.

The Chihuahuan Desert covers most of north-central Mexico. The question of this desert extending across West Texas and southern New Mexico into Arizona is a source of conflicting opinions since in Arizona it represents more a zone of transition between the Chihuahuan and Sonoran deserts than true Chihuahuan Desert (Schmidt, 1979). The region in the southeastern corner of Arizona, mostly in Cochise County, referred to as the Chihuahuan Desert is at elevations between 970 and 1,520 m (3,200 and 5,000 ft) but mostly above 1,060 m (3,500 ft). Although the precipitation in this area ranges from 200 to 360 mm (8 to 14 in), most of the area receives less than 250 mm (10 in). More than half the precipitation falls July through September.

Creosotebush, tarbush, whitethorn and sandpaperbush are among the important shrub species in the Chihuahuan Desert in Arizona. These species frequently grow in essentially pure stands, often on limestone, or in complex community mixtures with mariola, ocotillo, allthorn, shrubby senna, whitebrush, desert zinnia and little coldenia. The Chihuahuan desert scrub also is characterized by numerous herbaceous root perennials, and by numerous species of small cacti.

Chihuahuan desert scrub in Arizona is mostly in the Duncan, San Simon and San Pedro valleys where it grows in the soils of several of the Plate 1 TS, Thermic Semiarid Soils, mapping units.

Prairie dog, Cynonomys gunnisoni, *at mouth of burrow. (Photo courtesy of Arizona-Sonora Desert Museum)*

5

Animals and Soil in Arizona

Introduction

Arizona's animal life includes about 751 vertebrate species and more than 20,000 invertebrate species. Among the vertebrates are 64 species of fish, 22 species of amphibians, 94 species of reptiles, 434 species of birds and 137 species of mammals (Lowe, 1972). This diversity results from the extremes in climate (Plates 5 through 10), topography (Plates 2 and 4) and vegetation (Plates 11 and 12) that create many different environments. These environments range from the hot, dry deserts at low elevations in southern Yuma County through rich upland deserts, grasslands and woodlands at the mid-elevations, to cold, moist montane and alpine habitats such as those in the San Francisco Peaks and the White Mountains. In some places the division between habitats is quite distinct, particularly where relief is abrupt such as it is in the San Francisco Peaks. It was there in 1889 that C. Hart Merriam made observations that eventually led to the Life-Zone System for classifying plant and animal associations. Vegetative communities, representative animals and soil types within each life zone are illustrated in Figure 36.

Biotic provinces also have been used to describe the distribution of animals. Biotic provinces differ from life zones in that life zones are fully biogeographic systems whereas biotic provinces are founded primarily upon the distribution of animals (Lowe, 1972).

The biotic provinces of Arizona (Plate 17) were described and mapped by Dice (1943). Descriptions of the four biotic provinces as given by Lowe (1972) appear below.

1. The Sonoran biotic province is desert, and, in Arizona, it is intended to be essentially the area of the state that is within the Sonoran Desert.

2. The Mohavian biotic province is essentially the Mohave Desert of other authors (*sensu* Shreve, 1942).

3. The Apachian biotic province, according to Dice, is intended to represent the "grassy high plains, and the mountains included in them, of southeastern Arizona, southwestern New Mexico, northeastern Sonora, and northwestern Chihuahua." This is the Yaquian biotic province of others.

4. The Navahonian biotic province (Dice, 1943:39), "... is characterized by pinyon-juniper woodland." The "lowest life belt" of the Navahonian is characterized by "arid grassland," and the highest, the "alpine life belt," is made up of treeless areas above timberline.

Factors involved in delimiting the distribution of animals are complex and include topographic, hydrologic, pedogenic, climatic and biotic elements. The general relationships between some of these factors can be seen by comparing Plates 2 through 10 and 13, and Plates 11, 12, and 17. Specific information about the associations between animal populations and plant communities in Arizona has been collected and organized by Patton (1978) and is available as a computer printout or on microfiche (see Patton, 1979).

Many animals that inhabit these zones or provinces directly affect the characteristics of soil. Soil properties in turn can influence the distribution and abundance of animals living in (Kevan, 1962; Kuhnelt, 1976) and on (Allen, 1962;

Wallwork, 1982) the soil. The general roles that animals play in the soil community and the effects of soil on the distribution and abundance of animals are the subjects of this chapter.

General Role of Animals in the Soil Community

The precise ways that soil organisms interact with plant material, each other and the soil are difficult to describe. This is due to the lack of information about soil biochemistry, physiology and ecology. However, many soil animals depend either directly or indirectly on dead plant tissues as sources of energy. Thus, the decomposition of organic litter on the soil surface is of critical importance in soil ecology. The processes of decomposition are controlled largely by soil organisms. Consequently, the soil surface and immediate subsurface are regions of greatest biological activity.

In addition to their varying roles as decomposers, invertebrates such as protozoa, worms, insects and mites also contribute to the soil community by mixing, loosening and aerating soil (Evans, 1948; Englemann, 1961). Vertebrates influence soils primarily through their burrowing activities, and all organisms contribute to the soil by adding organic matter and chemicals via their bodies, excrement and food residues. See Figure 37 for representative soil animals.

Soil Fauna: Classification

Soil fauna have been classified by numerous authors including Kevan (1962), Schaller (1968) and Wallwork (1970). Five major groupings are widely accepted: classification based on body size; time spent in the soil; location or habitat in the soil profile; feeding strategies; and method of locomotion in the soil (Wallwork, 1970).

Soil fauna generally are small and have simplified appendages (Kuhnelt, 1976). Body size ranges from 0.0002 cm (0.00008 in) to more than 20 cm (8 in) and can be divided into microfauna, mesofauna and macrofauna (Figure 38). Microfauna range in body size from 0.0002 to 0.002 cm (0.00008 to to 0.0008 in) and consist of protozoa. Mesofauna range from slightly more than 0.002 to 1 cm (0.0008 to 0.4 in) and include mites, springtails, spiders, pseudoscorpions, pot-worms, insect larvae and the smaller millipedes and isopods. Macrofauna are at least 1 cm (0.4 in) or greater and include earthworms, the largest insects and arachnids, and the soil-dwelling vertebrates.

The amount of time organisms spend in the soil ecosystem also is used as a criterion for classification. Some organisms such as protozoans, nematodes, isopods and mites spend their entire lives in the soil, whereas other organisms like ground-nesting birds may be only tangentially associated with the soil community.

Life Zones*	Grand Canyon Formations	Vegetative Communities	Representative Animals	Soils and Elevation
Hudsonian, Canadian and Transition	Kaibab Plateau, Kaibab Limestone, Toroweap Limestone	Spruce-Alpine Fir and Montane Conifer Forests	mountain lion, mule deer, red squirrel, chipmunk, porcupine, Mexican vole, Chipping Sparrow, Yellow-bellied Sapsucker and Rocky Mountain Jay	cold to cool Frigid Subhumid Soils (FH). Approximately 3200 m (10700 ft) to 2100 m (6900 ft)
Upper Sonoran	Coconino Sandstone, Hermit Shale, Supai Group	Juniper-Pinyon and Encinal and Mexican Oak-Pine Woodlands, Chaparral, Grasslands and Great Basin Desert Scrub	mule deer, pocket gopher, desert cottontail, silky pocket mouse, kangaroo rat,, Acorn Woodpecker, Vesper Sparrow and Plain Titmouse	warm Mesic Subhumid and Semiarid Soils (MH, MS) to hot Mesic Arid Soils (MA) Approximately 1500 m (5000 ft) to 1000 m (3300 ft)
Lower Sonoran	Redwall Limestone, Muav Limestone & Dolomite, Bright Angle Shale, Tapeats Sandstone, Tonto Platform, Grand Canyon Super Group, Inner Gorge, Colorado River, Vishnu Precambrian Group	Mohave, Sonoran and Chihuahuan Desert Scrub	spotted skunk, prairie dog, cactus mouse, desert shrew, Cactus Wren, Gila Woodpecker and American Kestrel	very hot Hyperthermic and Thermic Arid and Semiarid Soils (HA, TA, TS). Approximately 30 m (100 ft) to 1060 m (3500 ft) depending on slope
	Riparian	Riparian Deciduous Woodland	river otter, beaver, badger, raccoon, Blue-gray Gnatcatcher, Lucy's Warbler and Blue Grosbeak	soil types and elevations range from those of the Lower Sonoran up into those of the Canadian Life zones

*Not shown in the figure is the Arctic-Alpine zone. This zone covers about 650 ha (1600 ac) in Arizona, mostly above 3085 m (11500 ft) on San Francisco Peaks near Flagstaff. The vegetative community is Alpine-Tundra.

FIGURE 36. Cross Section of the Grand Canyon and Associated Life Forms

A third way of classifying soil fauna is by their location in the soil profile. Soil animals generally fit into one of three profiles: the epigeon, or vegetation layer above the soil surface; the hemiedaphon, or organic layers; and the euedaphon, or mineral layers. Soils have numerous microhabitats and soil organisms can be classified according to their use of them.

Feeding strategies also are used in classification of soil fauna. Feeding classifications include carnivores, which feed on other fauna; phytophages, which feed on green plant and woody material; saprophages, which eat dead and decaying material; microphytic-feeders, which feed on fungal hyphae and spores, algae, lichens and bacteria; and miscellaneous feeders, which fit into two or more of the other categories.

An organism's method of locomotion through the soil provides yet another way of classifying soil animals. Distinctions are made between burrowing animals and those that move through the soil by making use of pore spaces, cavities or channels. Other classification schemes are used when applicable such as the burrowing behavior and activity patterns of desert animals briefly described below.

Of the relationships that exist between animals and the soils of Arizona, the interactions between fauna and desert soils are the most remarkable. Deserts are defined in numerous ways including kinds of vegetation and soil, and amount of precipitation. In this discussion, deserts are defined as being areas where annual evapotranspiration exceeds precipitation. By this definition, much of Arizona is desert and includes parts of the Great Basin, Chihuahuan, Mohave and Sonoran deserts.

Desert soils are sandy in texture and mineral in character since there is little cover vegetation to provide organic material to the soil (Wallwork, 1976). A classification of desert soils is provided by Wallwork (1982). It is in these desert soils that biological activity and productivity reach their lowest levels in Arizona. High temperatures and low amounts of precipitation limit the growing season of plants and the activities of many animals to short intervals during the year. In the lower Sonoran Desert, where annual precipitation may reach 200 mm (8 in) per year (Crosswhite and Crosswhite, 1982), a variety of perennial plants such as sage, saltbush, creosotebush, acacia and numerous cacti and succulents thrive. These plants provide food and refuge for a diverse group of desert animals.

Survival of desert animals depends on their ability to avoid extremely high temperatures. Soil temperatures beneath the desert floor are lower than those at the surface and provide relief from heat for desert fauna. Three basic faunal groups are common in deserts: small mammals, reptiles and arthropods. Individuals in these three groups frequently survive by burrowing during hot times of the day and being active during cooler periods (see Wallwork, 1982). Desert tortoises, lizards, rodents, foxes, and termites, bees and wasps all use this method. The burrowing behavior and the activity patterns of desert animals provide a simple means of classifying desert soil dwellers. They fall into four classes: 1) burrowers that are diurnal, 2) burrowers that are nocturnal, 3) nonburrowers that are active on the soil surface, and 4) nonburrowers that are active below the surface. The fourth classification includes some snakes and birds that make use of burrows made by other animals.

Several groups survive through tolerance; being adapted physiologically to require little water (see Wallwork, 1982). Reptiles and insects are ideally suited in this respect because their uric acid needs little water for elimination from the body. In the case of some small mammals such as the kangaroo rat, the urine is extremely concentrated. These animals rely almost entirely on metabolic water or the preformed water on their food for their moisture requirements (see Chew, 1965).

This discussion has merely highlighted soil fauna classification systems currently used. For detailed coverage of this aspect of soil ecology the reader is referred to Burges (1969), Hooper (1969), Kevan (1962), Schaller (1968), Sims (1969), Wallwork (1969, 1970, 1982) and Williams, Davies and Hall (1969).

Role of Animals in the Soil Community

Invertebrates

About 90 percent by weight and 99.9 percent by numbers of the animals of Arizona live in the soil and, for the most part, are so small that they are unnoticed (Hole, 1980; Jenny, 1980). Microscopic protozoa live in water films on soil particles and feed on bacteria and yeasts. Snails, slugs and elongate animals such as earthworms, flatworms and nematodes degrade organic matter. Among the most abundant arthropods are soil mites, springtails and various forms of insects including ants, termites, beetles and flies. Some centipedes, millipedes, spiders, scorpions, harvestmen and, in wet soils, crayfish are abundant in Arizona. The live weight of soil invertebrates in moist soil is about 3,335 kg per ha (3,000 lbs per ac), or about the weight of three horses (Jenny, 1980). When soil becomes dry, faunal biomass decreases.

Special features of soil that are fashioned by soil invertebrates include the following: 1) pea- or bean-size granules of soil in the form of worm casts; 2) thumb-size blocky soil peds shaped by cicada nymphs while tunneling through B horizons (Hugie and Passey, 1963); 3) pits and channels that are excavated by antlions, spiders, beetles, ants, termites, scorpions, worms and, in wet soils, crayfish; and 4) filled channels, or tubules, and chambers, or glaebules, packed with excreta, brood structures or edible plant and animal materials.

Although the precise ways soil invertebrates interact in soil communities are mostly unknown, important groups have been identified and their roles established.

Protozoa. Protozoa are represented in the soil mainly by rhizopods, ciliates and flagellates. Literally millions of protozoa inhabit a square meter of soil (slightly more than a square yard). These organisms generally are regarded as bacteria-feeders. Some ingest organic litter and fungi and even may be able to digest cellulose. Protozoa are decomposer organisms. Their actions may contribute significantly

ARIZONA SOILS

to turnover of available nutrients and to enhancement of biochemical activity in soils (Stout and Heal, 1967).

Nematodes. The feeding habits of soil nematodes vary considerably (Freckman and Mankau, 1977). Some inhabit decaying organic matter and ingest liquified components of decomposing animals and plants. Others feed on bacteria or fungi, while still others parasitize plants, beetles, worms and slugs. The feeding activity of nematodes generally does not contribute significantly to the decomposition of organic material or to the formation of soil humus, but they do provide an important food source for other members of the soil community (Wallwork, 1970).

FIGURE 37. Representative Soil Animals
   37a Termites
   37b Harvester Ants
   37c Spadefoot Toad
   37d Kangaroo Rat
   37e Shovel-Nosed Snake

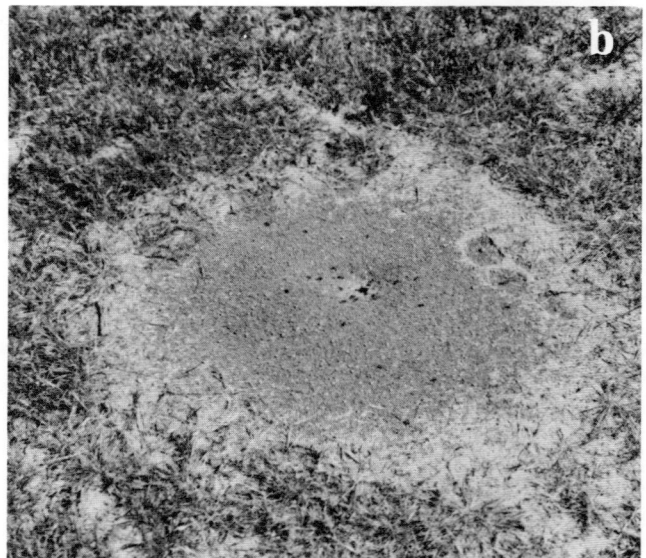

Worms. The best known of all soil animals are earthworms. They have a definite impact on the structure and properties of soils. Charles Darwin (1890) first examined the influence of earthworms on the decomposition of organic material. Subsequent investigations have examined the role of earthworms in the formation of organic-mineral complexes (Evans, 1948; Gerard, 1967; Satchell, 1967; and Thorp, 1949). Earthworms contribute to the soil community by ingesting and mixing decaying organic material and mineral soils. This action converts the bound nitrogen in organic complexes to ammonia, nitrites and nitrates that are more readily available to vegetation. Earthworms also influence soil drainage, fertility and stability (Wallwork, 1970) and promote the redistribution of organic debris.

Molluscs. Molluscs are represented in soil communities by slugs and snails. Land molluscs exhibit several types of feeding habits including herbivory, fungivory, predation and detritus feeding. This group probably influences soil most by feeding on surface vegetation, then moving into soil subsurface layers, thus incorporating organic material into the mineral structure of the soils.

Arthropods. Arthropods are another important and conspicuous part of the soil community. These organisms frequently dominate all other groups of the soil meso- and macrofauna, both in numbers of individuals and species. Shaller (1968) divides arthropods into crustaceans (wood lice), arachnids (scorpions, pseudoscorpions, harvestmen, soil spiders and mites), myriapods (millipedes and centipedes) and insects.

Crustaceans generally are not terrestrial and many have retained characteristics associated with aquatic life; most are not important soil organisms. However, the wood lice have established themselves as terrestrial forms and are abundant in a variety of soils ranging from humid litter in forests to the hot, dry soils of Arizona's deserts (Wallwork, 1982). They are omnivorous, feeding on dead plant material, feces and invertebrate carrion, and they play an important role in the decomposition of organic material.

Arachnids are predatory arthropods and frequently inhabit vegetation on the soil surface and loose leaf litter. The role of arachnids in the soil community has not been studied thoroughly, but they are important predators of insect populations and like all animals contribute organic matter to the soil when they die.

Millipedes and centipedes are common in many soils. Millipedes generally feed on plant detritus (Wallwork, 1982) to assist in the decomposition of organic matter, while centipedes are primarily predators. More information is required about both to determine the importance of their specific roles in soil ecology.

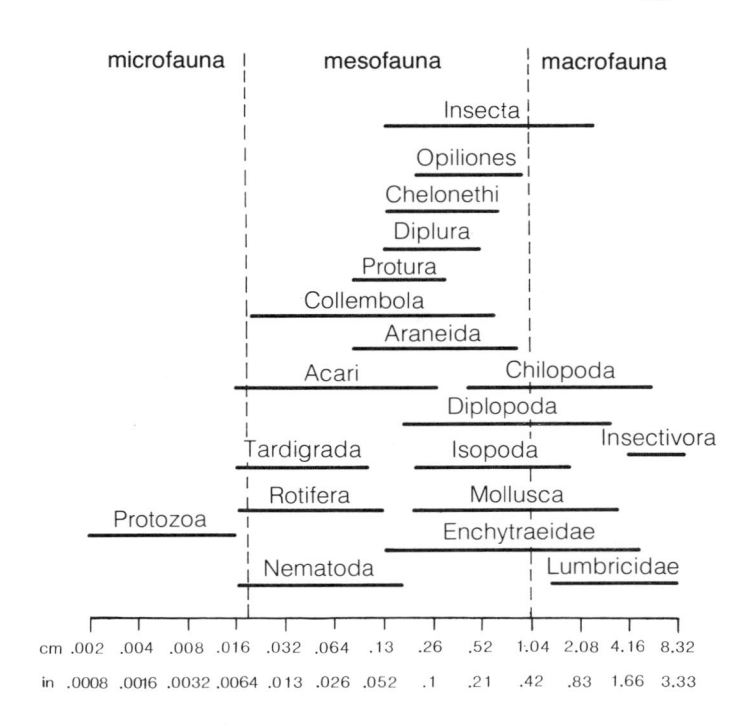

FIGURE 38. Soil Microfauna, Mesofauna and Macrofauna Classification (after J. A. Wallwork, 1970)

Numerous orders of insects are represented in soil fauna, but perhaps the most groups with respect to soil ecology are the termites and ants (Wallwork, 1982). Members of both groups construct numerous galleries in the soil, and many species transport large amounts of organic material from the surface to underground chambers; termites are particularly important in this respect (Schaefer and Whitford, 1981). These activities can contribute significantly to nutrient cycling.

Other species of insects may use the soil during part of their life cycles, larvae that overwinter in the soil, for instance. Most species, however, play minor roles in soil dynamics and should be considered passive members of soil communities.

Vertebrates

Larger vertebrates help shape the microtopography of soil landscapes. A moderate-size colony of prairie dogs may build clusters of mounds over an area of 1.2 ha (3 ac). Pocket gophers make conical mounds and thick, rope-like soil fillings in tunnels in basal layers of snow-banks. Wood or pack rats pile litter up to 1 m (3.3 ft) high and 2 m (6.6 ft) wide in a retreat or nest site.

Other vertebrates mix soils. Skunks, javelinas, coatis, whiptail lizards, roadrunners, Gambel's Quail and other animals dig and scratch through the upper soil layers in search of seeds, roots, tubers, insects, lizards and other small animals. Some snakes, such as the western shovel-nosed snake, and desert tortoises move or "swim" in sand. Spadefoot toads bury themselves in soil during dry periods and dig themselves out again when rains come.

Excavation of underground passageways and chambers affects the soil climate and alters soil horizons, in some instances to the point of obliterating argillic horizons. Pocket mice, ground squirrels, prairie dogs, skunks, pocket gophers, cottontail rabbits, Kit foxes, kangaroo rats, pack rats and wood rats make extensive systems of tunnels, shafts and chambers, and make dens under mounds in Fluvents. Badgers in pursuit of rodents enlarge burrows. Trampling by hooved animals collapses burrows and exposes soil material to wind erosion.

Animals also redistribute materials. Ground squirrels, rats, mice and gophers store plant materials, including seeds, in subsurface chambers. Bodily wastes of animals constitute local concentrations of nitrogen, phosphorus and potassium.

A great variety of rodents, birds, bats and other vertebrates, including coyotes, make their dens and nests in openings between masses or rocks. Mice and bats use crevices in the faces of high cliffs. Raptors build nests on ledges. In so doing, these animals introduce organic matter, some of which undoubtedly promotes weathering of bedrock and its conversion to new soil that supports vegetation. Bat and bird excreta are natural organic fertilizers, and large concentrations of these animals may have a significant influence on the chemical nature of soil.

The impact of horses, wild burros and, above all, cattle on soil landscapes has been enormous since the arrival of Europeans in Arizona. Loosening the sandy soils by overgrazing has accelerated both wind and water erosion. At some sites, sandy soil has blown short distances and collected in linear and oval deposits around mesquite trees. These deposits do not suppress growth of mesquite, but do provide an environment suitable for growth of new vegetation. These mounds are called "coppice dunes" and are Torripsamments (Gile and Grossman, 1979). The pattern of alternating bare and grassy patches and strips, then, may be ascribed to accelerated water erosion, resulting from overgrazing. Exposure of argillic soil horizons on the bare areas perpetuates movement of runoff and sediment into adjacent grassy areas where vegetative growth is fostered and soil is protected from erosion.

Although some vertebrates spend part of their time in the soil, they usually feed on the surface and their importance in the food web of soil ecosystems is often overlooked or deemed minimal. Because methods of study differ for vertebrates and invertebrates and because scientists tend to specialize, vertebrates are seldom included with invertebrates in investigations of soil fauna. Some vertebrates, however, do have an impact on soil ecosystems.

Vertebrates with Minimal Effect on Soils. "Periodic" vertebrates are those that associate with soils but have little impact on soil communities. These vertebrates include birds that nest in lagomorph or rodent dens; lizards that sleep in the ground; toads or frogs that lie dormant in soil when temperatures are high or that occasionally burrow in the soil in search of food; and foxes, badgers, coyotes, lagomorphs and desert tortoises all of which create dens in the soils.

The dens or chambers created by mammals and reptiles often become miniecosystems. When unoccupied by their creators, these underground chambers frequently are used by nonburrowing animals, such as beetles and frogs. The buildup of organic debris in the dens promotes growth of fungi, which, in turn, is eaten by insects and mites that become food for vertebrates. However, the overall effects of these chambers on soil communities probably are small.

Vertebrates with Substantial Impact on Soils. Many mammals have considerable influence on soil communities. The most important are the burrowing rodents including pocket gophers, kangaroo rats, ground squirrels and prairie dogs. Burrowing mammals raise soils from lower profiles to the surface where they are broken down, incorporated with organic matter and carried off by water and wind. Mixing deep and surface materials also may have significant effects on the texture and composition of soils at various levels (Koford, 1958).

Rodents also are responsible for moving large amounts of soil. Grinnell (1923) reported that pocket gophers moved more than 2.7 metric tons (3 tons) of soil per 2.6 km^2 (1 mi^2) during one winter. Prairie dogs also move soil; soil in the mounds excavated from 25 burrows may weigh as much as 27 to 36 metric tons (30 to 40 tons) (Thorp, 1949; Koford, 1958).

The net influence of vertebrates on soil composition is not easy to measure, but the following examples demonstrate important relationships. Badgers are strong diggers and can move large rocks. They have been known to change completely the soil surface from silt-loam to loam in some areas (Thorp, 1949).

Rodents and lagomorphs influence the soil by adding organic material. Feces alone is a significant contribution to soil communities. On the Santa Rita Experimental Range in Arizona, Vorhies and Taylor (1933) found an average of 16 kg per ha (14 lb per ac) of jackrabbit feces. If this value were extrapolated to include the entire 20,240 ha (50,000 ac) Santa Rita Experimental Range, fecal weight would be about 315 metric tons (350 tons), nearly 30 times the combined weight of the jackrabbits that lived on the range.

Soil chemical composition is altered by mammalian activities. Feces, urine and animal remains are rich in the salts of important soil chemicals. Greene and Reynard (1932) evaluated kangaroo rat dens on the Santa Rita Experimental Range and found increased quantities of soluble salts and nitrates in them. There was an average of 0.6 kg (3 lb) of nitrogen per hectare (1 ac) in kangaroo rat dens and 1.5 kg (8 lb) of stored food per hectare (1 ac).

Mammals also alter soil structure. Soil structure is determined by the size and arrangement of soil particles. Structure is important because it affects the ability of soils to absorb water and subsequently yield it to plants. Burrowing animals usually improve soil structure by loosening soil particles (Koford, 1958). Kangaroo rats in Arizona produce a measurable increase in the water-holding capacity of surface soils near their burrows (Greene and Murphy, 1932).

Not all mammal-soils associations are beneficial. Soil disturbance caused by burrowing animals can increase erosion and prevent natural revegetation. These changes can cause, in turn, the mortality of beneficial soil organisms such as earthworms. The extent to which mammals cause erosion is unclear. Wallwork (1970) maintained that the activities of burrowing animals can lead to soil erosion. But Koford (1958) maintained that although mammals may increase the speed of erosion after it is started, overgrazing by livestock, not burrowing activities of mammals, is most often the initial cause of excessive soil erosion.

Livestock and native ungulates affect soils by compacting them. Hungerford (1980) attempted to measure the effects of ungulate movement on soil by establishing soil stability classes. He based these classes on the amount of vegetational cover, litter or rock on the soil surface. Other indicators of soil stability included the number of seedling perennial plants, observed soil movement, amount of litter against rocks or plants, presence of rills and gullies without perennial vegetation. Animal trailing, grazing, playing, fighting and walking to and from water also were measured. Impact of animals on soils and vegetation was assessed using a site disturbance index (SDI). Hungerford (1980) calculated the SDI as shown below.

$$SDI = \frac{M(WRA)}{S}$$

Where
- M = Moisture vulnerability. This ranges from 1.0, dry or frozen soil, to 4.0, saturated soils.
- W = The mean force exerted on the soil by an average hoofprint expressed as psi.
- R = The daily range of an animal. Cattle are used as the norm with their range being represented as 1.0. Ranges lesser or greater than cattle would be ±1.0. (Less than cattle, then, would be 0.0 and more than cattle would be 2.0.)
- A = Relative activity. This value is estimated for each month of the year. Breeding, fighting and playing are some behavioral elements that cause more impact by one species than another. Cattle have a value of 1.0 so other ungulates have relative activity values of ±1.0, if they are more or less active than cattle.
- S = Range shift. A migration through and off the site would have impact during the movement, yet it would preclude any action adding to the movement for the subsequent months and would therefore be for only a portion of the time interval; the value of S would be 0.0. If no shift occurred during the time interval, S = 1.0.

Hungerford's SDI for mule deer, elk, cattle and horses in Carson National Forest, New Mexico, are in Table 5. Use of the SDI can quantify the physical impact of large herbivores on soil communities to ascertain when management practices should begin. The SDI is a comparative measure of the impact of one class of animal upon a site when compared with another animal. The effect herbivores have on soil communities may be nearly as important as the amount of vegetation they consume.

Effects of Soil on Animals

Soil properties that most affect the distribution and abundance of animals include soil structure, texture, moisture, aeration and chemical composition (Kevan, 1962; Kuhnelt, 1976). These properties do not influence animals independently, but act in concert with each other and with biotic factors such as the presence of food, symbiots and predators. The following discussion briefly outlines the role of soil as it influences animals that live in and on soils.

Animals That Live in the Soil

Soil Structure and Texture. The structure and texture of soil affect the distribution of burrowing animals. Compact soils or very rocky soils may reduce the rate of burrowing by earthworms, for instance (Guild, 1955), or preclude burrowing altogether. Soils that are too fine may not be suitable for burrowing, except for those animals with special adaptations (Kuhnelt, 1976). These generalities hold not only for animals that spend all their lives in the soil, but also for vertebrates such as burrowing rodents that spend varying amounts of time underground. For example, Koford (1958) found that most prairie dog towns were on deep alluvial soils of medium to fine texture, whereas few were on shallow, sandy or rocky soils.

Another aspect of soil structure, the size and number of spaces between soil particles, influences the species composition and verticle distribution of nonburrowing animals. A clear, positive correlation exists between the average size of pore space in soils and the animals that inhabit them (Kuhnelt, 1958). Furthermore, the size and number of spaces in soils affect soil moisture and carbon dioxide content. Thus, soil structure may indirectly influence animal distribution in other ways described below.

Moisture. The species composition of animals that live in the microcaverns of the soil is influenced not only by the size of the microcaverns, but by the moisture content of these small cavities. Generally, the smallest of the soil organisms are the most susceptible to desiccation. They often are found deep in soils in the most narrow crevices. These small spaces hold water for the longest time due to surface tension. The thin film of water that forms around soil particles provides habitat for a number of small organisms (Kevan, 1962; Kuhnelt, 1976). Lack of moisture may limit the number of soil animals, as illustrated by the limited biological activity in the soils of Arizona deserts (Wallwork, 1976).

Aeration. Soil aeration is difficult to separate from soil moisture because they generally correlate inversely (Kevan, 1962). The resistance of soil organisms to high carbon dioxide levels is extremely variable, and little is known about aeration requirements of many soil animals. But some species such as nematodes (Wallace, 1956) apparently depend on specific levels of oxygen for successful emergence.

Chemical Composition. The pH of soils depends to a large extent upon the soil parent material, but the kind of vegetation on the surface and the level of aerobic and anaerobic decomposition processes also influence soil pH (Kevan, 1962; Wallwork, 1976). Soil organisms vary considerably in their preferences for soil pH, but most avoid very acid soils (Kevan, 1962).

Animals That Live on the Surface

Perhaps the most significant influence that soil has on the distribution and number of animals that live aboveground is through the relationships between soil and vegetation. In the broadest sense, the distribution of vegetation on the surface of the Earth is controlled primarily by climatic conditions. Temperature and precipitation are the most important climatic factors (Ricklefs, 1979). However, superimposed upon these general vegetative patterns are numerous other factors, both biotic and abiotic, that influence plant communities. Soil conditions are among the more important of these factors and play a significant role in determining the characteristics of vegetation on a particular site.

The importance of the structure, composition and general vigor of vegetation in determining the distribution and health of animal populations is difficult to overstate. The general configuration of vegetation is thought to be an important proximate factor in habitat selection by birds (Hilden, 1965; James, 1971). For some mammals, such as elk and deer, the structure of vegetation determines the suitability of a site for thermal or hiding cover (Thomas et al, 1979). The species composition and vigor of vegetation are of obvious importance to animals that feed directly on plants, but also are important to predators. In the latter group, selection of a particular species of plant or vegetation configuration when foraging may occur because the number of prey is greater there, or because the configuration allows the predator to search more easily for and capture prey (e.g. in birds; Holmes and Robinson, 1981). In the Southwest, vegetation even may be the primary source of water for some birds and mammals such as Gambel's Quail and bighorn sheep. Ultimately then, vegetation can provide most of the elements necessary for survival: food, shelter and water. And changes in soil conditions that produce changes in vegetation potentially have tremendous impacts on animal populations.

Allen (1962) recognized that soils influence animals indirectly through vegetation, and emphasized the correlations that exist between soil fertility and the density and health of animal populations. His examples from the eastern United States clearly showed that white-tailed deer, rabbits, raccoons, muskrats, wild turkeys, pheasants and Bobwhite Quail were in better condition on fertile soil than on poor soil. Similar examples can be found in the Southwest. For instance, the most dense populations of Gambel's Quail inhabit areas of residual soils of decomposed granite, or of floodplain soils of river bottoms (Johnsgard, 1976). Both types of soils support the relatively luxuriant and diverse vegetation preferred by Gambel's Quail.

TABLE 5.

Summary of Site Disturbance Indices for Deer, Elk, Cattle and
Horses in the Carson National Forest, New Mexico
(Hungerford, 1980)

	July	Aug	Sept	Oct	Nov	Dec	Jan	Feb	Mar	Apr	May	June	Annual
Mule Deer	0	18	18	0	37	37	37	37	28	37	18	—	267
Elk	29	29	0	0	87	0	0	87	58	29	29	0	348
Cattle	48	24	24	0	0	0	0	0	0	0	0	24	144
Horses	52	104	104	52	52	104	104	104	157	157	104	52	1,146

Plate 2

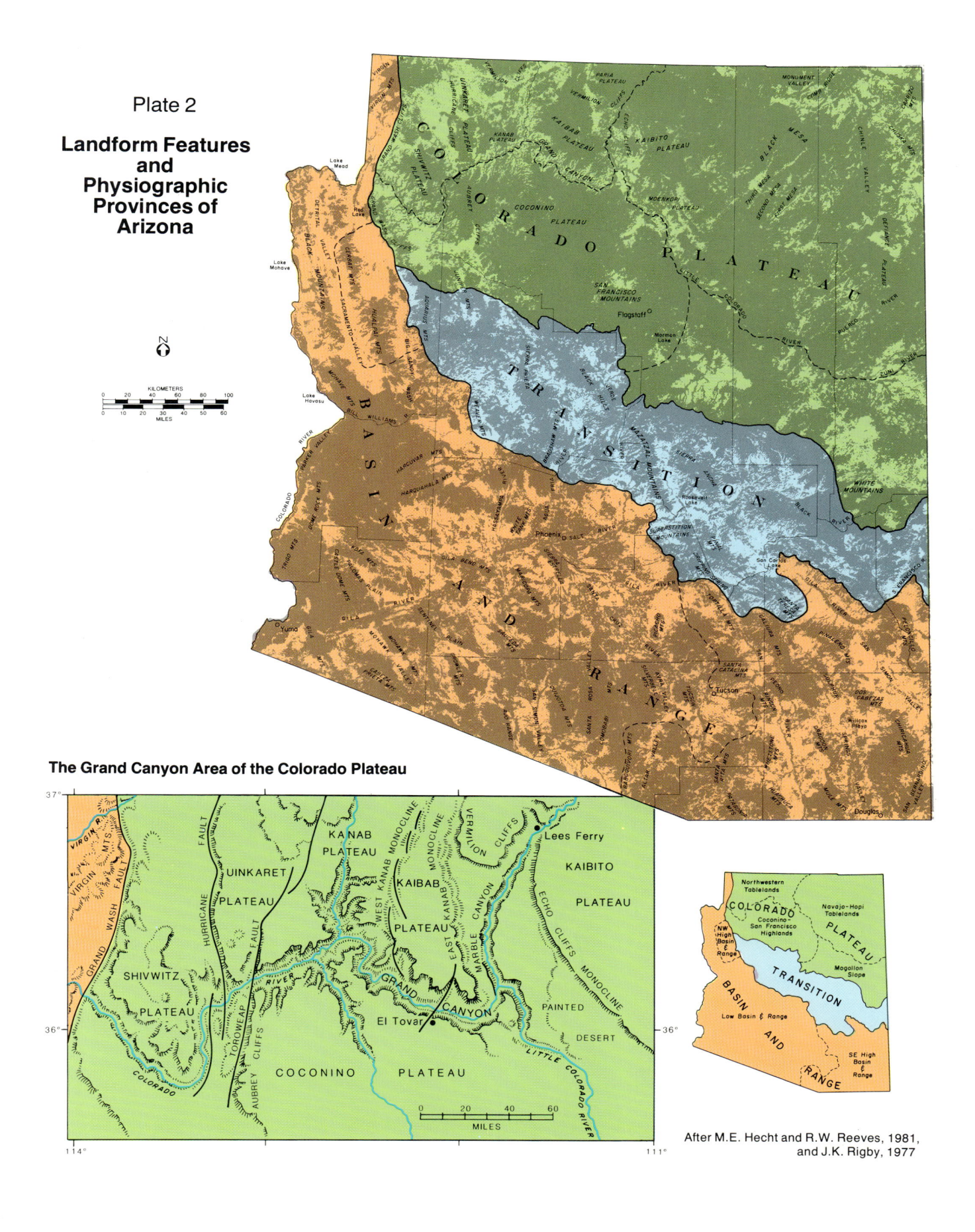

Landform Features and Physiographic Provinces of Arizona

The Grand Canyon Area of the Colorado Plateau

After M.E. Hecht and R.W. Reeves, 1981, and J.K. Rigby, 1977

Plate 3

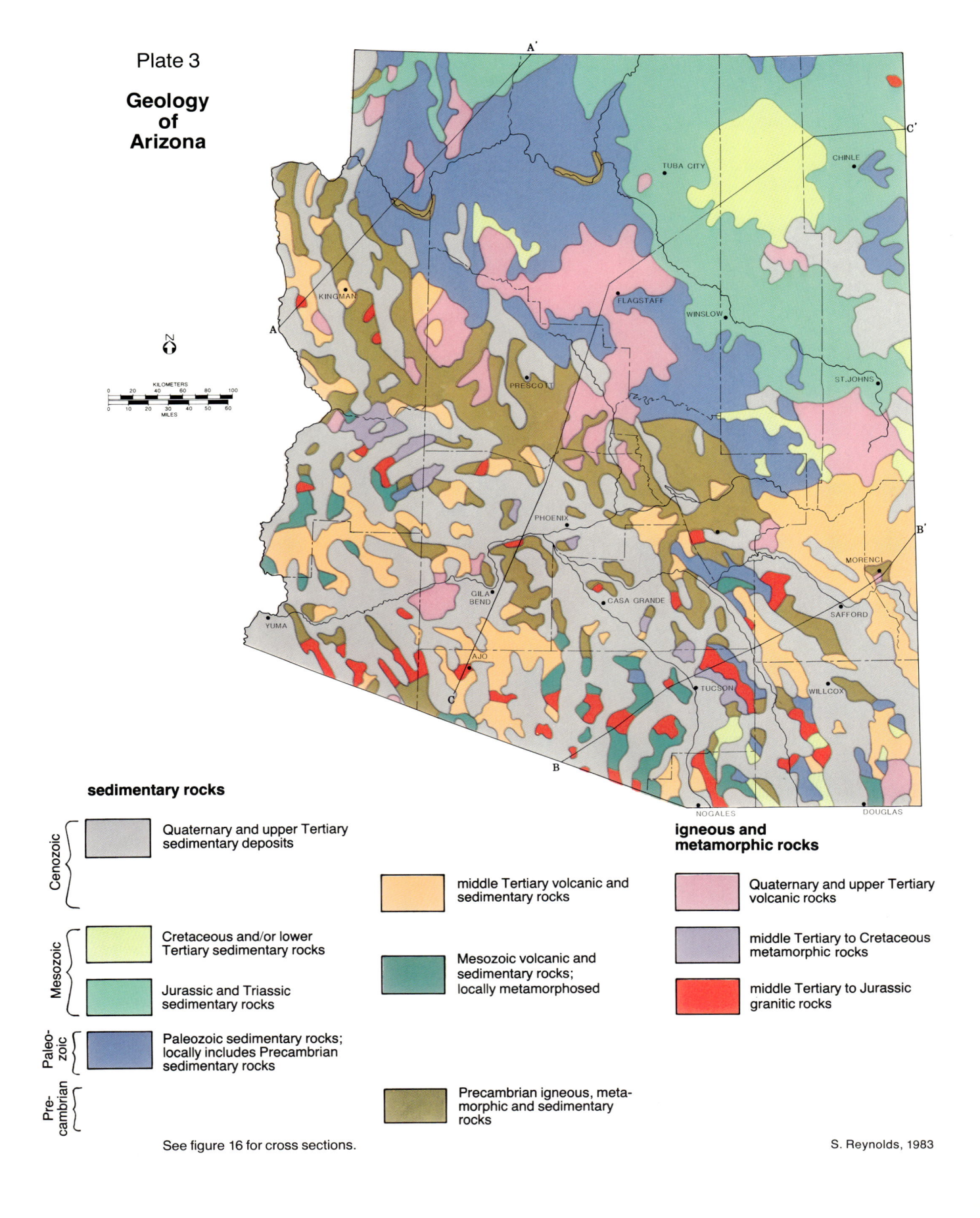

Geology of Arizona

sedimentary rocks

Cenozoic
- Quaternary and upper Tertiary sedimentary deposits
- middle Tertiary volcanic and sedimentary rocks

Mesozoic
- Cretaceous and/or lower Tertiary sedimentary rocks
- Jurassic and Triassic sedimentary rocks
- Mesozoic volcanic and sedimentary rocks; locally metamorphosed

Paleozoic
- Paleozoic sedimentary rocks; locally includes Precambrian sedimentary rocks

Precambrian
- Precambrian igneous, metamorphic and sedimentary rocks

igneous and metamorphic rocks
- Quaternary and upper Tertiary volcanic rocks
- middle Tertiary to Cretaceous metamorphic rocks
- middle Tertiary to Jurassic granitic rocks

See figure 16 for cross sections.

S. Reynolds, 1983

Plate 4

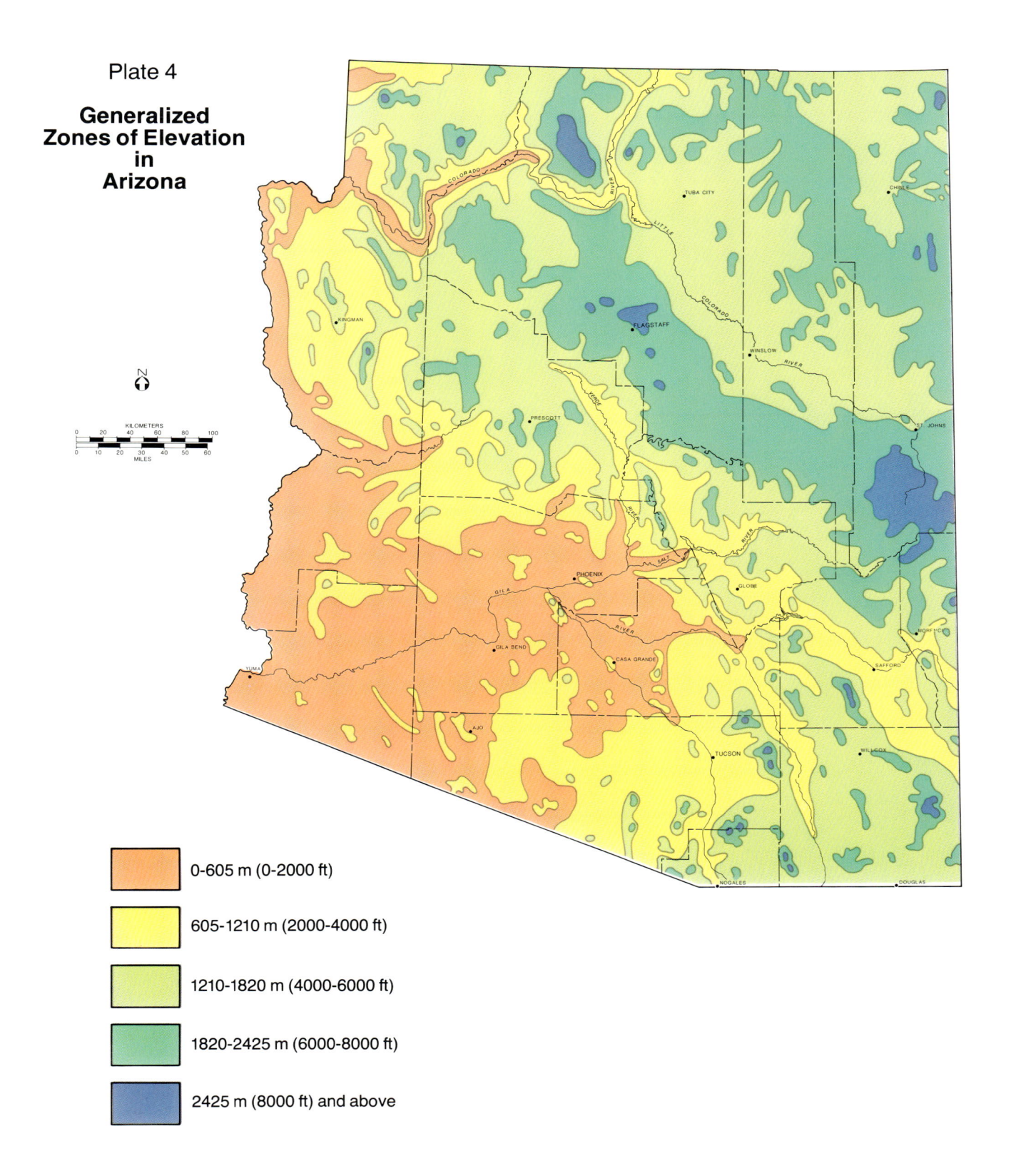

Generalized Zones of Elevation in Arizona

- 0-605 m (0-2000 ft)
- 605-1210 m (2000-4000 ft)
- 1210-1820 m (4000-6000 ft)
- 1820-2425 m (6000-8000 ft)
- 2425 m (8000 ft) and above

After S. Bahre, 1966

Plate 5

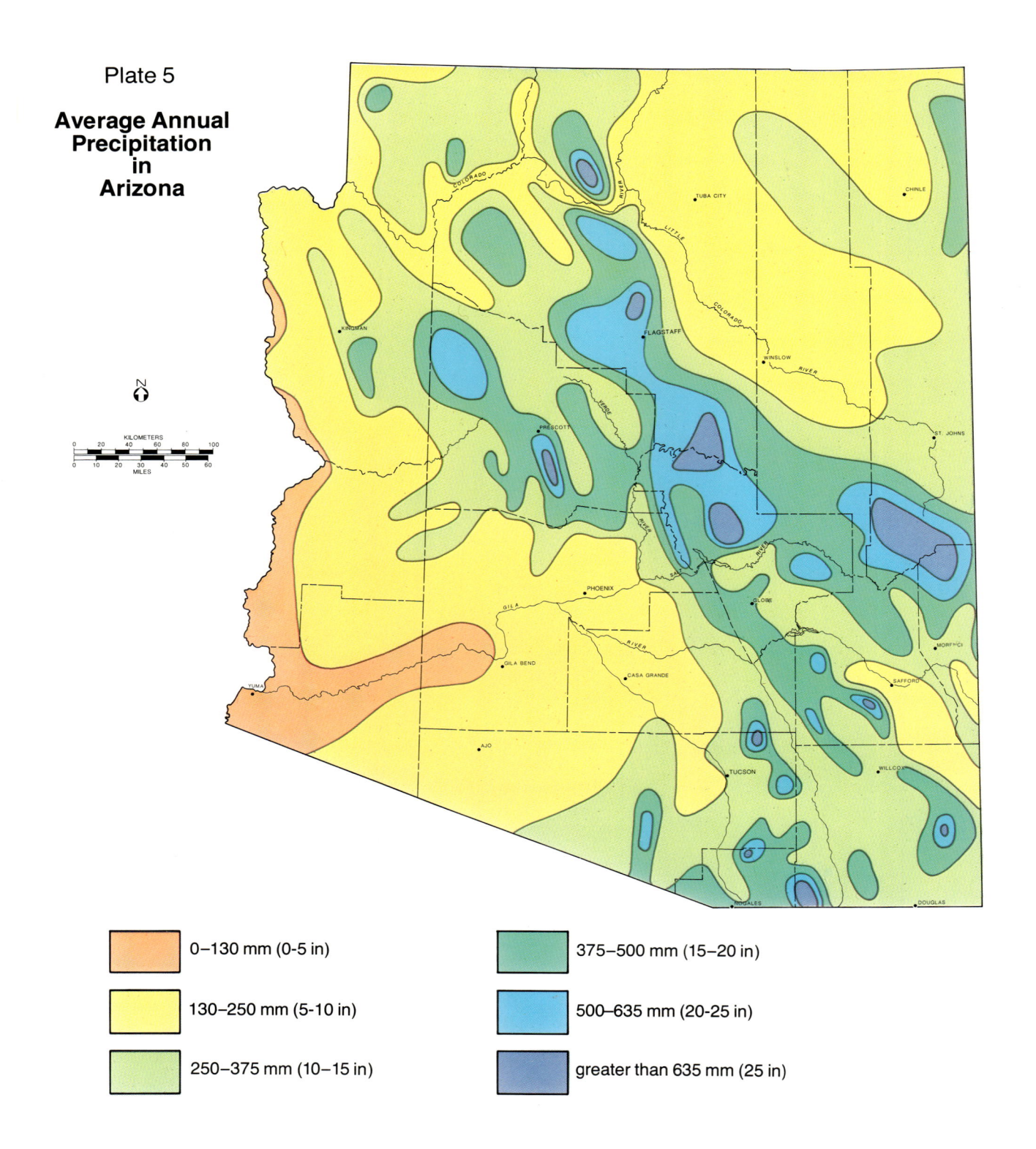

Average Annual Precipitation in Arizona

- 0–130 mm (0-5 in)
- 130–250 mm (5-10 in)
- 250–375 mm (10–15 in)
- 375–500 mm (15–20 in)
- 500–635 mm (20-25 in)
- greater than 635 mm (25 in)

After W.D. Sellers and R.H. Hill, 1974

Plate 6

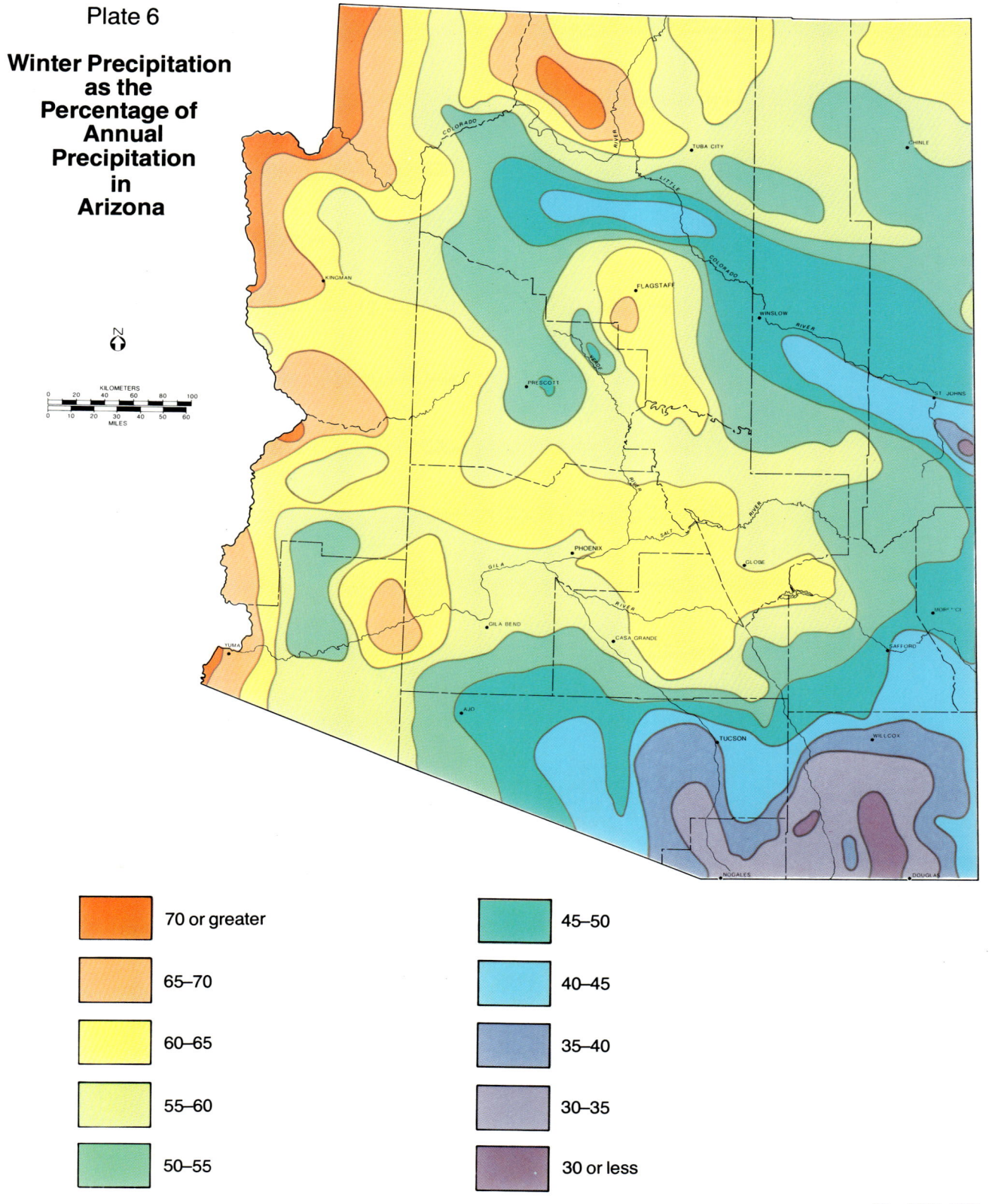

Winter Precipitation as the Percentage of Annual Precipitation in Arizona

	70 or greater		45–50
	65–70		40–45
	60–65		35–40
	55–60		30–35
	50–55		30 or less

After ARIS, 1975

Plate 7

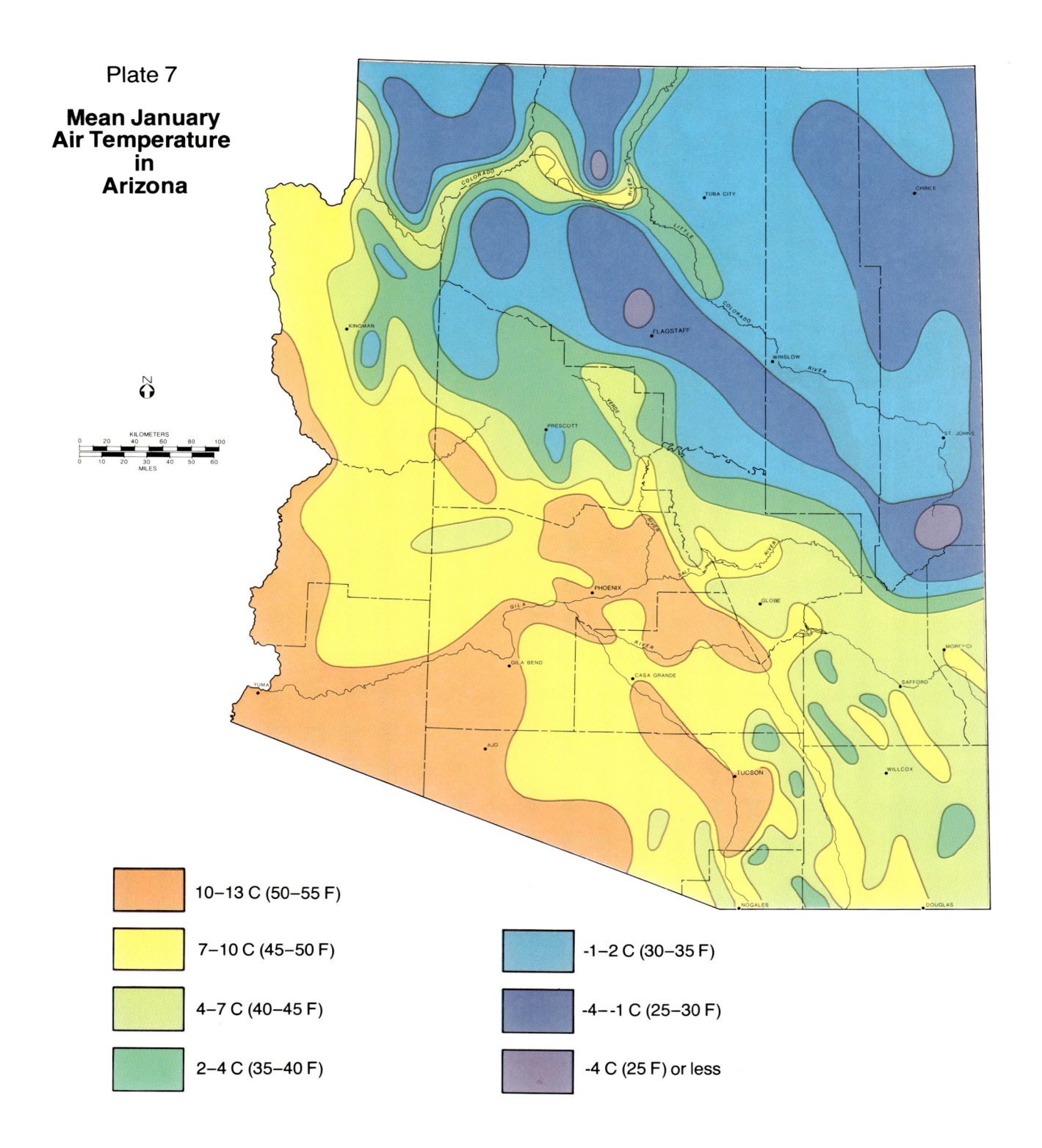

Mean January Air Temperature in Arizona

Legend:
- 10–13 C (50–55 F)
- 7–10 C (45–50 F)
- 4–7 C (40–45 F)
- 2–4 C (35–40 F)
- -1–2 C (30–35 F)
- -4–-1 C (25–30 F)
- -4 C (25 F) or less

After M.E. Hecht and R.W. Reeves, 1981

Plate 8

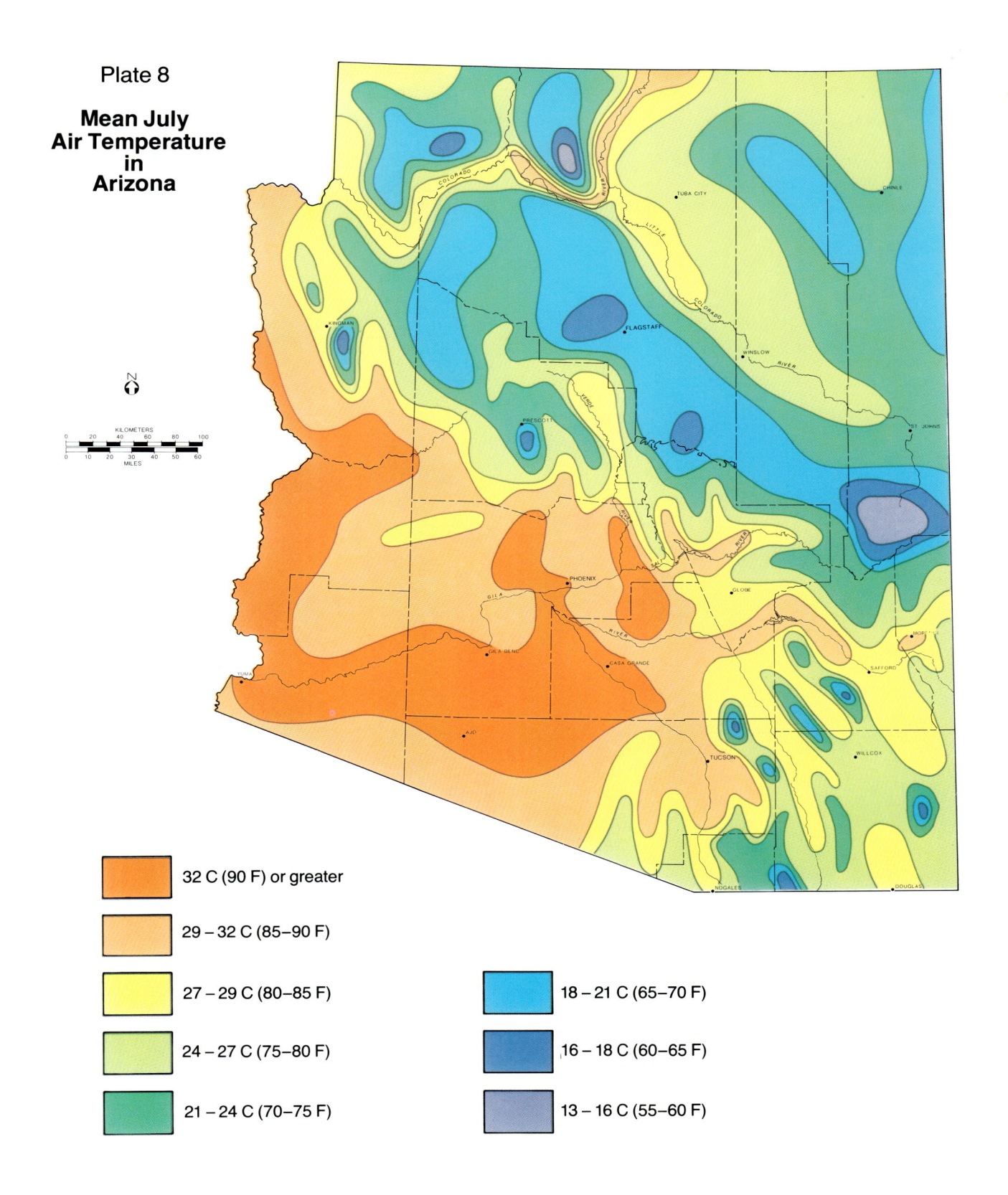

Mean July Air Temperature in Arizona

Legend:
- 32 C (90 F) or greater
- 29 – 32 C (85–90 F)
- 27 – 29 C (80–85 F)
- 24 – 27 C (75–80 F)
- 21 – 24 C (70–75 F)
- 18 – 21 C (65–70 F)
- 16 – 18 C (60–65 F)
- 13 – 16 C (55–60 F)

After M.E. Hecht and R.W. Reeves, 1981

Plate 9

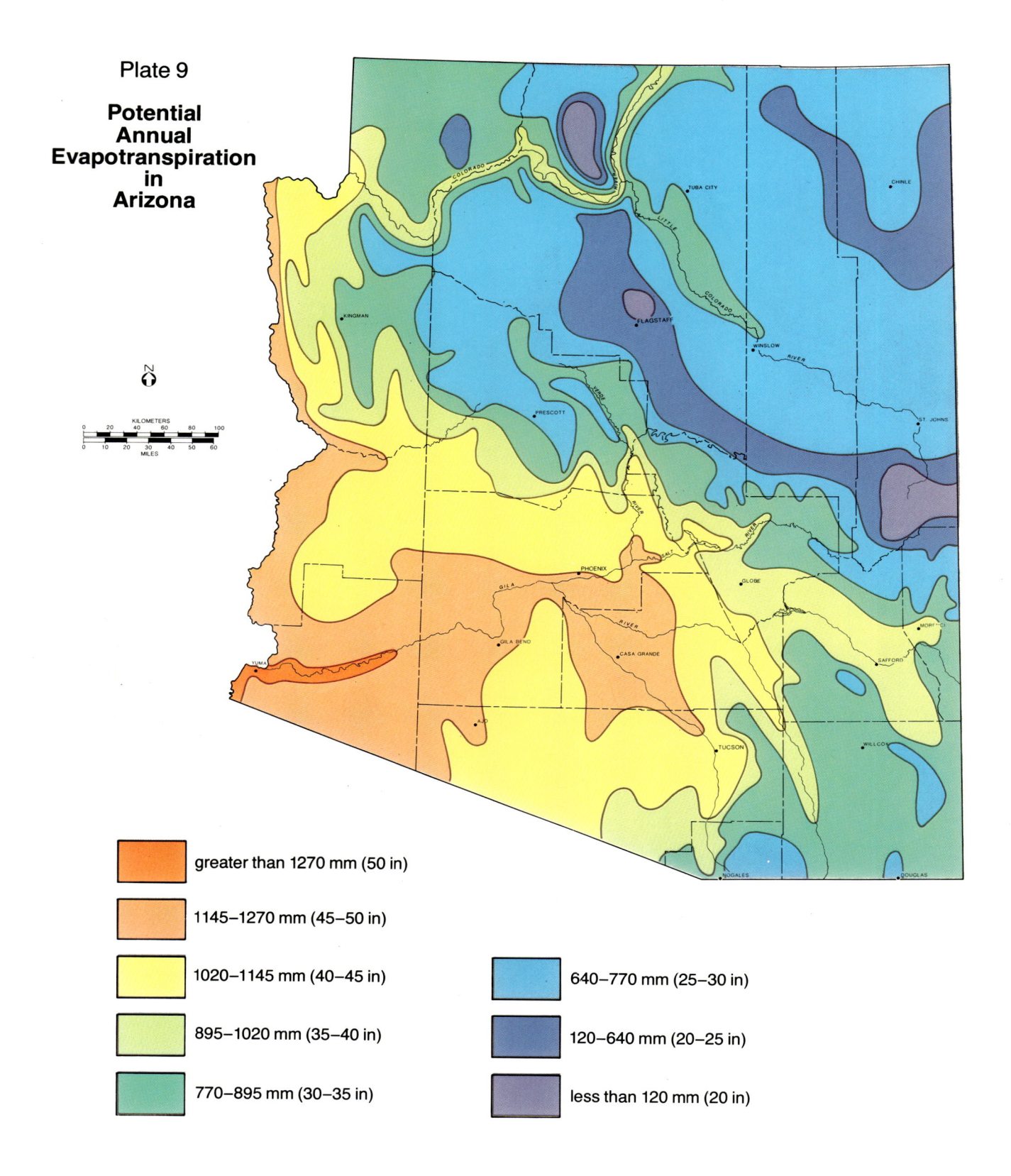

Potential Annual Evapotranspiration in Arizona

- greater than 1270 mm (50 in)
- 1145–1270 mm (45–50 in)
- 1020–1145 mm (40–45 in)
- 895–1020 mm (35–40 in)
- 770–895 mm (30–35 in)
- 640–770 mm (25–30 in)
- 120–640 mm (20–25 in)
- less than 120 mm (20 in)

After S.W. Buol, 1964

Plate 10

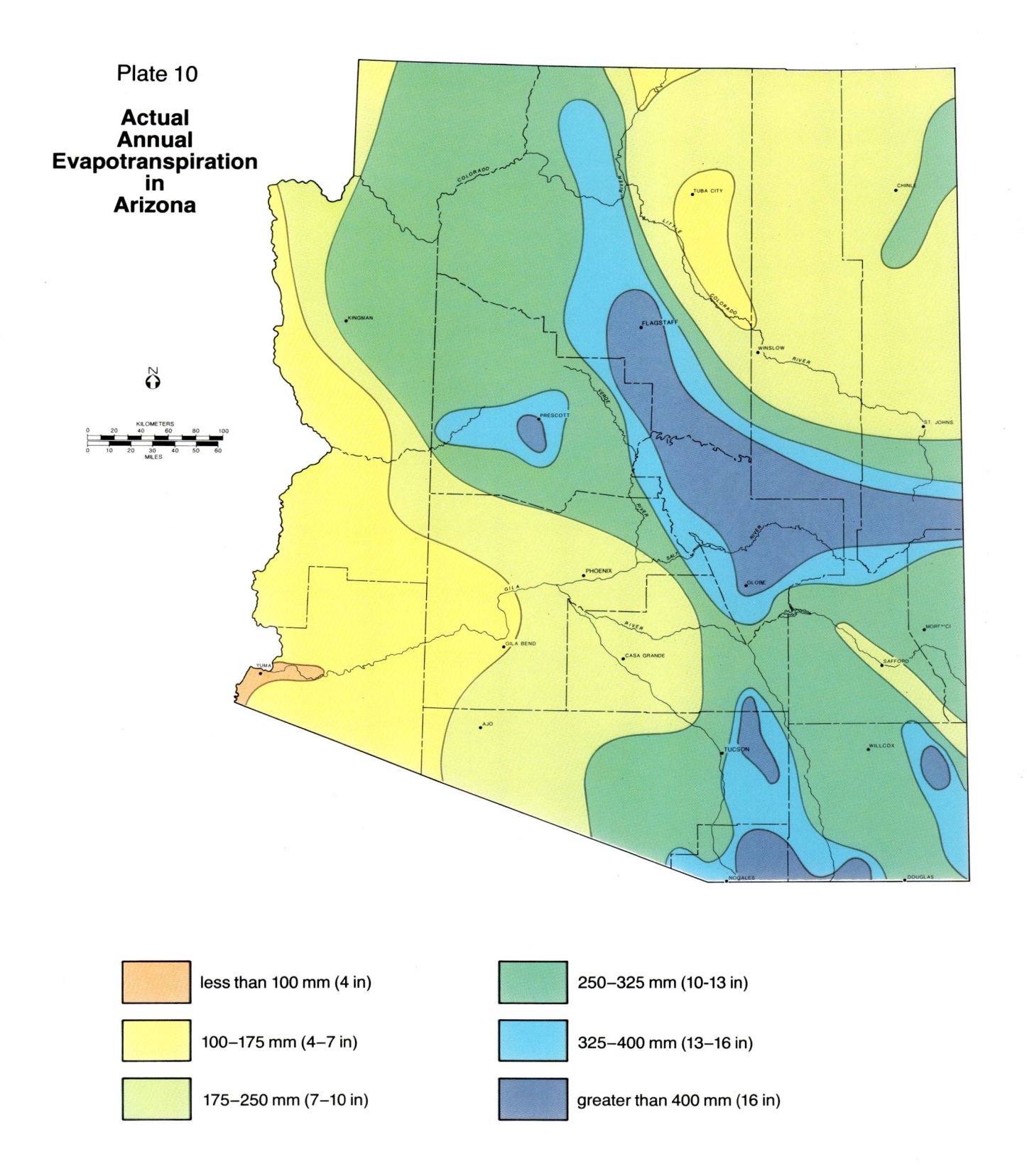

Actual Annual Evapotranspiration in Arizona

▨ less than 100 mm (4 in)	▨ 250–325 mm (10-13 in)
▨ 100–175 mm (4–7 in)	▨ 325–400 mm (13-16 in)
▨ 175–250 mm (7–10 in)	▨ greater than 400 mm (16 in)

After S.W. Buol, 1964

Plate 11

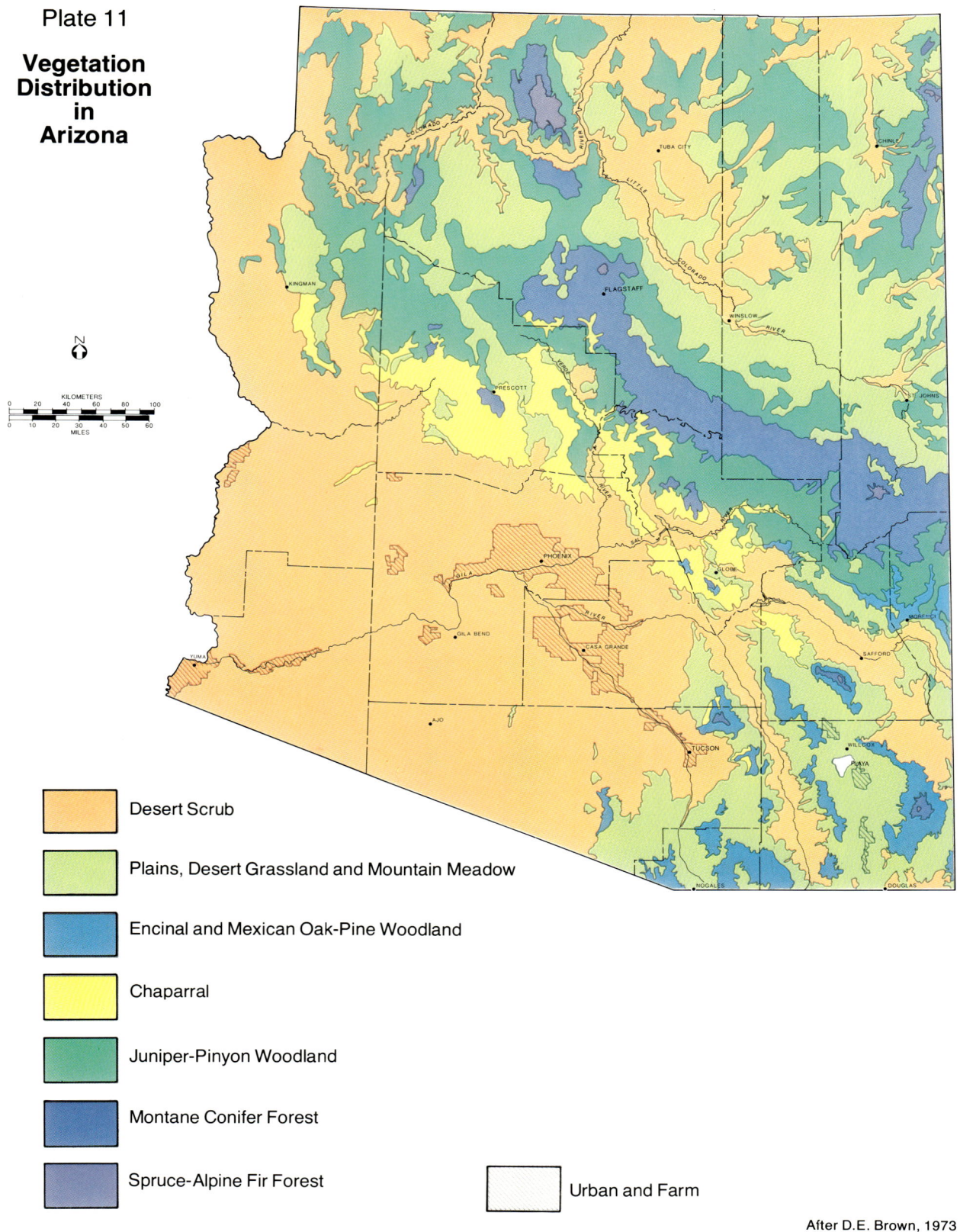

Vegetation Distribution in Arizona

- Desert Scrub
- Plains, Desert Grassland and Mountain Meadow
- Encinal and Mexican Oak-Pine Woodland
- Chaparral
- Juniper-Pinyon Woodland
- Montane Conifer Forest
- Spruce-Alpine Fir Forest
- Urban and Farm

After D.E. Brown, 1973

Plate 12

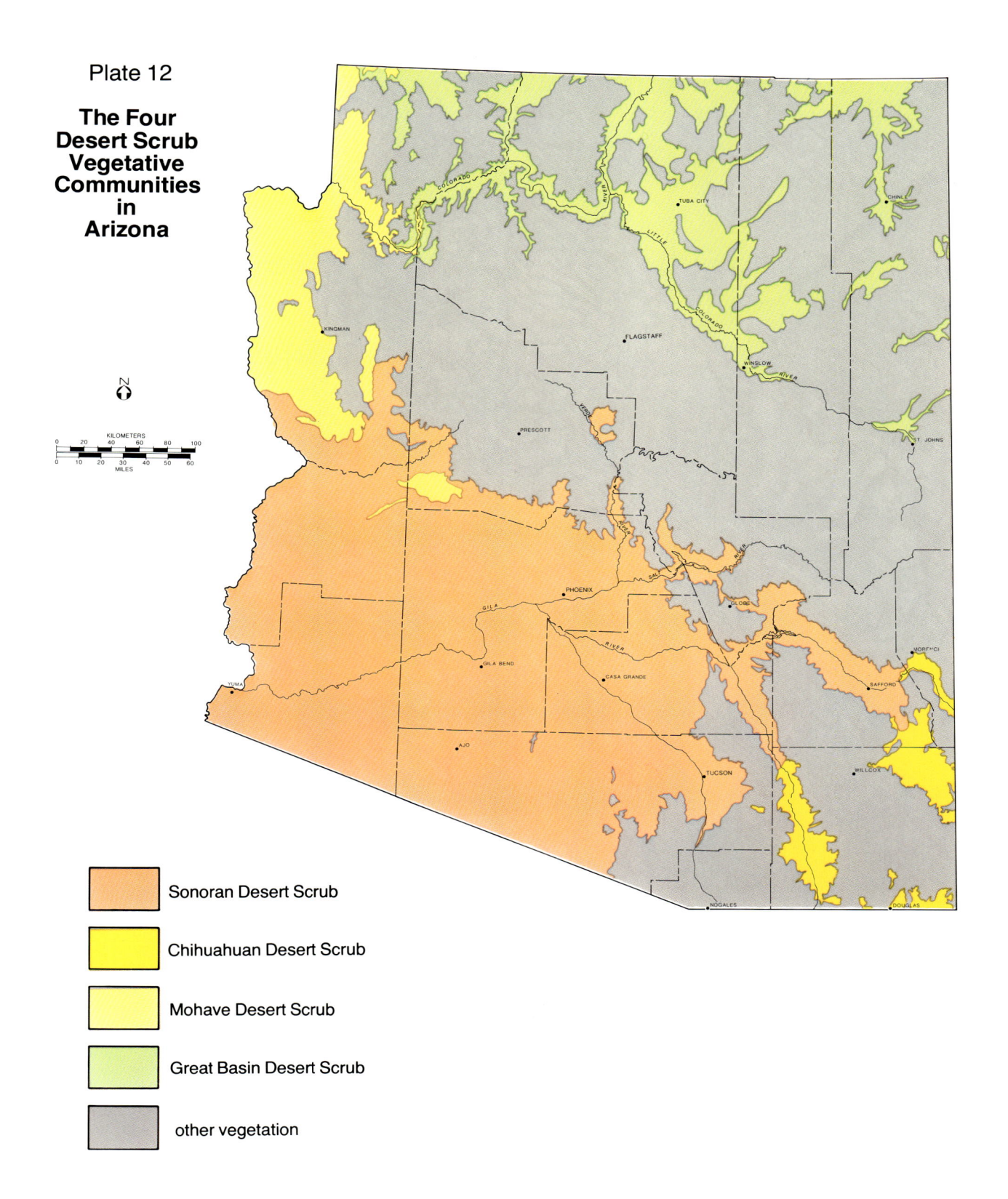

The Four Desert Scrub Vegetative Communities in Arizona

- Sonoran Desert Scrub
- Chihuahuan Desert Scrub
- Mohave Desert Scrub
- Great Basin Desert Scrub
- other vegetation

After D.E. Brown, 1973

Plate 13

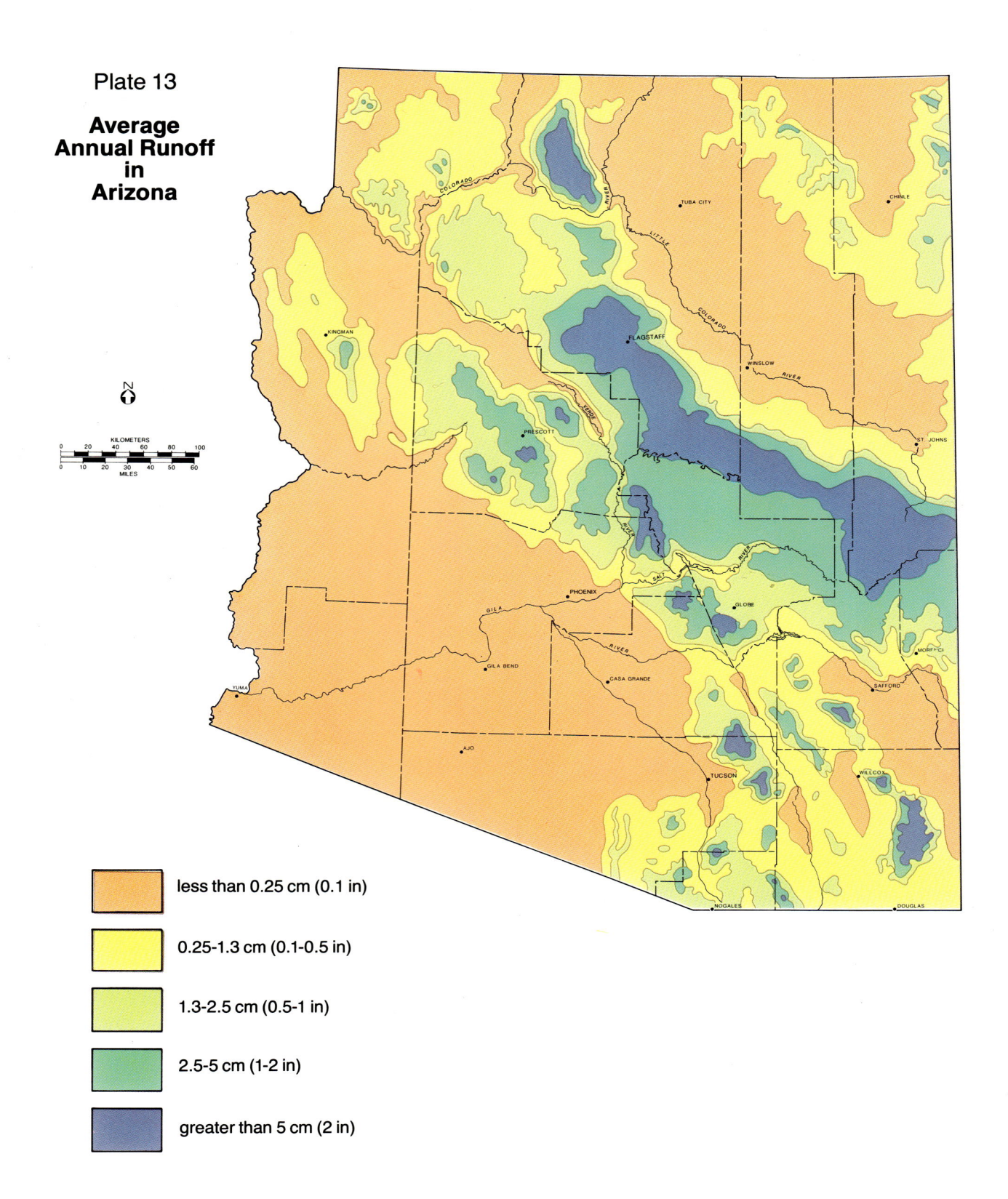

Average Annual Runoff in Arizona

less than 0.25 cm (0.1 in)

0.25-1.3 cm (0.1-0.5 in)

1.3-2.5 cm (0.5-1 in)

2.5-5 cm (1-2 in)

greater than 5 cm (2 in)

after J. H. Dorroh in Spencer, 1966

Plate 14

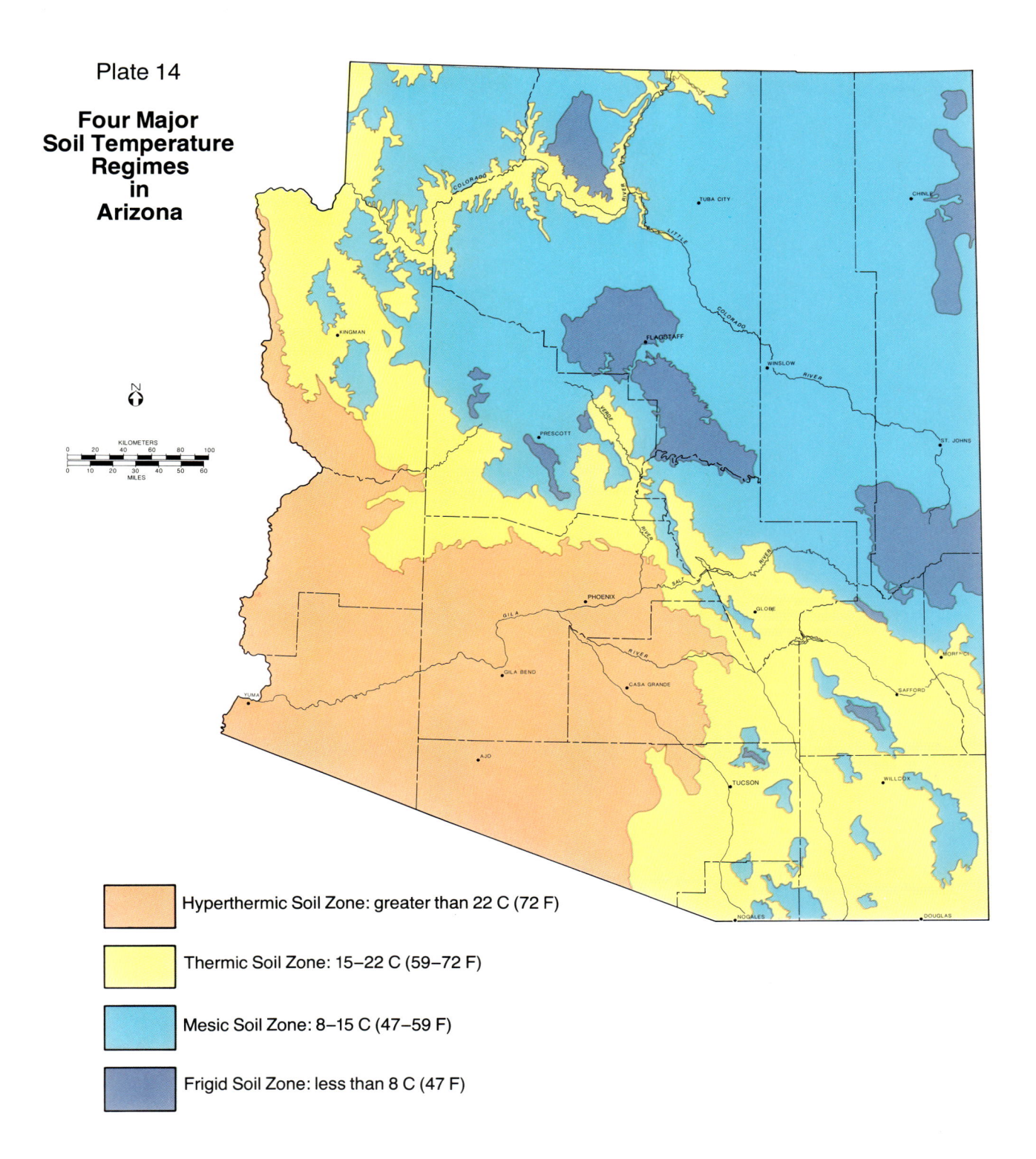

Four Major Soil Temperature Regimes in Arizona

Hyperthermic Soil Zone: greater than 22 C (72 F)

Thermic Soil Zone: 15–22 C (59–72 F)

Mesic Soil Zone: 8–15 C (47–59 F)

Frigid Soil Zone: less than 8 C (47 F)

D.M. Hendricks, 1982

Plate 15
Horizons That Could Occur in a Soil Profile

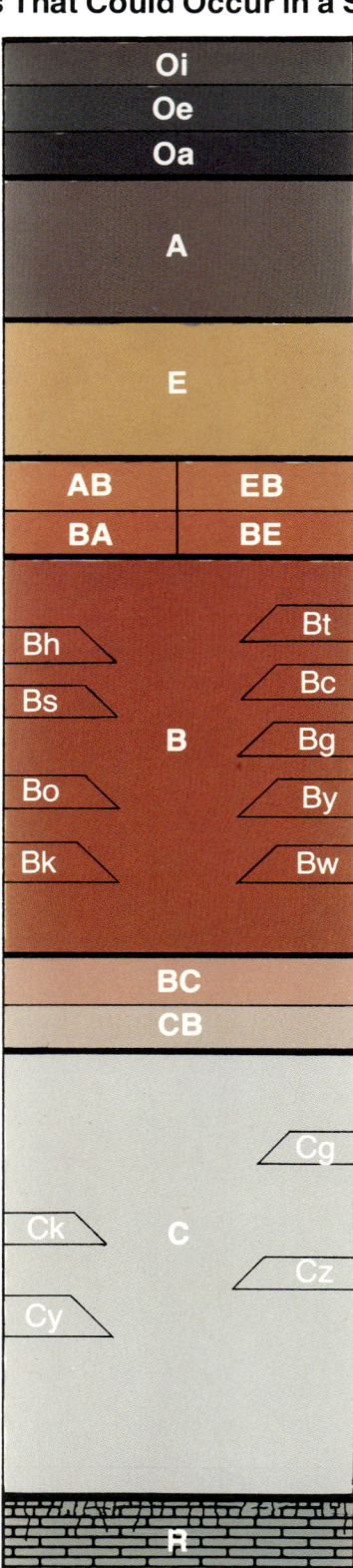

Organic Horizons
- **Oi** — Slightly decomposed organic matter.
- **Oe** — Intermediately decomposed organic matter.
- **Oa** — Highly decomposed organic matter.

Mineral Horizons of Maximum Biological Activity
- **A** — Surface mineral horizon that has an accumulation of well-decomposed organic matter that coats the mineral particles and darkens the soil mass. With plowing or other disturbances the Ap notation is used.

Horizons of Eluviation (removal of materials dissolved or suspended in water)
- **E** — Subsurface horizon that has lost organic matter, clay, iron or aluminum through eluviation with concentration of resistant sand and silt-sized particles.

- **AB**, **EB** — Transitional from the A or E to the B but more like the A or E horizon.
- **BA**, **BE** — Transitional from the A or E to the B but more like the B than the A or E horizon.

Mineral Horizons of Illuviation (accumulation of dissolved or suspended material from above and/or alteration of the parent material)

B (Bh, Bs, Bo, Bk, Bt, Bc, Bg, By, Bw) — Mineral horizon that is characterized by one or more of the following:

1. Illuvial accumulation of clay (Bt), organic matter (Bh), carbonates (Bk), silica (Bq), gypsum (By), iron and aluminum oxides (Bs);
2. Residual concentration of iron and aluminum oxides (Bo);
3. Development of structure and/or coatings of iron and aluminum oxides that give darker, stronger or redder colors (Bw); and
4. Evidence of carbonate removal.

- **BC** — Transitional from B to C but more like the B horizon.
- **CB** — Transitional from B to C but more like the C horizon.

Mineral Horizon (excluding bedrock, which is little affected by soil forming processes)

- **C** (Cg, Ck, Cy, Cz) — Mineral horizon, other than bedrock, that may or may not be similar to presumed parent material. Has been little affected by soil-forming processes but may be otherwise weathered. Numerical prefixes are used to designate C horizons unlike presumed parent material as 2C, 3C, etc. This designation also is used with other horizons.

  Cg = C horizon with intense gleying or reduction of iron compounds.
  Ck = C horizon with accumulation of carbonates such as CaCO3.
  Cy = C horizon with accumulation of gypsum.
  Cz = C horizon with accumulation of soluble salt.

Hard Bedrock
- **R** — Underlying consolidated bedrock.

D.M. Hendricks, 1982

Plate 16

**Actual Soil Profile
and
Associated Vegetation**

The soil is classified as a fine, montmorillonitic Mollic Entroboralf. It is approximately 11 km (7 mi) north-northwest of Jacobs Lake at an elevation of approximately 2130 m (7000 ft) on the Kaibab Plateau. It is in Soil Mapping Unit FH4, but a series name has not yet been designated for this soil.

A horizon

gravelly loam texture
platy structure
(moist)—very dark brown

Bt horizon

clay texture
prismatic and blocky structure
(moist)—yellowish red

C horizon

weathered Kaibab Limestone
clay loam texture
(moist)—yellow

R horizon

Kaibab Limestone
(moist)—very pale brown

Plate 17

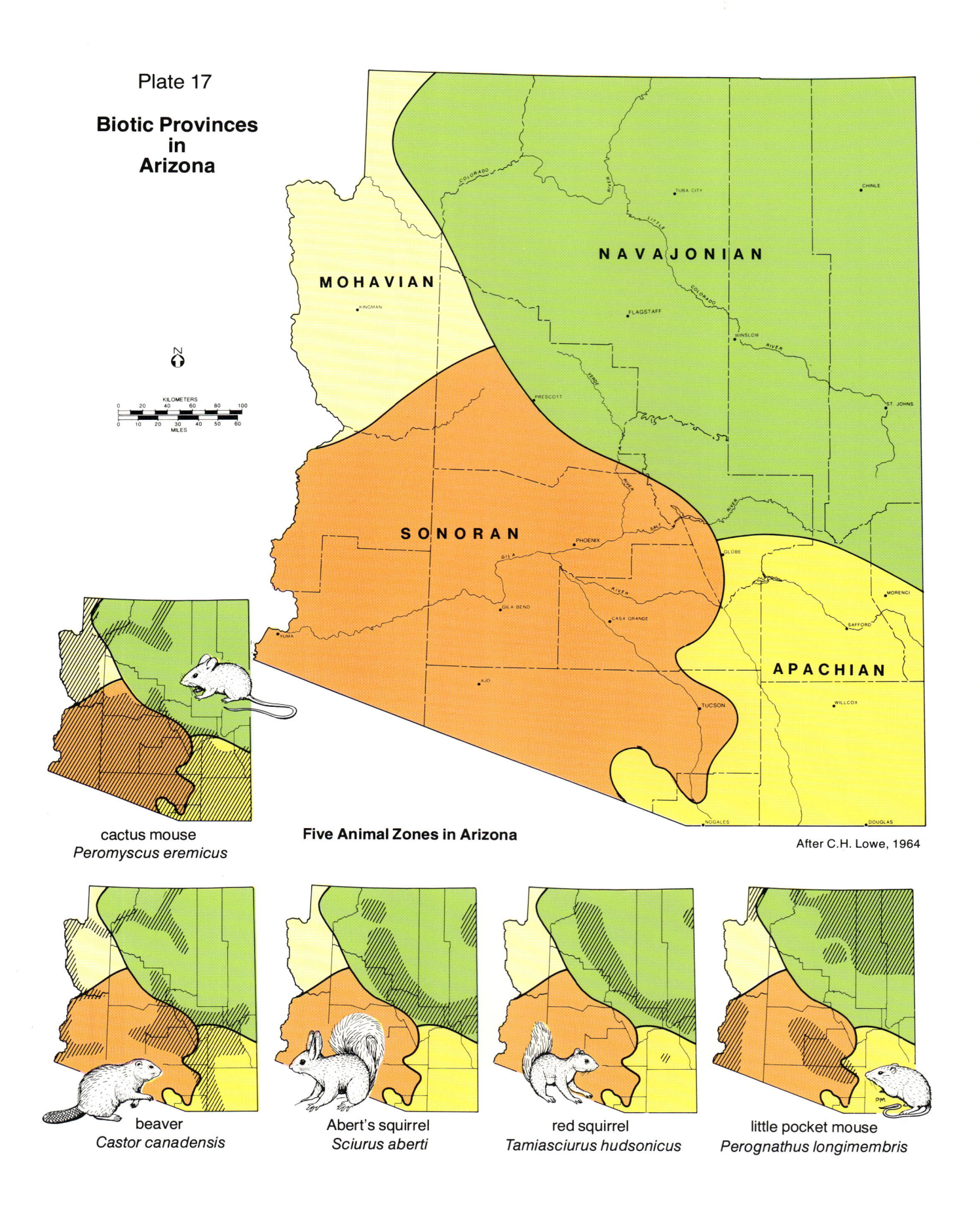

Biotic Provinces in Arizona

MOHAVIAN
NAVAJONIAN
SONORAN
APACHIAN

Five Animal Zones in Arizona

After C.H. Lowe, 1964

cactus mouse
Peromyscus eremicus

beaver
Castor canadensis

Abert's squirrel
Sciurus aberti

red squirrel
Tamiasciurus hudsonicus

little pocket mouse
Perognathus longimembris

6

The United States Soil Classification System and Its Application in Arizona

Introduction

Classification is fundamentally important to any science. Not only is it a means to impose order on diversity between and within objects and concepts, but classification also provides the avenue through which research can be addressed in a rigorously systematic manner. Classifications also have more practical applications. Classification of soils, for instance, is indispensible to the soil survey program of mapping the soils of Arizona. Soil surveys, in turn, can be used to apply the principle functions of soil science to agriculture, forestry and engineering to predict soil behavior under defined use and management or manipulation.

The soil classification system now used in the United States was developed by USDA Soil Conservation Service Soil Survey staff between 1951 and 1975. The system, published as *Soil Taxonomy* (Soil Survey Staff, 1975), was designed to classify all the world's soils because expanding soil survey programs demanded more precise definitions of soil properties than were possible with previous soil classification systems. An improved conceptual frame of reference also was needed so that research data could be more readily communicated, tested and applied between soils of one area to soils of another area where conditions of soil formation or genesis were similar (Simonson, 1962; Smith, 1963; Aandahl, 1965).

The soil classification system described in *Soil Taxonomy* (Soil Survey Staff, 1975) focuses on soil properties that for the most part can be measured quantitatively rather than on soil-forming processes or factors. Yet, the system certainly can not exclude soil genesis since many soil properties quantified have considerable significance in soil genesis. General objectives of the new system are to make the characteristics of various soils easier to remember, to make clearer the relationships among soils and between soils and other elements of the environment, and to provide a basis for developing principles of soil genesis and soil behavior that have prediction value.

The classification system as described in *Soil Taxonomy* (Soil Survey Staff, 1975) is too extensive to relate wholly in this publication. Nonetheless, the rest of the text in this chapter introduces the reader to a few of the basic elements of the system and describes its application in Arizona.

Overview of Soil Taxonomy

Diagnostic Horizons

Soil classes in *Soil Taxonomy* (Soil Survey Staff, 1975) are defined by properties that can be measured quantitatively. Some properties used to classify soils are soil depth, moisture, temperature, texture, structure, cation exchange capacity, base saturation, clay mineralogy, organic matter content and salt content.

Certain soil horizons referred to as diagnostic horizons are the primary building blocks of the *Soil Taxonomy* (Soil Survey Staff, 1975) system. The diagnostic horizons that commonly are found in Arizona soils are listed in Table 6. Presence or absence of certain diagnostic horizons, which can be attributed to conditions of soil formation, is an im-

TABLE 6.

Diagnostic Horizons Common in Arizona Soils

Diagnostic Horizons	Characteristics
Surface Horizons	
mollic epipedon	Surface horizon of accumulation of organic matter; dark colored
ochric epipedon	Surface horizon of limited accumulation of organic matter; light colored
Subsurface Horizons	
argillic	Horizon of clay accumulation
natric	Horizona of clay and sodium accumulation
cambic	Horizon of pedogenic alteration usually expressed by soil structure or removal of calcium carbonate
calcic	Horizon of pronounced carbonate accumulation
gypsic	Horizon of pronounced gypsum accumulation
salic	Horizon of pronounced soluble salt accumulation
petrocalcic	Hard indurated horizon with calcium carbonate as the dominant cementing material
duripan	Hard indurated horizon with silica as the dominant cementing material
albic	Light colored horizon in which clay and organic matter have been significantly removed by leaching

portant criterion in defining many of the classes.* Although the processes by which diagnostic horizons formed are not always understood, the properties possessed by them are those that are significant to the behavior and management of soils.

Epipedons are diagnostic horizons that have formed at the surface and occur nowhere else in the soil, unless the soil is buried under fluvial, aeolian or volcanic deposits. They are defined mostly in terms of soil color, content of organic matter and base saturation; i.e., relative amounts of bases such as calcium, magnesium, sodium and potassium as compared with hydrogen. Of the six epipedons currently recognized in the United States, only two are important in Arizona: the mollic and ochric. The mollic epipedon generally is darker and higher in organic matter than the ochric epipedon.

The remaining nine diagnostic horizons form below the soil surface, although they may become exposed if surface horizons are removed. For the most part, subsurface diagnostic horizons develop from materials leached and accumulated from upper horizons, or they exhibit special features that are useful in differentiating soils. Argillic horizons, for example, are accumulations of clay derived from overlying horizons. Cambic horizons, on the other hand, are only slightly altered by soil-forming processes from the parent material.

Categories

Soil Taxonomy (Soil Survey Staff, 1975) has six categories as illustrated in Figure 39. These are, from top to bottom, order, suborder, great group, subgroup, family and series. The highest categories have the fewest classes and criteria separating classes, while the lowest categories have the most classes and criteria. The "soil type," a soil series subdivision, used in previous classification systems is not a category in *Soil Taxonomy* (Soil Survey Staff, 1975), but is a phase within a mapping unit.

Ten classes are in the order level. Criteria used to differentiate orders are highly generalized and based more or less on the kinds and degrees of soil-forming processes. Mostly these criteria include properties that reflect major differences in the genesis of soils.

A suborder category is a subdivision of an order within which genetic homogeneity is emphasized. Soil characteristics used to distinguish suborders within an order vary from order to order. For example, soil moisture and temperature are the important factors that differentiate the suborders in the order Alfisols. The presence or absence of an argillic horizon, on the other hand, distinguishes the two suborders of the order Aridisols. Forty-seven suborders are recognized in the United States.

The great group category is a subdivision of a suborder. They are distinguished one from another by kind and sequence of soil horizons. All soils belonging to one of the suborders of Aridisols have argillic horizons. They also may have additional diagnostic horizons such as a petrocalcic as well as several others. Soils having these additional horizons are placed in separate great groups. About 185 great groups are recognized in the United States.

Great group categories are divided into three kinds of subgroups: typic, intergrade and extragrade. A typic subgroup represents the basic concept of the great group from which it derives. An intergrade subgroup contains soils of one great group, but have some properties characteristic of soils in another great group or class. These properties are not developed or expressed well enough to include the soils within the great group toward which they grade. Extragrade subgroup soils have aberrant properties that do not intergrade to any known soil. There are about 1,000 kinds of subgroups in the United States.

A soil family category is a group of soils within a subgroup that has similar physical and chemical properties that affect response to management and manipulation. The principal characteristics used to differentiate soil families are texture, mineralogy and temperature. Family textural classes, in general, distinguish between clayey, loamy and sandy soils. For some soils the criteria also specify the amount of silt, sand and coarse fragments such as gravel, cobbles and rocks.

---

*The smallest natural body that can be defined as a thing complete in itself is an individual. A class is a group of individuals that have been grouped together on the basis of certain selected characteristics. A class is distinguished from all other classes by differences in these characteristics. In a hierarchical classification system classes of a lower category are grouped together according to common properties to form a class of a higher category.

These criteria are important in determining the agricultural and engineering uses of soils. Application of mineralogy class criteria to clayey soils primarily is to group soils of similar clay mineralogy. The kinds of clay minerals in soils may have a strong influence on use, management and behavior, especially engineering behavior. Mineralogy classes for sandy and silty soils primarily separate soils that have weatherable minerals from those that have non-weatherable minerals. This tends to group soils having a common natural fertility potential because weatherable minerals provide nutrients. Soil temperature classes also have practical value in assessing engineering and agricultural applications. Family temperature classes are distinguished by soil temperatures measured at a 50 cm (20 in) depth. About 4,500 soil families are recognized in the United States.

The soil series is the narrowest category in *Soil Taxonomy* (Soil Survey Staff, 1975). Its general concept essentially is the same as the soil classification system that *Soil Taxonomy* (Soil Survey Staff, 1975) superceded (Simonson, 1964). Some series have been redefined, some subdivided and some newly established. These changes have produced more precise definitions of soil series and narrower ranges of properties used in those definitions. And more specific statements can be made about soils than were possible in the past. This, in turn, enhances the value of research and permits more effective application of existing knowledge to use of soil resources because information is more easily and accurately communicated. More than 14,000 soil series are recognized in the United States.

Nomenclature

When a person first encounters *Soil Taxonomy* (Soil Survey Staff, 1975), one of the features most startling is the nomenclature. It is not unusual for such a person to throw up his or her hands in despair at the seemingly incoherent, barbaric and meaningless terms. Some critics of the system reserve their most vehement criticism of *Soil Taxonomy* (Soil Survey Staff, 1975) for its nomenclature. However, once a person becomes familiar with the construction of the nomenclature, he or she soon recognizes the utility and the advantages of the system.

The nomenclature was developed so that each class had a name that was mnemonic, that is, to help memory, and that would connote some properties of the soils of each class. The name also places a class in the system so that a person can recognize both the category of the class and the classes of the higher categories to which it belongs. Incidentally the system of nomenclature was developed primarily not by soil scientists but by classical linguists, Professors J.L. Heller, University of Illinois, and A.L. Leemans, State University of Ghent, Belgium.

Class names were coined from Greek and Latin roots, for the most part, that are familiar because of their use in many common words. Order names end in "sols" (Table 7). A formative element from each order is the ending of names of the suborders, great groups and subgroups. The order formative elements are in Table 8 (See also Appendix D).

Names of suborders have two syllables. The last syllable is

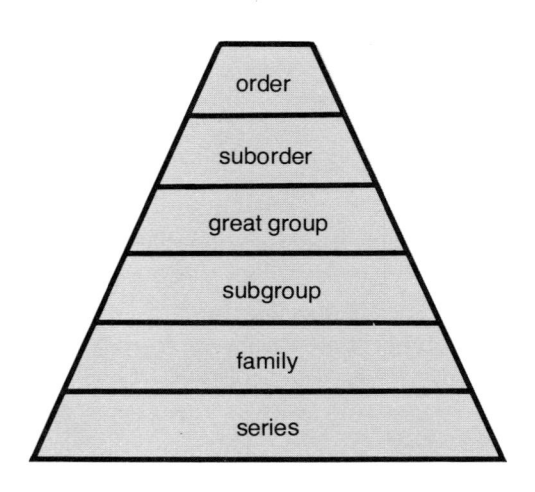

FIGURE 39. *The Hierarchical Soil Classification System*

the formative element from the order. The first syllable is the suborder formative element that suggests certain characteristics about the soil. A few examples and meanings of suborder formative elements are in Table 9. One subgroup is the Argids. These are Aridisols, dry soils, with argillic horizons (arg + id = Argids).

Continuing in the same manner, great group names are formed simply by putting a great group formative element in front of the suborder name. The prefix is connotative and suggests the unique features of the particular great group named. The great group Durargids, for example, include Aridisols with argillic horizons and duripans (dur + Argid = Durargids).

Names of subgroups are binomial. Each binomial has an adjective before the name of the great group to which the subgroup belongs. The term "Typic" is added to a great group name to form the subgroup name for soils that are typical or modal of the great group. Thus, Typic Durargids is the name of the modal subgroup of the Durargids great group. An intergrade subgroup also carries the name of its great group but it is modified by the adjectival form of the name of the class or classes towards which it grades. The adjective is formed by adding "ic" to the class name. For example, Durargids that have some properties of the Haploxerolls great group belong to the Haploxerollic Durargid subgroup. Extragrade subgroup names are formed by adding to the great group name a formative element that describes the extragradational feature. Abruptic Durargids, for example, include soils that have properties of Durargids but have, in addition, an abrupt transition between the A and B horizons characterized by a large difference in percentage of clay.

Family names are fairly common terms that denote in more detail the features of a particular subgroup. A fine, montmorillonitic, thermic soil family is one that is fine textured, contains montmorillonite as the dominant mineral in the clay and has a warm, or thermic mean annual soil temperature.

Soil series are abstract names sometimes taken from some local geographic feature near the site where the series was first established. Examples include the Gila, Graham, Mohave, Tubac and Moenkopie series. Other names are variants of these place names and still others simply are coined.

Each soil series used in the United States is classified with *Soil Taxonomy* (Soil Survey Staff, 1975) nomenclature. Thus, the Suncity series is described as shown below.

Order	Suborder	Great Group	Subgroup	Family	Series
Aridisols	Argids	Durargids	Typic Durargids	fine-loamy, mixed, hyperthermic	Suncity

TABLE 7.

Names and Important Properties of the Orders

Name	Important Properties
Alfisols	Mineral soils relatively low in organic matter with relatively high base saturation. Contains horizon of illuvial clay. Moisture is available to mature a crop.
Aridisols	Mineral soils relatively low in organic matter. Contain developed soil horizons. Moisture is inadequate to mature a crop without irrigation in most years.
Entisols	Mineral soils lacking developed soil horizons. Moisture content varies.
Histosols	Soils composed mostly of organic matter. Moisture content varies.
Inceptisols	Mineral soils containing some developed horizons other than one of illuvial clay. Moisture is available to mature a crop.
Mollisols	Mineral soils with thick, dark surface horizons relatively high in organic matter and with high base saturation.
Oxisols	Mineral soils with no weatherable minerals. High in iron and aluminum oxides. Contain no illuvial horizons.
Spodosols	Soils that contain an illuvial horizon of amorphous aluminum and organic matter, with or without amorphous iron. Usually moist or well leached.
Ultisols	Mineral soils with an illuvial clay horizon. Has low base saturation. Generally found in humid climates.
Vertisols	Clayey soils with deep wide cracks at some time in most years. Moisture content varies.

Soil Individual

The major difficulty in classifying soils is that soils are a continuum in which properties may change gradually with distance. Unlike the classification of plants and animals, where an individual ponderosa pine tree or Abert's squirrel is readily recognized, defining the basic soil entity or entities that are to be grouped into classes is a problem. In an attempt to solve this problem the basic diagnostic entity is defined in *Soil Taxonomy* (Soil Survey Staff, 1975) as being the smallest volume within the landscape needed to sample and describe the soil. This description must include the nature and arrangement of soil horizons and the variability of other properties. The term "pedon" is applied to this small volume of soil. A soil individual, called a "polypedon," is composed of a group of contiguous pedons that belong to the same soil series. Through the soil series designation, the polypedon links soil bodies as they exist in nature with *Soil Taxonomy* (Soil Survey Staff, 1975) (Johnson, 1963) (See figure 40).

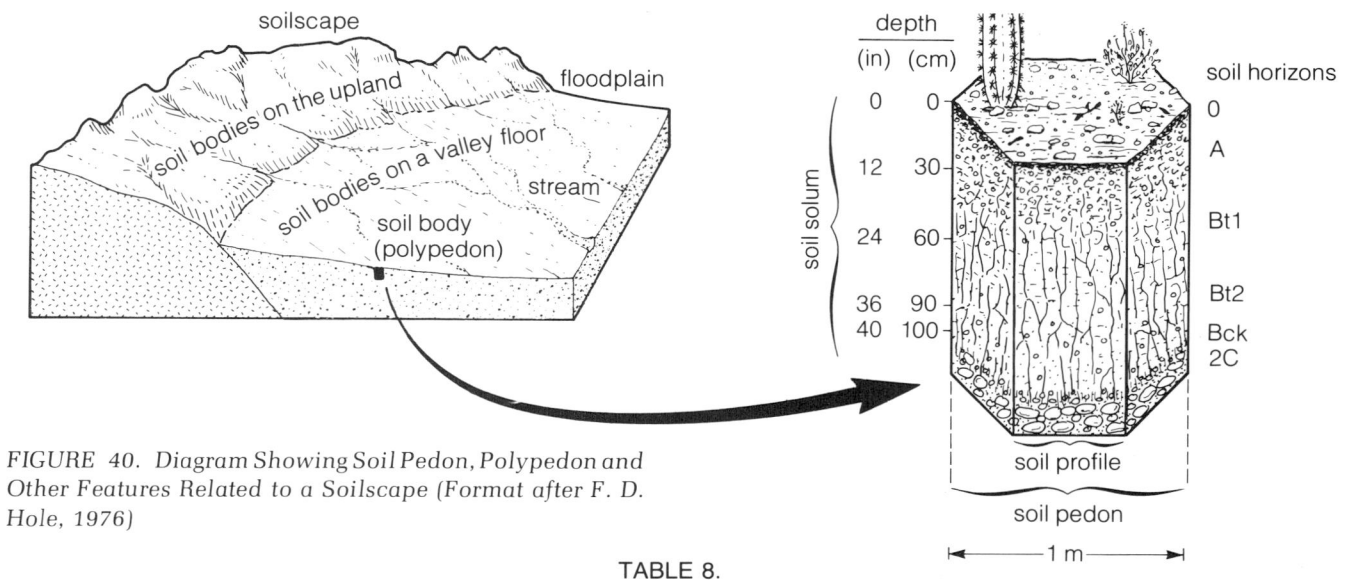

FIGURE 40. Diagram Showing Soil Pedon, Polypedon and Other Features Related to a Soilscape (Format after F. D. Hole, 1976)

TABLE 8.

Soil Order Names and Formative Elements

Order	Formative Element	Derivation*	Mnemonicon
Alfisols	alf	(nonsense syllable)	Pedalfer
Aridisols	id	L.—aridus, dry	arid
Entisols	ent	(nonsense syllable)	recent
Histosols	ist	Gr.—histos, tissue	histology
Inceptisols	ept	L.—inceptum, beginning	inception
Mollisols	oll	L.—mollis, soft	mollify
Oxisols	ox	Fr.—oxide, oxide	oxide
Spodosols	od	Gr.—spodos, wood ashes	Podzol; odd
Utisols	ult	L.—ultimus, last	ultimate
Vertisols	ert	L.—verto, turn	invert

*L. = Latin   G. = Greek   Fr. = French

Source: Soil Survey Staff, 1975.

TABLE 9.

Examples of Formative Elements of Suborder Names*

Formative Element	Meaning
aqu	A soil that is very wet or that has been artificially drained.
arg	A soil that has an illuvial horizon of clay.
fluv	A soil that is composed of recent alluvium.
orth	A soil that is the most representative.
psamm	A soil that has sandy texture, sand or loamy sand.
torr	A soil that is too dry to mature a crop without irrigation.
ud	A soil that is moist but not wet.
ust	A soil that is dry for long periods but moist in a growing season for 90 days or more.
xer	A soil that is moist in winter and dry in summer.

*Formative elements are used in more than one suborder.

Application to Arizona

Orders and Suborders

Six of the 10 soil orders are officially recognized as being in Arizona. The presence of these six soil orders, Alfisols, Aridisols, Entisols, Inceptisols, Mollisols and Vertisols, reflects the wide range of soil-forming conditions in Arizona. In addition to these six orders about 65 ha (160 ac) of Histosols have been described in the floodplain of the Little Colorado River south of Joseph City (USDA Soil Conservation Service, 1976). Of the three remaining soil orders, Spodosols and Ultisols usually form in climates more humid than Arizona. Oxisols are not found outside of tropical regions with the exception of certain relict or exhumed Oxisols such as those in the Sacramento Valley-Sierra Nevada foothills in California (Singer and Nkedi-Kizza, 1980).

Alfisols. Alfisols are soils with light-colored surface layers and clayey subsurface (argillic) horizons. Alfisols, like the Mollisols, occur at higher elevations than Aridisols and are scattered throughout the semiarid and subhumid regions of Arizona. They generally are fairly old soils since probably more than 10,000 years are required to form their argillic horizons. They are mostly in the forested or wooded regions. Boralfs and Ustalfs are the two suborders of Alfisols recognized in Arizona.

Boralfs occur in the cool and cold mountain regions that primarily have coniferous vegetation. They have mean annual soil temperatures of less than 8 C (47 F) at a depth of 50 cm (20 in). Boralfs often are associated with Borolls, but they contain lower amounts of organic matter than Borolls.

Ustalfs occur in warmer and generally drier climates than Boralfs. They are scattered throughout the semiarid and subhumid regions of Arizona in association with Ustolls. Ustalfs usually are reddish and have some accumulations of carbonates in or below the subsoil. They generally lack diagnostic horizons other than ochric epipedons and argillic horizons. Ustalfs with high sodium content occur to a limited extent in the Willcox Playa vicinity.

Aridisols. Aridisols are the developed soils of dry regions. They have light-colored surface layers (ochric epipedons), generally have low amounts of organic matter and have at least one diagnostic subhorizon. Calcium carbonate usually is in some or all parts of the soil. Some soils are high in soluble salts. The two suborders of Aridisols, Orthids and Argids, are widely distributed in arid and semiarid Arizona.

Orthids show little or no textural change with depth. Most are calcareous throughout and many have a distinct accumulation layer of carbonates (calcic horizons). Some Orthids have hardpans cemented by carbonates (petrocalcic horizons) or by silica (duripans). Soils with layers of gypsum accumulation (gypsic horizons) occur to a limited extent in northern and eastern Coconino, Apache and Navajo counties and in the San Simon Valley and possibly elsewhere in Arizona.

Argids have an accumulation of translocated clay in the subsurface layers (argillic horizons). Argids usually are on older landscapes. Soil genesis research findings indicate that at least 10,000 years are required for Argids to form (Soil Survey Staff, 1975). In fact, many soil scientists believe that argillic horizons in Aridisols developed in the past under a more humid climate. Argids may have other diagnostic horizons as well. Calcic horizons are the most common, but a few Argids have petrocalcic horizons or duripans. Sodium has accumulated in some Argids near the Willcox Playa, in the Gila River Valley near Casa Grande and in a few other areas.

Entisols. Entisols are soils that show little or no evidence of horizon development. They may have thin surface horizons with some accumulation of organic matter, but they lack enough alteration of parent materials to form other horizons. Entisols generally are in young landscapes where time has not been sufficient for soils to develop. Some Entisols may occur on older landscapes if they are composed of materials resistant to weathering, or if the climate has been too dry for appreciable soil formation.

Entisols are widely distributed throughout all climates in Arizona. Fluvents, Orthents and Psamments are the three suborders of Entisols in the state.

Fluvents formed in recently deposited alluvium in floodplains and near stream channels on alluvial fans or piedmont slopes. Fluvents lack horizon development because of flooding at fairly frequent intervals, leaving too little time for significant soil formation between alluvial depositions. Fluvents are a significant proportion of the irrigated agricultural land in Arizona.

Most Orthents in Arizona are shallow over rock, usually less than 50 cm (20 in). They typically occur on steep slopes where soil material is removed so fast that time is insufficient for significant horizon development. Other Orthents include deeper and older soils that lack horizon development because of dry environment.

Psamments are sandy Entisols. Soil textures are uniformly coarse, either sand or loamy sand. They include soils of stabilized sand dunes such as in Yuma, Navajo and Apache counties. Psamments also formed in sandy alluvium.

Inceptisols. Inceptisols are relatively young soils that lack horizons of illuvial clay accumulation (argillic horizons). They differ from Entisols because of weak to moderate profile horizonation. The horizons of Inceptisols result mostly from slight to moderate alteration of the parent material. These alterations may be expressed by soil structure development, carbonate removal and hydrolytic weathering to produce clay, form iron oxide minerals and accumulate organic matter. Inceptisols are limited in Arizona primarily to subhumid regions. Only the Ochrepts suborder of Inceptisols is officially recognized in Arizona, but soils of the Andepts suborder have been identified (Hendricks and Davis, 1979).

Ochrepts recently were recognized officially in the soil survey of central Coconino County (Taylor, 1982). Ochrepts also are recognized in several of the national forests (personal communication, Owen Carlton, 1983). Most Ochrepts identified in Arizona are shallow and characterized by weakly developed B horizons (cambic horizons). Because they occur mostly on relatively young geomorphic surfaces, time has limited horizon development.

Andepts are associated with pyroclastic materials such as volcanic ash and cinders. Andepts characteristically have low bulk densities and appreciable amounts of allophane in the clay fraction. They also commonly have higher amounts of organic matter than other soils in the same climatic regime. In Arizona they have been identified on Green's Peak (see Figure 15), a cinder cone in southern Apache County (Hendricks and Davis, 1979). Andepts probably also occur elsewhere in the subhumid and to a limited extent in the semiarid regions of the state. They may have formed in those regions from late Pleistocene to early Holocene cinders and volcanic ash deposits.

Mollisols. Mollisols have thick, dark-colored surface horizons (mollic epipedons). Mollisols in Arizona occur at higher elevations under semiarid and subhumid climates in landscapes covered with grass or grass-tree mixtures, although in some areas they are mostly tree covered. Aquolls, Borolls and Ustolls are the three suborders of Mollisols in Arizona.

Limited Aquolls are in the Willcox Playa vicinity. These soils formed because of poor drainage and are generally high in sodium.

Borolls are cool and cold Mollisols in Arizona's higher elevations that have a mean annual soil temperature of less than 8 C (47 F) at a depth of 50 cm (20 in). Borolls include dark-colored forest and mountain meadow soils and are most common along the Mogollon Rim and in the higher mountain ranges throughout Arizona. Some Borolls have argillic horizons. Others have formed from calcareous parent materials and have calcic horizons.

Ustolls occur under warmer and drier conditions at elevations below Borolls. They are fairly widely distributed throughout Arizona under conditions a little cooler and more moist than those of Aridisols. Some Ustolls have argillic horizons. Others have calcic horizons and a few have petrocalcic horizons.

Vertisols. Vertisols are clayey soils that have deep wide cracks at some time during the year and that are associated primarily with volcanic (basalt) rocks in Arizona. They occur mostly in semiarid to subhumid regions in Yavapai, Coconino, Apache and Navajo counties and in the semiarid San Bernardino Valley of southeastern Cochise County.

Vertisols generally are clayey in all horizons. The clay is composed dominantly of montmorillonite. This mineral causes the soil to undergo considerable shrinking and swelling during drying and wetting cycles. As a result, the soils crack and move both horizontally and vertically with changes in moisture content to produce a mixing action. This action yields uniform textures with depth and causes stones and other rock fragments to move to the soil surface. The soil movement makes Vertisols troublesome for engineering uses and can seriously affect the growth of trees. Building foundations on Vertisols may crack and fences, power lines, highways and trees often become misaligned or variously tilted.

The Usterts suborder represents most Vertisols in Arizona. In addition there are a few Torrerts. Torrerts occur in a more arid climate than Usterts.

Soil Families

Soil families were established based on criteria significant to soil use, management and behavior. Although nine families are defined in *Soil Taxonomy* (Soil Survey Staff, 1975), only three, soil temperature, particle size and mineralogy, are discussed here. These three are the most widely used in soil family groupings. The other six are applied only to a limited number of soils.

Soil Temperature Families. Four soil temperature families are recognized in Arizona: hyperthermic, thermic, mesic and frigid. They correspond to the four soil temperature regimes defined and discussed in the chapter on climate. *Soil Taxonomy* (Soil Survey Staff, 1975) also contains four other temperature families: isohyperthermic, isothermic, isomesic and isofrigid. These families experience less difference between mean winter and summer soil temperatures.

Soil temperature has a strong influence on plant growth. For example, commercial citrus production is limited primarily to hyperthermic soils and cotton production is restricted to hyperthermic and thermic soils. The boundary between the pinyon-juniper woodland and the montane coniferous forest zones in the Southwest generally follows the mesic-frigid boundary, although exceptions are known.

Particle Size Families. Particle size refers to the grain size distribution of the whole soil and includes stones and gravel as well as sand, silt and clay particles. All 11 particle size family classes defined in *Soil Taxonomy* (Soil Survey Staff, 1975) occur in Arizona soils because of the wide range of soil-forming conditions. Their presence reflects the great diversity of soil parent materials and, to a lesser extent, that of climate in Arizona.

Each particle size family is defined by properties that tend to group soils in relation to use and management. For example, soils high in silt are susceptible to erosion. Many soils in southern Arizona floodplains are high in silt. The erodibility of these soils may have been a factor in downcutting some streams during the past 80 to 100 years. Soils high in silt thus are placed in families different from more stable soils low in silt. Plastic and nonplastic soils are distinguished by percent clay content. Particle size families that are plastic have more than 18 percent clay and those that are nonplastic have less than 18 percent clay content. Clayey soil families have more than 35 percent clay and tend to create difficulties in tillage, seedling survival and, in some instances, engineering behavior. Skeletal families, those containing significant amounts of coarse fragments such as gravels, cobbles and stones, usually create problems in soil-plant relations and tillage operations. But they may be potential sources of gravel for construction.

Mineralogy Families. Mineralogy families are defined mostly by the mineral composition of selected size fractions of the soil. With a few exceptions the clayey soils in Arizona are classified either as montmorillonitic or mixed. Montmorillonitic soils, the Vertisols for example, present problems in use, management and behavior not encountered in other soils.

Most medium and coarse-textured soils are in mixed mineralogy families. This indicates that the content of quartz and other resistant minerals is less than 90 percent of the sand fraction and that no other mineral dominates. Consequently, Arizona soils generally have a high inherent fertility, but are deficient in nitrogen, which is not derived from mineral weathering. Although not officially recognized in Arizona, there may be some soils that belong to a siliceous family. These soils contain more than 90 percent quartz and other resistant minerals and have low inherent fertility.

Soils exceptionally high in carbonates or gypsum are classified as carbonatic or gypsic. Soils high in carbonates tend to cause chlorosis in plants and, from an engineering standpoint, ameliorate some properties of noncarbonate clays, such as amount of shrink-swell. High quantities of gypsum may cause rapid corrosion of uncoated steel pipes and concrete structures. Gypsiferous soils materials also are undesirable for foundations and for use in hydraulic structures such as canal and irrigation embankments.

7

Description of the Mapping Units of the Arizona General Soil Map

Basis and Development of the Units

Arizona has a great diversity of soils that generally are distributed in an intricate pattern throughout the state. These patterns are produced by the relatively extreme differences of climate, vegetation, lithology and physiography that are compressed within the state's boundaries. One objective of the Arizona General Soil Map (Plate 1) was to organize this intricate geographic soil distribution pattern so that it is comprehensible. The map then is the basis for organizing and extending knowledge about soils pertinent to use or potential use and to scientific study.

Mapping units used in the Arizona General Soil Map (Plate 1) are soil associations. A soil association consists of a set of geographic bodies that are segments of the soil mantle covering the land surface (Simonson, 1971). Each association consists of polypedons of two or more soils that occur together in a characteristic and repetitious manner (Figure 40). Every delineated body of one soil association has the same major component kinds of soils occurring together in a similar pattern so that the patterns and proportions of major soils are alike within limits among the delineated bodies identified as a single soil association.

County soil maps were used in part to establish soil associations and prepare the map. General soils maps were available for each of the 15 counties but were prepared by different persons over a period of time. Few, if any, attempts were made to correlate soil associations in one county with those in another county at their common boundary. Consequently the mapping units of the counties were not uniform. Levels of generalization varied from county to county due partly to the availability or nonavailability of completed or ongoing standard soil surveys. Variability also was due in part to surveys conducted in response to requests about using soils for different specific purposes. The county general soils maps, then, were adjusted along county boundaries so that similar mapping units in bordering counties could be given similar names. It was necessary also to combine, rename and adjust units to stay within the number of units recommended for general soil maps (Simonson, 1971).

Little was known about the soils in some fairly large areas of the state. A number of field checks were made of these areas, but some were inaccessible due to remoteness and lack of access roads, or to exclusion by the U.S. Department of Defense from bombing ranges and proving grounds. All available information of these areas was culled from sources such as geologic, climatic and topographic maps as well as from general knowledge to describe their soil associations.

Most of the soil associations of the general soil map were named after the dominant soil series that compose them. Some associations were named using subgroup or great group category names including those of the alluvial floodplain soils, of the shallow, rocky soils on low desert mountains and of a few relatively inaccessible areas. A few units were named after both the soil series and subgroup class names. Some associations contain "not soils" in which case the name of a miscellaneous land area was used, rock outcrop, for instance.

Soil Taxonomy (Soil Survey Staff, 1975) prescribes grouping soils by the soil temperature zones in which they occur. Doing so produced four reasonable regions: hyperthermic, thermic, mesic and frigid. Then, the addition of precipitation zones arid, semiarid and subhumid produced three more reasonable units. When the soil units developed from the county general soil maps were modified and grouped into the temperature-precipitation zones, the result was a total of 64 soil associations in seven temperature-precipitation units (Table 4).

It is emphasized that the Arizona General Soil Map is not a guide for the user to specific soil uses in specific areas. This map is a broad guide to area planners engaged in general planning efforts. Proper use of this map is to select areas apparently suitable for specific uses for more detailed investigations. The final determination of suitability of a specific area for a specific use has to be determined by on-site investigations.

Reliability of the mapping units varies considerably. Descriptions of units identified by great group or subgroup names, such as HA1, TS2 and MA5, are very general. Descriptions of those soil associations in remote and inaccessible areas also have a low degree of reliability. Still, such descriptions may be useful for general planning purposes. Soil associations in areas previously described in detailed soil maps have a high degree of reliability for general soil map purposes. Examples of these units are HA2, HA3 and TS11.

Mapping Unit Descriptions

Explanation for block diagrams

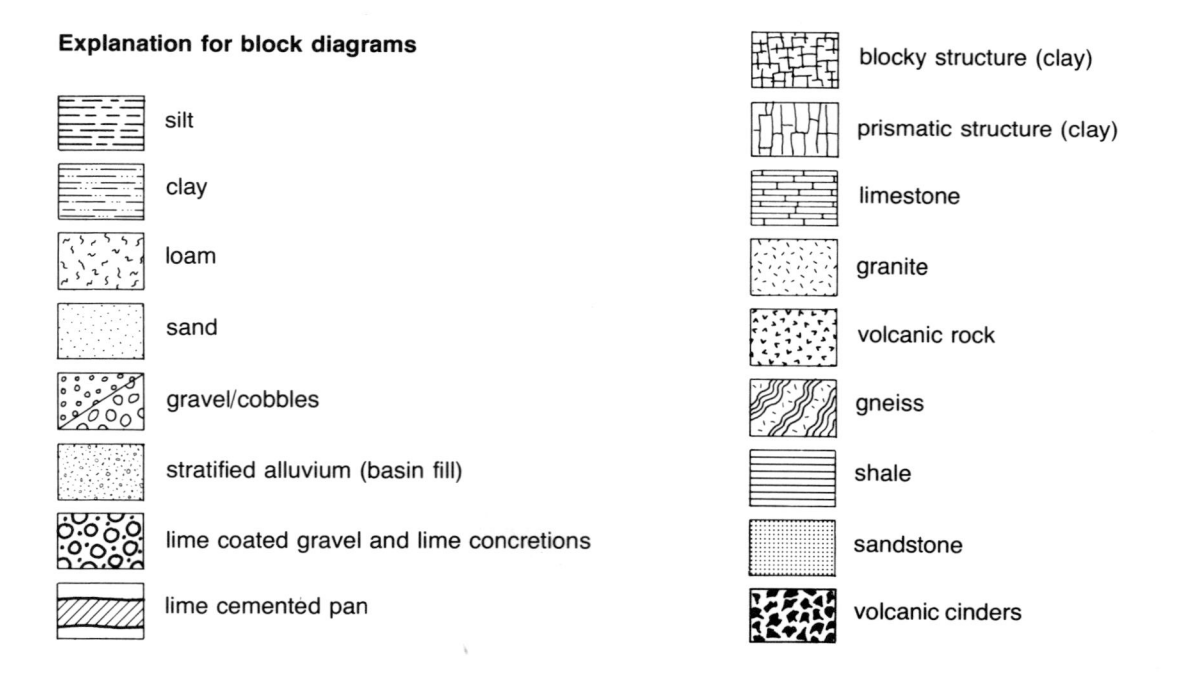

- silt
- clay
- loam
- sand
- gravel/cobbles
- stratified alluvium (basin fill)
- lime coated gravel and lime concretions
- lime cemented pan
- blocky structure (clay)
- prismatic structure (clay)
- limestone
- granite
- volcanic rock
- gneiss
- shale
- sandstone
- volcanic cinders

FIGURE 41. *Explanation of Soil and Rock Symbols in Soilscapes and Profiles*

Hyperthermic Arid
Soils

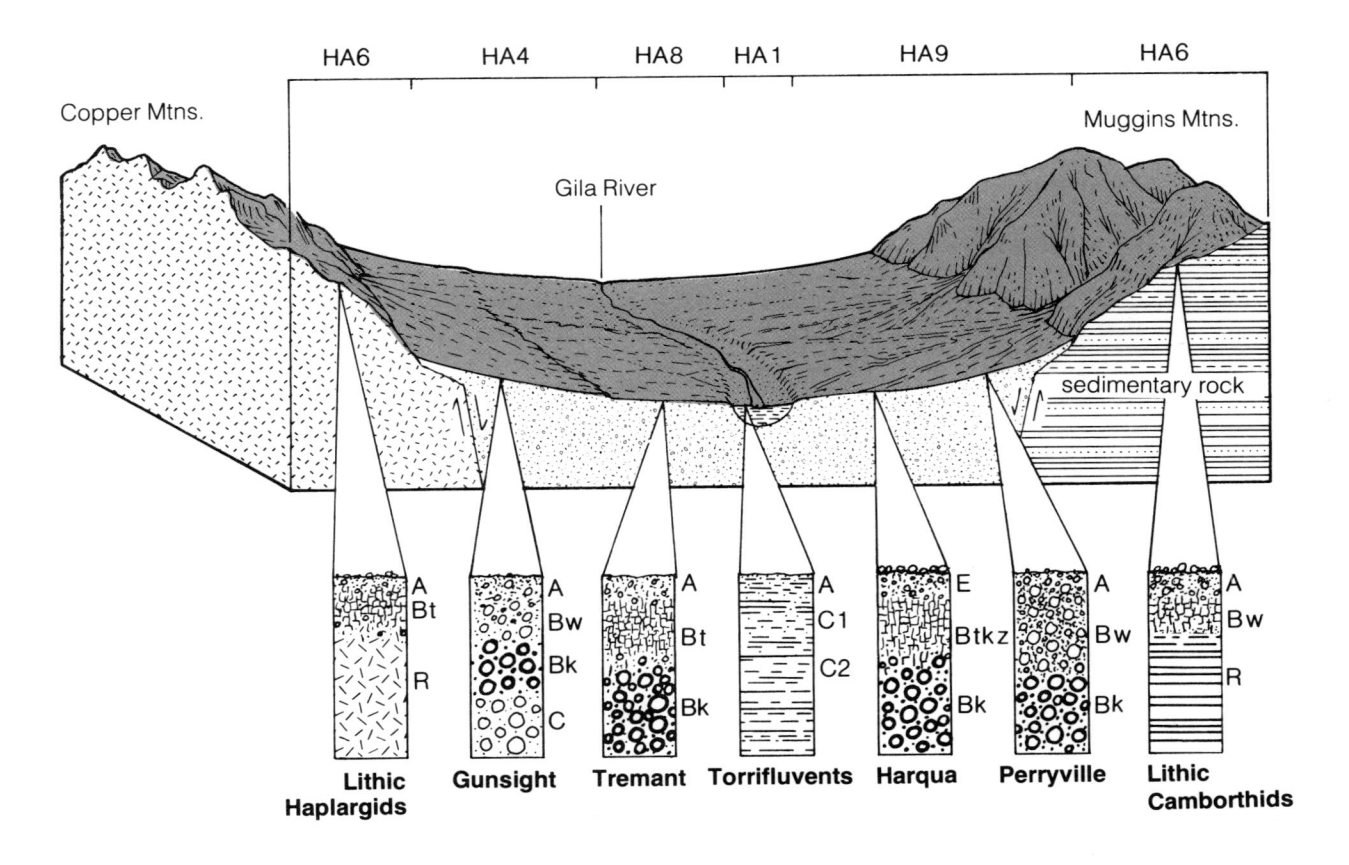

FIGURE 42. Representative Hyperthermic Arid Soils Soilscape and Profiles (D. M. Hendricks)

HA
Hyperthermic Arid Soils

FIGURE 43. *Geographic Distribution of Hyperthermic Arid (HA) Soils in Arizona (D. M. Hendricks)*

- HA1   Torrifluvents Association
- HA2   Casa Grande-Mohall-La Palma Association
- HA3   Mohall-Vecont-Pinamt Association
- HA4   Gunsight-Rillito-Pinal Association
- HA5   Laveen-Rillito Association
- HA6   Lithic Camborthids-Rock Outcrop-Lithic Haplargids Association
- HA7   Laveen-Carrizo-Antho Association
- HA8   Tremant-Coolidge-Mohall Association
- HA9   Harqua-Perryville-Gunsight Association
- HA10 Superstition-Rositas Association

HA Hyperthermic Arid Soils

Hyperthermic Arid Soils have mean annual soil temperatures of 22 C (72 F) or higher. The difference between mean summer and mean winter temperatures is greater than 5 C (9 F) at a depth of 50 cm (20 in) or at the soil-bedrock interface in shallow soils. These soils receive less than 250 mm (10 in) mean annual precipitation. Hyperthermic Arid Soils are at the lower elevations in western and southwestern Arizona. They cover about 8,198,860 ha (20,259,100 ac), 27 percent of Arizona.

HA1 Torrifluvents Association

Deep, stratified, coarse to fine-textured, nearly level to gently sloping soils on floodplains and lower alluvial fans.

Soil Classification Typic Torrifluvents

Percent Slope none to 5

Elevation 30 to 760 m (100 to 2,500 ft)

Mean Annual Precipitation 80 to 250 mm (3 to 10 in)

Winter Precipitation as the Percentage of Annual Precipitation 40 to 70

Mean January Air Temperature 7 to 10 C (45 to 50 F)

Mean July Air Temperature 27 to 32 C (80 to 90 F)

Mean Annual Soil Temperature 22 to 27 C (72 to 80 F)

Frost-Free Days 255 to 320

Area and Percent of State 1,928,400 ha (4,765,000 ac), 6.5

Land Uses irrigated cropland and pasture, rangeland and urban

This association consists of well-drained to somewhat excessively drained soils formed in sandy to clayey recent mixed alluvium on the floodplains and adjacent lower alluvial fans of the lower Santa Cruz, Gila, Salt and Colorado rivers and their major tributaries.

The major Torrifluvents in this map unit include Gilman soils, 20 percent; Antho soils, 20 percent; Vint soils, 15 percent; and Valencia soils, 15 percent. Other similar soils that may be dominant in some areas include Estrella, Avondale, Glenbar, Trix, Holtville, Kofa, Indio, Pimer, Laguna and Gadsden. They constitute about 20 percent of this association. Minor coarse-textured soils and medium and moderately fine-textured soils that are moderately deep over coarse-textured material are about 5 percent of this unit. Included are Carrizo, Brios, Agualt and Maripo and riverwash.

All of the soils in this association are subject to seasonal, brief flooding unless protected. Runoff is slow and the hazard of erosion is usually slight except along entrenched streams where soils are subject to bank cutting, piping and gullying.

About 30 percent of the major irrigated croplands in Arizona are in this association. The principal crops are cotton, grain sorghum, small grains, alfalfa, sugar beets, pasture grasses, vegetables and citrus.

Desert rangeland has low carrying capacity. The native vegetation is mesquite, catclaw, cresosotebush, cacti, bursage, ironwood, arrowweed, saltbush and annual grasses and weeds. Riparian vegetation along major streams includes cottonwood, paloverde, desertwillow and tamarisk, which provide shade for livestock and habitat for many desert wildlife species.

Phoenix and several towns and metropolitan areas are on these soils. Flooding potential is the major limitation of these soils. The permeability of the Valencia soil and the included Estrella, Avondale, Glenbar, Trix, Pimer and Gadsden soils is too limited for septic tank disposal fields. The excessive permeability of the Vint, Antho and included Carrizo, Brios, Agualt and Maripo soils prohibits their use for water retention structures.

HA2 Casa Grande-Mohall-La Palma Association

Deep and moderately deep, moderately fine-textured, nearly level soils on valley plains.

Soil Classification
 Typic Natrargids
 Typic Haplargids
 Typic Durargids

Percent Slope none to 2

Elevation 240 to 490 m (800 to 1,600 ft)

Mean Annual Precipitation 150 to 250 mm (6 to 10 in)

Winter Precipitation as the Percentage of Annual Precipitation 50 to 60

Mean January Air Temperature 7 to 10 C (45 to 50 F)

Mean July Air Temperature 29 to 32 C (85 to 90 F)

Mean Annual Soil Temperature 22 to 27 C (72 to 80 F)

Frost-Free Days 250 to 300

Area and Percent of State 225,420 ha (557,000 ac), 0.8

Land Uses irrigated cropland, desert rangeland, wildlife habitat and building sites

This association of well-drained soils formed in mixed old alluvium is on valley plains and lower slopes.

Casa Grande soils make up about 35 percent of the map unit, Mohall soils 30 percent, La Palma soils 10 percent, Harqua soils 10 percent and minor soils 15 percent. The minor soils are mostly Tremant, Toltec, Vecont, Laveen and Gilman.

Low amounts of forage for livestock and wildlife are in this unit. The native vegetation is creosotebush, mesquite, paloverde, bursage, cacti and annual weeds and grasses. Supplemental grazing is provided in wet years by the annual grasses and forbs.

Most of the irrigated cropland is on Mohall soils and Casa Grande and Harqua soils that have been reclaimed by leaching. The principal crops are cotton, small grains, grain sorghum, alfalfa and pasture grasses.

Nonirrigated soils have low potential for use as rangeland wildlife habitat. Irrigated soils have good potential for use as openland wildlife habitat.

Factors limiting urban development on these soils are slow permeability, high lime zones in the subsoils, moderate depth to hardpan in La Palma soils, and high salinity and sodium in Casa Grande, La Palma and Harqua soils.

HA3 Mohall-Vecont-Pinamt Association

Deep, moderately fine and fine-textured and gravelly, moderately fine-textured, nearly level to gently sloping soils on valley plains.

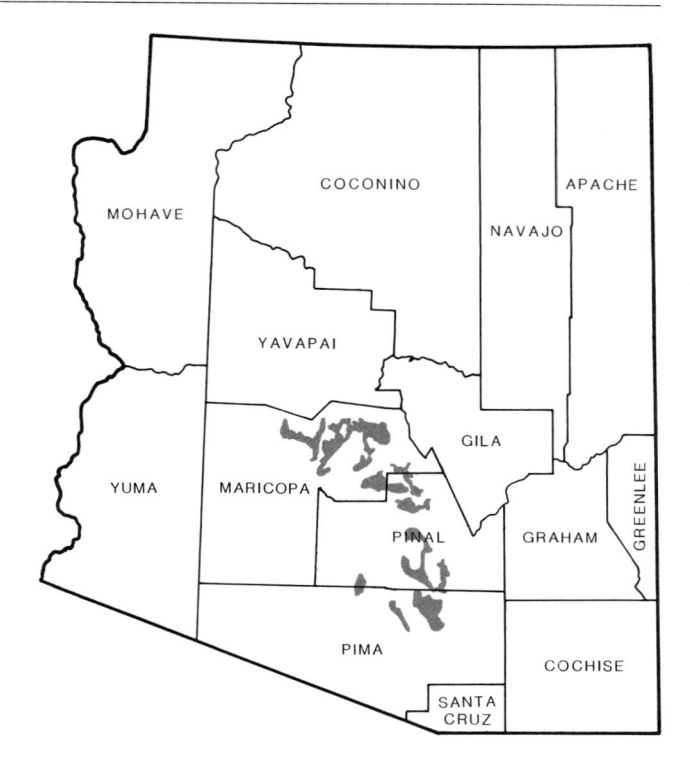

Soil Classification Typic Haplargids

Percent Slope mostly less than 2; some to 8

Elevation 240 to 760 m (800 to 2,500 ft)

Mean Annual Precipitation 150 to 280 mm (6 to 11 in)

Winter Precipitation as the Percentage of Annual Precipitation 45 to 60

Mean January Air Temperature 7 to 10 C (45 to 50 F)

Mean July Air Temperature 29 to 32 C (85 to 90 F)

Mean Annual Soil Temperature 22 to 27 C (72 to 80 F)

Frost-Free Days 230 to 300

Area and Percent of State 350,070 ha (865,000 ac), 1.2

Land Uses desert rangeland and wildlife habitat; some irrigated cropland and homesites

This association of well-drained soils formed in mixed old alluvium on broad valley plains.

Mohall soils make up about 25 percent of the association. Vecont, Pinamt and Tremant soils each represent 20 percent, and minor soils 15 percent. The minor soils include small, intermixed areas of Laveen, Ebon, Contine and Rillito, and narrow bodies of Torrifluvents along the drainageways.

Most of the native vegetation is paloverde, mesquite, creosotebush, ironwood, cacti, bursage and annual weeds and grasses.

Low amounts of forage for livestock and wildlife grow in the unit. Supplemental grazing is provided in wet seasons by annual grasses and forbs.

Mohall soils are well suited for growing arid-adapted crops when irrigation water is available. Slow intake and permeability of Vecont soils require careful water management. Tremant and Pinamt soils usually are not cultivated.

Factors limiting the potential of these soils for homesites are the moderately slow or slow permeability for use as septic tank absorption fields, high shrink-swell in Vecont soils and moderate shrink-swell in Mohall and Tremant soils. The gravelly, high-lime substratum material of the Tremant and Pinamt soils is poor for use as topsoil or landfill cover.

HA4 Gunsight-Rillito-Pinal Association

Deep and shallow, limy, gravelly, medium and moderately coarse-textured, nearly level to strongly sloping soils on alluvial surfaces and valley plains.

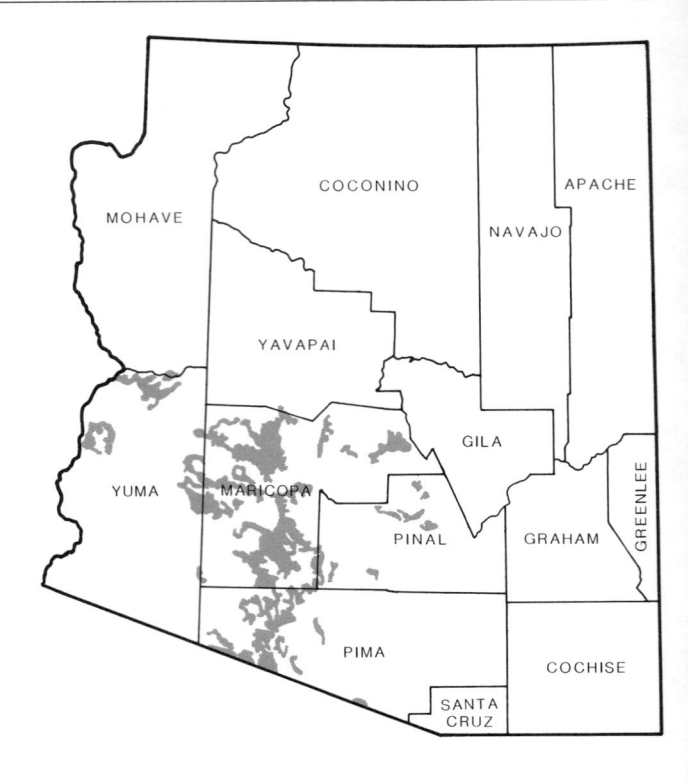

Soil Classification
  Typic Calciorthids
  Typic Durorthids
  Typic Paleorthids

Percent Slope mostly none to 5; some to 15 or more

Elevation 120 to 760 m (400 to 2,400 ft)

Mean Annual Precipitation 100 to 250 mm (4 to 10 in)

Winter Precipitation as the Percentage of Annual Precipitation 55 to 65

Mean January Air Temperature 7 to 10 C (45 to 50 F)

Mean July Air Temperature 29 to 32 C (85 to 90 F)

Mean Annual Soil Temperature 22 to 27 C (72 to 80 F)

Frost-Free Days 255 to 320

Area and Percent of State 832,470 ha (2,057,000 ac), 2.8

Land Uses mostly desert rangeland and wildlife habitat; some homesites

This association of well-drained soils is on broad, shallowly dissected alluvial fans and valley slopes. The soils formed in calcareous, old mixed alluvium derived from volcanic rocks, schist, limestone and granite.

Gunsight and Rillito soils each constitute about 30 percent of this association. Pinal soils cover about 15 percent, Cavelt soils 10 percent and other minor soil inclusions 15 percent. The principal minor soils are Ajo, Cipriano, Laveen, Perryville, Ligurta, Cristobal and Harqua along with small areas of rock outcrop and Torrifluvents in the drainageways. Areas of these soils are in the Organ Pipe National Monument.

The soils produce little forage for livestock and wildlife. Native vegetation consists of widely spaced paloverde, mesquite, cacti, creosotebush, bursage, ironwood, saltbush and annuals. Limited forage is provided by annual grasses and forbs following rainy periods. Larger vegetation along drainageways provides the best wildlife habitat.

Factors limiting these soils for homesite and community uses are high lime and excessive gravel contents. Shallow depth to hardpan and low water capacity in the Pinal and Cavelt soils restrict plant growth. Excavations require the use of heavy equipment in most places. The soils are fairly well suited to support for low buildings without basements. Gunsight and Rillito soils are suitable for use as septic tank absorption fields but have excessive seepage for use as water retention structures.

HA5 Laveen-Rillito Association

Deep, medium and moderately coarse-textured, nearly level to gently sloping, limy soils on low alluvial surfaces and valley plains.

Soil Classification Typic Calciorthids

Percent Slope mostly none to 3

Elevation 180 to 550 m (600 to 1,800 ft)

Mean Annual Precipitation 130 to 250 mm (5 to 10 in)

Winter Precipitation as the Percentage of Annual Precipitation 60 to 65

Mean January Air Temperature 7 to 10 C (45 to 50 F)

Mean July Air Temperature 29 to 32 C (85 to 90 C)

Mean Annual Soil Temperature 22 to 27 C (72 to 80 F)

Frost-Free Days 250 to 300

Area and Percent of State 179,080 ha (442,500 ac), 0.6

Land Uses irrigated cropland, desert rangeland, wildlife habitat and communities

This association of limy, well-drained soils formed in calcareous, old alluvium derived from limestone and other rocks.

Laveen soils make up about 40 percent of the association, Rillito soils 35 percent and minor soils 25 percent. The minor soils are mostly Mohall, Coolidge, Tremant, Perryville, Antho and Gilman.

Irrigated Laveen soils have good and Rillito soils have fair potential for cropland use. Crops are alfalfa, cotton, small grains, grain sorghum, safflower and sugar beets. Lime sensitive plants may show some chlorosis on these soils. Rillito soils require more frequent irrigation due to lower available water capacity.

Nonirrigated areas produce limited forage for livestock and wildlife. Native vegetation consists of sparse stands of creosotebush, saltbush, bursage, mesquite, paloverde, ironwood, cacti and annual weeds, forbs and grasses. Annuals that grow following rainy periods provide supplemental seasonal grazing.

Where irrigated, these soils are good openland wildlife habitat for doves, quail, rabbits, songbirds, rodents, foxes, badgers, coyotes and snakes.

The soils in this association generally have only slight limitations for most community uses such as homesites, septic tank absorption fields, sanitary landfill and excavations. They are somewhat dusty for use as playgrounds, camping sites and picnic areas. Water retention structures such as sewage lagoons and earthen ponds may seep excessively. Substratum materials may cause chlorosis in plants if used as topsoil.

HA6 Lithic Camborthids-Rock Outcrop-Lithic Haplargids Association

Shallow, very gravelly and cobbly, moderately coarse to moderately fine-textured, gently sloping to very steep soils and rock outcrop on hills and mountains.

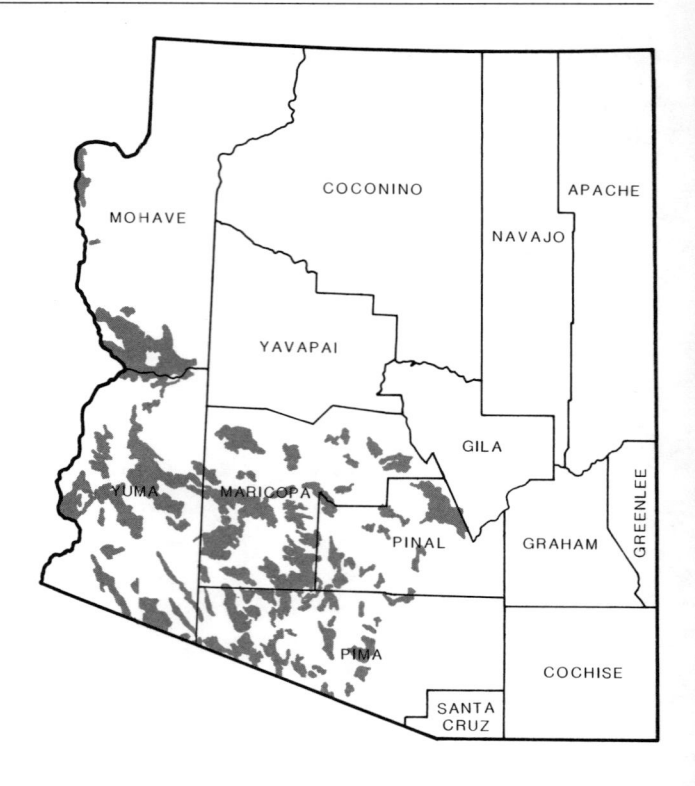

Soil Classification
Lithic Camborthids
Typic Durorthids
Lithic Haplargids

Percent Slope 2 to 60 or more

Elevation 90 to 1,490 m (300 to 4,900 ft)

Mean Annual Precipitation 100 to 250 mm (4 to 10 in)

Winter Precipitation as the Percentage of Annual Precipitation 50 to 65

Mean January Air Temperature 7 to 10 C (45 to 50 F)

Mean July Air Temperature 29 to 32 C (85 to 90 F)

Mean Annual Soil Temperature 22 to 27 C (72 to 80 F)

Frost-Free Days 250 to 300

Area and Percent of State 3,094,130 ha (7,645,500 ac), 10.5

Land Uses desert rangeland, wildlife habitat, game refuge, military proving grounds and bombing range and recreation; also heavy equipment proving grounds and city and county parks

This association consists of well-drained, shallow soils and rock outcrop on hills and low mountains. The soils formed in materials weathered residually from granitic rocks, schists, volcanic tuffs and conglomerates, basalt and some shale and sandstone.

Lithic Camborthids make up about 20 percent of the association, Typic Durorthids 15 percent, rock outcrop 35 percent, Lithic Haplargids 15 percent and minor soils 15 percent. Lomitas soils are representative of the Lithic Camborthids. Cherioni soils are the dominant Lithic Durorthids and Gachado soils the Lithic Haplargids. The minor soils are mostly Gunsight, Harqua, Rillito, Ligurta, Cristobal and Pinamt. Torrifluvents and Torriorthents are along drainageways.

Rock outcrop tops the mountain ranges as near-vertical ledges, escarpments and pinnacles, and occurs scattered throughout the shallow soils at the lower elevations.

Dominant native vegetation is creosotebush, paloverde, ironwood, ocotillo, bursage, jojoba, range ratany, saguaro and other cacti, bush muhly, big galleta, tridens, Rothrock and black grama, fluffgrass and annuals.

These soils accommodate little livestock grazing due to rockiness and steepness of the terrain and too low rainfall. Parts of the unit are prime habitat for bighorn sheep.

Factors limiting these soils for homesite and other community uses are shallow depth to rock and slopes of more than 8 percent. Rock fragments on the surface limit use for playgrounds, campgrounds and picnic areas. However, some smoother areas can be used for these purposes or the loose rocks may be removed in others. Many areas, such as the Ajo Mountains, are scenic and unique.

HA7 Laveen-Carrizo-Antho Association

Deep, medium-textured, limy and gravelly, moderately coarse and coarse-textured, nearly level to moderately sloping soils on floodplains and dissected alluvial surfaces.

Soil Classification
  Typic Calciorthids
  Typic Torriorthents
  Typic Torrifluvents

Percent Slope mostly none to 8; some steep slopes

Elevation 150 to 610 m (500 to 2,000 ft)

Mean Annual Precipitation 100 to 200 mm (4 to 8 in)

Winter Precipitation as the Percentage of Annual Precipitation 65 to 70

Mean January Air Temperature 7 to 10 C (45 to 50 F)

Mean July Air Temperature 29 to 32 C (85 to 90 F)

Mean Annual Soil Temperature 22 to 27 C (72 to 80 F)

Frost-Free Days 250 to 300

Area and Percent of State 154,390 ha (381,500 ac), 0.5

Land Uses rangeland, wildlife habitat, homesites and recreation

This association consists of well-drained and excessively drained soils on dissected old alluvial fans and sandy floodplains. They formed in transported alluvium derived from mixed igneous and sedimentary rocks.

Laveen soils make up about 30 percent of this association, Carrizo soils 25 percent, Antho soils 25 percent and minor soils 20 percent. The minor soils are mostly Rillito, Laguna, Brios, Vint, Gilman, Ligurta, Cristobal and Cavelt, and riverwash.

The native vegetation is a sparse growth of creosotebush, bursage, paloverde, catclaw, filaree, cacti, threeawn and annual weeds and grasses.

These soils produce little forage for livestock and wildlife. Seasonal grazing is possible for short periods following rainy periods. Riparian vegetation along some drainageways furnishes cover and some food for wildlife.

Laveen and Antho soils are good for homesite and other community uses, if protected from flooding. They may be somewhat dusty for use as playgrounds and campgrounds. Carrizo soils have very rapid permeability. They are subject to flooding and excessive seepage prohibits their use for sanitary facilities. They are good potential sources of sand and gravel.

HA8 Tremant-Coolidge-Mohall Association

Deep, moderately coarse and gravelly, moderately fine-textured, nearly level and gently sloping soils on low fan surfaces and valley plains.

Soil Classification
  Typic Haplargids
  Typic Calciorthids

Percent Slope mostly none to 2; some to 5

Elevation 120 to 760 m (400 to 2,500 ft)

Mean Annual Precipitation 80 to 250 mm (3 to 10 in)

Winter Precipitation as the Percentage of Annual Precipitation 50 to 65

Mean January Air Temperature 7 to 10 C (45 to 50 F)

Mean July Air Temperature 29 to 32 C (85 to 90 F)

Mean Annual Soil Temperature 22 to 27 C (72 to 80 F)

Frost-Free Days 250 to 300

Area and Percent of State 890,340 ha (2,200,000 ac), 3.0

Land Uses desert rangeland, wildlife habitat, irrigated cropland, homesites and military reservations

This association consists of well-drained soils in broad valleys and adjacent lower alluvial fans. The soils formed in old mixed alluvium derived from igneous and calcareous sedimentary rocks.

Tremant soils make up about 40 percent of this association, Coolidge soils 15 percent, Laveen soils 15 percent and Mohall soils 15 percent. Included in the mapping unit are large areas of Valencia, Rillito and Pinamt soils. Also included in the unit are a few low hills and mountains that have rocky, shallow soils and areas of moderately coarse-textured recent stratified alluvial soils in drainageways. The minor soils make up about 15 percent of the association.

Nonirrigated areas provide some forage for livestock and wildlife. The vegetation consists of creosotebush, scattered paloverde, mesquite and ironwood, saltbush, bursage, cacti and annuals. Limited grazing of annual grasses is possible following wet seasons. Wildlife habitat is mostly along drainageways.

Only small areas are irrigated cropland. They produce alfalfa, small grain, grain sorghum and other arid-adapted crops. Tremant and Coolidge soils are somewhat droughty.

Factors limiting these soils for homesite and community uses are the moderately slow permeability and moderate shrink-swell of the Tremant and Mohall soils. Laveen and Coolidge soils generally are well suited for building sites and sanitary facilities. The limy substrata of all of these soils are poor for topsoil and landfill cover. Denuded areas are dusty.

HA9 Harqua-Perryville-Gunsight Association

Deep, gravelly, moderately fine-textured, and gravelly, limy, medium-textured, nearly level to moderately sloping soils on old fan surfaces.

Soil Classification
  Typic Haplargids
  Typic Calciorthids

Percent Slope mostly 1 to 8; some to 15

Elevation 90 to 460 m (300 to 1,500 ft)

Mean Annual Precipitation 80 to 200 mm (3 to 8 in)

Winter Precipitation as the Percentage of Annual Precipitation 55 to 65

Mean January Air Temperature 7 to 10 C (45 to 50 F)

Mean July Air Temperature 29 to 32 C (85 to 90 F)

Mean Annual Soil Temperature 22 to 27 C (72 to 80 F)

Frost-Free Days 250 to 300

Area and Percent of State 394,180 ha (974,000 ac), 1.3

Land Uses wildlife habitat and military proving grounds; some livestock grazing and building sites

This association consists of well-drained, gravelly soils in broad intermountain valleys. They formed in old alluvium derived from volcanic and calcareous sedimentary rocks. Many areas are saline and sodic.

Harqua soils make up about 35 percent of the mapping unit, Perryville soils 25 percent, Gunsight soils 20 percent and minor soils 20 percent. The minor soils are mostly Coolidge, Rillito, Ligurta and Cristobal and small, rocky areas of Lomitas and Cherioni.

Parts of the unit are in the Kofa and Cibola National Wildlife refuges. The dominant native vegetation is a sparse cover of creosotebush, saltbush, bursage, cholla and other cacti, and annual grasses and weeds.

The major limitations of these soils for community uses are moderate to high gravel content of all soils, high salt and alkali content and moderate shrink-swell in Harqua soils and high lime content of the Perryville and Gunsight soils. Perryville and Gunsight soils are suitable sites for low buildings and local roads, and for use as roadfill. The soils are poor for topsoil due to high lime and gravel content. Dust may be a problem in denuded and disturbed areas.

HA10 Superstition-Rosita Association

Deep, coarse-textured, nearly level and undulating soils on terraces.

Soil Classification
 Typic Calciorthids
 Typic Torripsamments

Percent Slope mostly none to 2; some to 15

Elevation 30 to 305 m (100 to 1,000 ft)

Mean Annual Precipitation 80 to 150 mm (3 to 6 in)

Winter Precipitation as the Percentage of Annual Precipitation 60 to 65

Mean January Air Temperature 7 to 10 C (45 to 50 F)

Mean July Air Temperature 29 to 32 C (85 to 90 F)

Mean Annual Soil Temperature 23 to 26 C (74 to 79 F)

Frost-Free Days 280 to 320

Area and Percent of State 150,390 ha (371,600 ac), 0.5

Land Uses irrigated cropland, urban and military

This association consists of somewhat excessively drained soils.

Superstition soils make up about 60 percent of the association and Rositas soils 20 percent. The remaining 20 percent includes intermixed areas of Coolidge, Rillito and Harqua soils and a few small areas of rocky buttes and sand dunes.

The natural vegetation consists of widely spaced creosotebush, big galleta, white bursage, turkshead and annual grasses and forbs.

The irrigated crops grown are mostly alfalfa, citrus and a few dates. Numerous frequent irrigations are required for these crops and much of the water applied percolates to the aquifer. Where they are used, new methods of irrigation such as drip or bubbler irrigation have reduced water losses. Cover crops and windbreaks usually are necessary while establishing crops to prevent soil loss by wind erosion.

Factors limiting the potential of these soils for community uses are droughtiness, blowing soil, excessive seepage, loose sandy sufaces and soil caving in excavations. These limitations can be partially overcome by special measures such as windbreaks, drip irrigation and soil stabilization.

Thermic Arid Soils

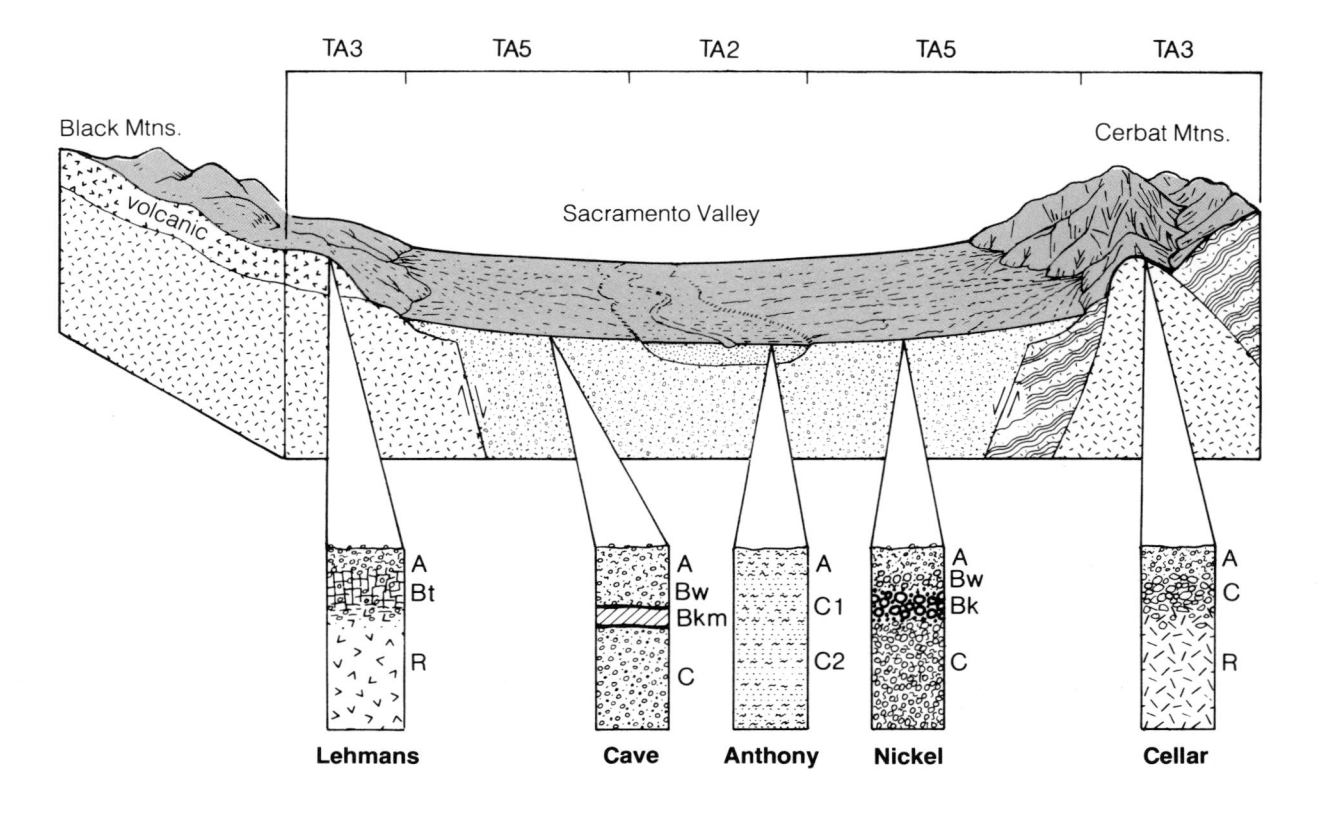

FIGURE 44. Representative Thermic Arid Soils Soilscape and Profiles (D. M. Hendricks)

TA Thermic Arid Soils

FIGURE 45. Geographic Distribution of Thermic Arid (TA) Soils in Arizona (D. M. Hendricks)
- TA1 Torriorthents-Camborthids-Rock Outcrop Association
- TA2 Anthony-Vinton-Agua Association
- TA3 Lithic Torriorthents-Rock Outcrop-Lithic Haplargids Association
- TA4 Latene-Anthony-Tres Hermanos Association
- TA5 Paleorthids-Calciorthids-Torriorthents Association

TA Thermic Arid Soils

Thermic Arid Soils have mean annual soil temperatures of 15 to 22 C (59 to 72 F). The difference between mean summer and mean winter temperatures is greater than 5 C (9 F) at 50 cm (20 in) or at the soil-bedrock interface in shallow soils. The mean annual precipitation associated with these soils falls between 130 to 250 mm (5 to 10 in). These soils are primarily at low to intermediate elevations in northwestern Arizona as well as along and in the Grand Canyon. Thermic Arid Soils cover about 2,263,240 ha (5,592,400 ac), or 8 percent of Arizona.

TA1 Torriorthents-Camborthids-Rock Outcrop Association

Shallow and moderately deep soils and rock outcrop of the canyons, cliffs and mesas.

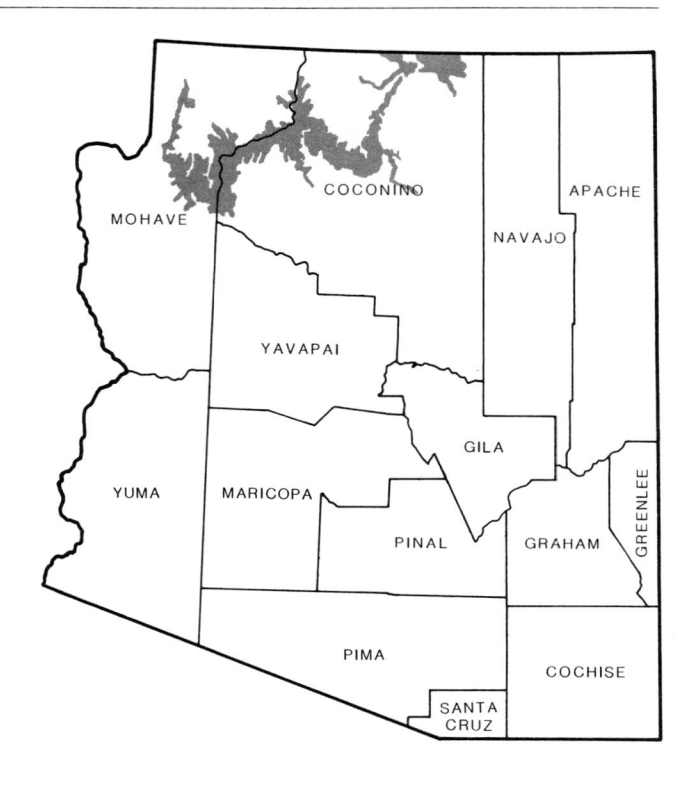

Soil Classification
Torriorthents
Camborthids

Percent Slope 5 to vertical, some to 1,820 m (6,000 ft)

Elevation 360 to 2,430 m (1,200 to 8,000 ft)

Mean Annual Precipitation mostly 150 to 250 mm (6 to 10 in); range is 150 to 500 mm (6 to 20 in)

Winter Precipitation as the Percentage of Annual Precipitation 60 to 65

Mean January Air Temperature -1 to 7 C (30 to 45 F)

Mean July Air Temperature 21 to 27 C (70 to 80 F)

Mean Annual Soil Temperature mostly 15 to 22 C (59 to 72 F); range is 8 10 24 C (47 to 75 F)

Frost-Free Days 90 to 340

Area and Percent of State 779,690 ha (1,926,600 ac), 2.7

Land Uses recreation, wildlife habitat and some cropland

This association consists primarily of the Grand Canyon area and the major tributaries to the Colorado River. These are shallow and moderately deep, moderately sloping to extremely steep, gravelly, cobbly and stony, moderately coarse to moderately fine-textured soils developed in colluvial and on residual materials such as limestone, sandstone and shale bedrock.

Torriorthents make up about 65 percent of this association, Camborthids about 15 percent and rock outcrop about 15 percent. About 5 percent of the mapping unit is Ustorthents, recent alluvial soils along the tributary drainageways and the Colorado River, very steep talus materials and water, including the Colorado River and the Arizona portions of Lake Mead and Lake Powell.

The contrasting hues of reds, grays and whites and the vastness of the relief formed by millions of years of geologic erosion make this area one of unparalleled scenic grandeur. Recreation activities include hiking, camping, boating and fishing. Most of the area is in the Lake Mead National Recreational Area and the Grand Canyon National Monument, both administered by the National Park Service. Several campgrounds and resorts are in the area. The Havasupai Indians farm a small area near the bottom of the Grand Canyon. Vegetation ranges from desert shrubs in the hot, arid inner canyon to juniper and pine at the higher elevations.

Factors limiting the potential of these soils for development of homesites or campgrounds are the shallow depths to bedrock and the steep slopes.

TA2 Anthony-Vinton-Agua Association

Deep, medium to coarse-textured, nearly level to gently sloping soils on floodplains and low alluvial fans.

Soil Classification Typic Torrifluvents

Percent Slope none to 5

Elevation 610 to 1,220 m (2,000 to 4,000 ft)

Mean Annual Precipitation 130 to 250 mm (5 to 10 in)

Winter Precipitation as the Percentage of Annual Percipitation 60 to 70

Mean January Air Temperature 4 to 7 C (40 to 45 F)

Mean July Air Temperature 24 to 29 C (75 to 85 F)

Mean Annual Soil Temperature 15 to 22 C (59 to 72 F)

Frost-Free Days 200 to 275

Area and Percent of State 448,210 ha (1,107,500 ac), 1.5

Land Uses rangeland, wildlife habitat and homesites; some irrigated cropland and pasture along Big Sandy River

This association consists of well-drained soils on the floodplains of intermountain valleys. The soils formed in recent mixed alluvium derived from granite and other rocks.

Anthony soils make up about 35 percent of the association, Vinton soils 30 percent and Agua soils 20 percent. Included in the mapping unit are Gila, Glendale, Tobler and Harrisburg soils in drainageways, and small areas of Rillito and Nickel on low fan surfaces. These minor soils make up about 15 percent of the unit.

All of these soils are subject to brief flooding during wet seasons unless protected.

Vegetation is creosotebush, paloverde, mesquite, cacti, catclaw and annual grasses and weeds. Joshua trees grow north of Red Lake and cottonwood and tamarisk grow along Big Sandy River.

The rangeland generally has low carrying capacity due to the low rainfall and the dominance of low-palatability brushy species. Areas that receive runoff from adjacent uplands support vegetation for seasonal grazing. Riparian vegetation along major streams provides good habitat for various wildlife species.

Possible flooding is a severe limitation for homesite and many other community uses. If protected from flooding, these soils are suitable for use as homesites and sanitary facilities. Excessive seepage, however, prohibits their use for water impoundment structures. These soils are somewhat droughty for growing lawns and landscape plants.

TA3 Lithic Torriorthents-Rock Outcrop-Lithic Haplargids Association

Shallow, gravelly and cobbly, moderately sloping to very steep soils and rock outcrop on hills and mountains.

Soil Classification
  Lithic Torriorthents
  Lithic Haplargids

Percent Slope mostly 30 to 50; range is 5 to 70 or more

Elevation 670 to 1,730 m (2,200 to 5,700 ft)

Mean Annual Precipitation 130 to 250 mm (5 to 10 in)

Winter Precipitation as the Percentage of Annual Precipitation 60 to 70

Mean January Air Temperature 2 to 10 C (35 to 50 F)

Mean July Air Temperature 18 to 32 C (65 to 90 F)

Mean Annual Soil Temperature 17 to 22 C (62 to 72 F)

Frost-Free Days 200 to 275

Area and Percent of State 443,510 ha (1,095,900 ac), 1.5

Land Uses rangeland, wildlife habitat and recreation; some building sites and mines

This association consists of well-drained, shallow soils and rock outcrop on hills and low mountains in hot, arid areas. The soils formed in residuum weathered from granite, gneiss, schist and volcanic rocks, including tuffs.

Lithic Torriorthents (Cellar and House Mountain soils) make up about 35 percent of the association, rock outcrop 30 percent, Lithic Haplargids (Chiricahua and Lehmans soils) 20 percent and minor soils 15 percent. The minor soils are mostly small areas of Lampshire, Moano, Continental, Pinaleno and Tres Hermanos, and narrow bands of moderately coarse and coarse-textured, stratified soils along drainageways.

Dominant native vegetation is paloverde, mesquite, creosotebush, ocotillo, blackbrush, cacti, threeawn, galleta, bush muhly and blue, slender, sideoats and black grama.

Steepness and stoniness of the soils in this association limit livestock grazing to the smoother, less sloping areas. White- and blacktail deer and bighorn sheep are the primary big-game animals of these areas.

Most homesites and recreation areas are on soils adjacent to Lake Mead. The major soils of the association are severely limited for use as building sites due to excessive slopes and shallowness to bedrock. Chiricahua and Lehmans soils also have high shrink-swell.

TA4 Latene-Anthony-Tres Hermanos Association

Deep, moderately coarse to moderately fine-textured, nearly level to moderately steep soils on alluvial fan surfaces and valley plains.

Soil Classification
  Typic Calciorthids
  Typic Torrifluvents
  Typic Haplargids

Percent Slope mostly none to 15; some to 30

Elevation 610 to 1,060 m (2,000 to 3,500 ft)

Mean Annual Precipitation 150 to 250 mm (6 to 10 in)

Winter Precipitation as the Percentage of Annual Precipitation 60 to 65

Mean January Air Temperature 7 to 10 C (45 to 50 F)

Mean July Air Temperature 29 to 32 C (85 to 90 F)

Mean Annual Soil Temperature 17 to 22 C (62 to 72 F)

Frost-Free Days 210 to 275

Area and Percent of State 340,270 ha (840,800 ac), 1.2

Land Uses rangeland and wildlife habitat; some homesites and irrigated cropland

This association consists of well-drained soils on old alluvial fans and valley plains. They formed in mixed alluvium from igneous and calcareous sedimentary rocks.

Latene and Anthony soils each constitute about 30 percent of this association and Tres Hermanos soils 20 percent. Small areas of Cave, Continental, Gila, Mohave, Nickel, Pinaleno and Whitlock soils make up about 20 percent of the mapping unit.

Most of the native vegetation is bursage, creosotebush, galleta, snakeweed, paloverde, cacti and annual grasses and forbs. Forage production is low due to limited rainfall, but annual grasses and forbs provide some grazing following wet seasons. The best wildlife habitat is along drainageways.

These soils have fair to good potential for use as homesites if the following limitations are overcome: Anthony soils are subject to flooding and are moderately droughty; Latene and Tres Hermanos soils have high lime content on the substrate that is poor for topsoil; and some areas of Tres Hermanos soils have excessive slope.

TA5 Paleorthids-Calciorthids-Torriorthents Association

Shallow and deep, gravelly, medium to coarse-textured, limy, gently sloping to moderately steep soils on valley slopes and hills.

Soil Classification
  Paleorthids
  Calciorthids
  Torriorthents

Percent Slope 1 to 30; a few near-vertical escarpments

Elevation 610 to 1,220 m (2,000 to 4,000 ft)

Mean Annual Precipitation 150 to 300 mm (6 to 12 in)

Winter Precipitation as the Percentage of Annual Precipitation 60 to 70

Mean January Air Temperature 2 to 10 C (35 to 50 F)

Mean July Air Temperature 24 to 32 C (75 to 90 F)

Mean Annual Soil Temperature 15 to 22 C (59 to 72 F)

Frost-Free Days 200 to 275

Area and Percent of State 251,560 ha (621,600 ac), 0.9

Land Uses rangeland and wildlife habitat; some homesites and recreation near Lake Mead and irrigated cropland along Virgin River

This association consists of well-drained, calcareous soils on valley slopes, mesas and ridges. The soils formed in calcareous, gravelly alluvium derived from sedimentary and volcanic rocks.

Paleorthids make up about 30 percent of the association, Calciorthids 30 percent, Torriorthents 15 percent and minor soils 25 percent. Cave and Morman Mesa soils are the dominant Paleorthids, Nickel the dominant Calciorthid and Arizo the dominant Torriorthent. The minor soils include Arado, Winkel, Tonopah, Flattop, Bitter Spring, Moapa and St. Thomas. Also included are small areas of rock outcrop and narrow bodies of Torrifluvents along drainageways.

The native vegetation is mostly a sparse cover of creosotebush, blackbrush, bursage, Mormon-tea, desert almond, big galleta, Indian ricegrass, cacti, Mohave yucca and annual grasses and forbs.

These soils have low potential for forage production due to low rainfall and shallow soils. The best wildlife habitat is along drainageways.

Factors limiting the potential for use as homesites are shallow depths of the Paleorthids; high lime content of the Paleorthids and Calciorthids; and flooding hazard and droughtiness of the Torriorthents.

Thermic Semiarid Soils

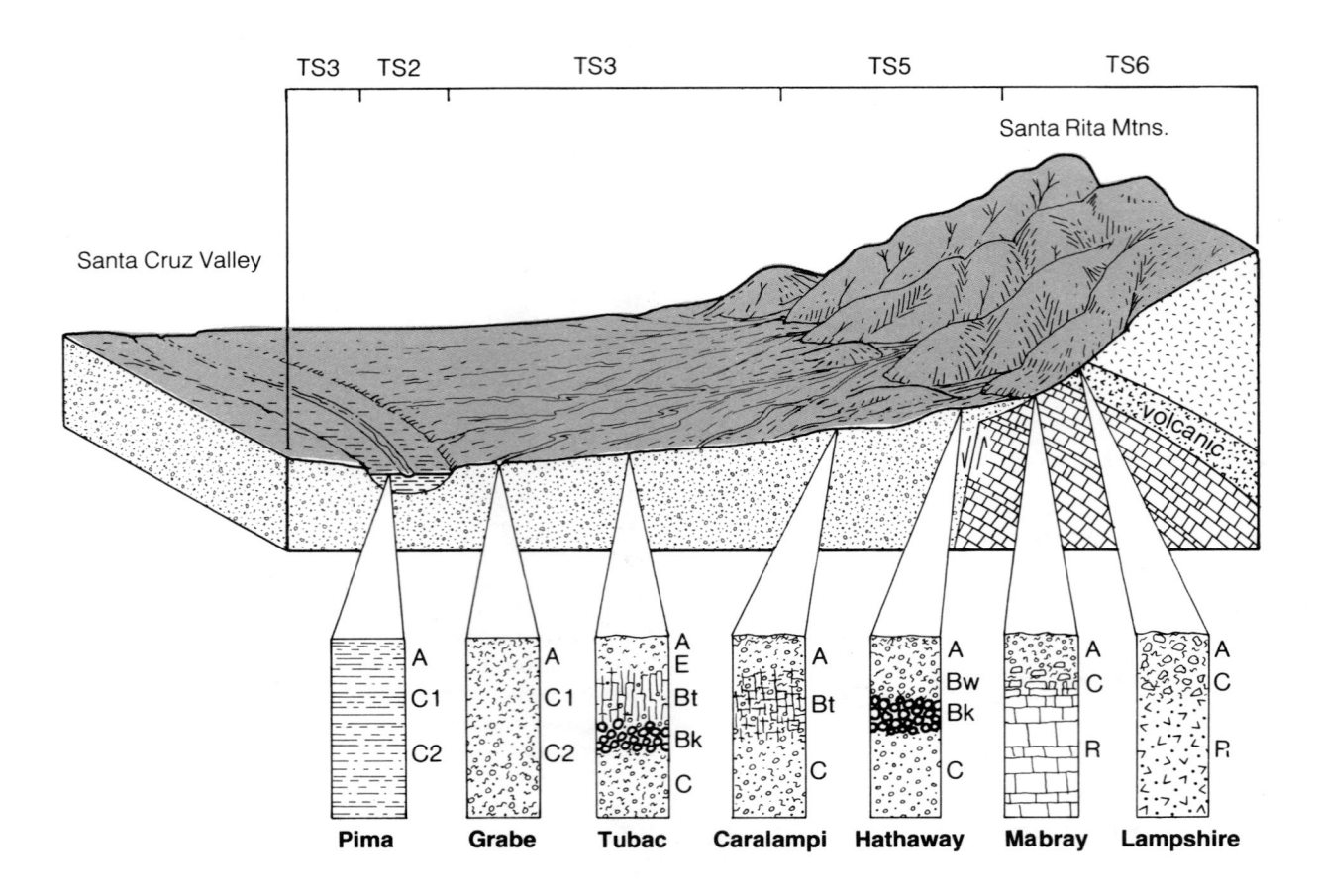

FIGURE 46. Representative Thermic Semiarid Soils Soilscape and Profiles (D. M. Hendricks)

TS
Thermic Semiarid Soils

FIGURE 47. Geographic Distribution of Thermic Semiarid (TS) Soils in Arizona (D. M. Hendricks)

- TS1 Torrifluvents-Torripsamments Association
- TS2 Torrifluvents Association
- TS3 Tubac-Sonoita-Grabe Association
- TS4 White House-Bernardino-Hathaway Association
- TS5 Caralampi-Hathaway Association
- TS6 Lithic Torriorthents-Lithic Haplustolls-Rock Outcrop Association
- TS7 White House-Caralampi Association
- TS8 Caralampi-White House Association
- TS9 Latene-Nickel-Pinaleno Association
- TS10 Chiricahua-Cellar Association
- TS11 Gothard-Crot-Stewart Association
- TS12 Continental-Latene-Pinaleno Association
- TS13 Karro-Gothard Association
- TS14 Nickel-Latene-Cave Association
- TS15 Bonita-Graham-Rimrock Association
- TS16 Penthouse-Latene-Cornville Association
- TS17 Signal-Grabe Association
- TS18 Graham-Lampshire-House Mountain Association
- TS19 Anthony-Sonoita Association

TS Thermic Semiarid Soils

Thermic Semiarid Soils have mean annual soil temperatures of 15 to 22 C (59 to 72 F). The difference between mean summer and mean winter temperatures is greater than 5 C (9 F) at a depth of 50 cm (20 in) or in shallow soils at the soil-bedrock interface. These soils receive 250 to 410 mm (10 to 16 in) annual precipitation. Elevations of Thermic Semiarid Soils range from low to intermediate. They cover about 5,966,410 ha (14,742,800 ac), or 20 percent of Arizona.

TS1 Torrifluvents-Torripsamments Association

Deep, moderately coarse and coarse-textured, nearly level to strongly sloping soils on floodplains and low alluvial fans.

Soil Classification
  Torrifluvents
  Torripsamments

Percent Slope none to 5

Elevation 1,030 to 1,220 m (3,400 to 4,000 ft)

Mean Annual Precipitation 200 to 300 mm (8 to 12 in)

Winter Precipitation as the Percentage of Annual Precipitation 40 to 45

Mean January Air Temperature 4 to 7 C (40 to 45 F)

Mean July Air Temperature 27 to 29 C (80 to 85 F)

Mean Annual Soil Temperature 16 to 21 C (61 to 69 F)

Frost-Free Days 210 to 250

Area and Percent of State 19,830 ha (49,000 ac), less than 0.1

Land Uses rangeland and wildlife habitat

This association consists of well-drained and somewhat excessively drained soils on floodplains and alluvial fans. They formed in recent sandy alluvium derived from igneous and sedimentary rocks.

Coarse, loamy Torrifluvents make up about 40 percent of the association, Torripsamments 40 percent and other fine, loamy Torrifluvents and sandy or gravelly Torriorthents 20 percent.

The vegetation is low-growing, scattered mesquite, saltbush, creosotebush, Mohave yucca, bush muhly, sand dropseed, blue and Rothrock grama and annual grasses and forbs.

The soils have low potential for forage production due to low rainfall and droughtiness of the soils. Many grasses and forbs, however, respond rapidly to any precipitation that falls during the growing season. Care should be taken to avoid overgrazing or disturbing these soils due to creating the hazard of blowing soil.

Factors limiting the potential of these soils for development of homesites are the flood hazards of the Torrifluvents and the very sandy texture of the Torripsamments.

TS2 Torrifluvents Association

Deep, moderately coarse to moderately fine-textured, nearly level to gently sloping soils on floodplains and alluvial fans.

Soil Classification Typic Torrifluvents

Percent Slope none to 3

Elevation 670 to 1,220 m (2,200 to 4,000 ft)

Mean Annual Precipitation 230 to 300 mm (9 to 12 in)

Winter Precipitation as the Percentage of Annual Precipitation 40 to 50

Mean January Air Temperature 4 to 10 C (40 to 50 F)

Mean July Air Temperature 24 to 29 C (75 to 85 F)

Mean Annual Soil Temperature 16 to 22 C (60 to 72 F)

Frost-Free Days 200 to 270

Area and Percent of State 142,860 ha (353,000 ac), 0.5

Land Uses irrigated cropland, rangeland, wildlife habitat, urban and recreation

This association consists of well-drained soils formed on recent mixed alluvium on the floodplains of the Santa Cruz, Upper Gila and San Pedro rivers and their tributaries.

Torrifluvents make up about 95 percent of this association and Torriorthents and other soils make up about 5 percent. The major Torrifluvents in this unit are Grabe, Pima and Anthony soils, each of which make up about 25 percent. About 20 percent consists of Comoro, Gila, Vinton, Guest and Glendale soils. Torriorthents and other included soils are mostly Arizo and Brazito soils and riverwash. Any of these soils may be dominant at a given site.

All of these soils are subject to flooding unless protected. Flooding is usually very brief and local. Areas along entrenched streams are subject to soil piping, gullying and bank cutting.

Cropland and community uses in this association require protection from flooding. Irrigated crops include cotton, alfalfa, grain sorghum, small grains, vegetables and pecans. Natural vegetation includes mesquite, saltbush, catclaw, creosotebush, arrowweed, desertwillow, bush muhly, Arizona cottontop, threeawn and annuals. These soils have medium potential for use as rangeland under good management. They have high potential for openland wildlife habitat in irrigated areas and fair potential for rangeland wildlife habitat where nonirrigated.

TS3 Tubac-Sonoita-Grabe Association

Deep, moderately coarse to fine-textured, nearly level to strongly sloping soils of the uplands and drainageways.

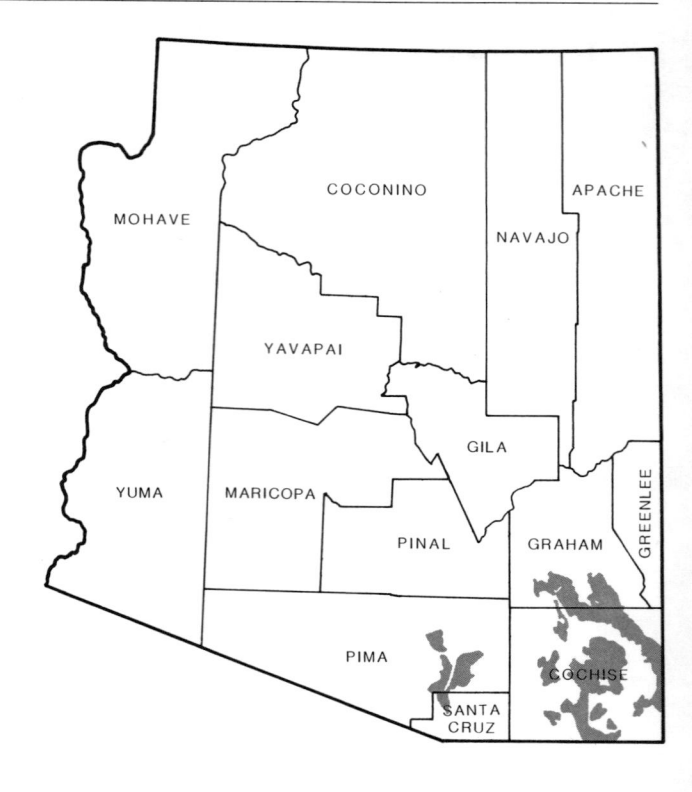

Soil Classification
  Typic Paleargids
  Typic Haplargids

Percent Slope mostly none to 8; some to 15

Elevation 910 to 1,520 m (3,000 to 5,000 ft)

Mean Annual Precipitation 250 to 400 mm (10 to 16 in)

Winter Precipitation as the Percentage of Annual Precipitation 40 to 50

Mean January Air Temperature 4 to 10 C (40 to 50 F)

Mean July Air Temperature 24 to 29 C (75 to 85 F)

Mean Annual Soil Temperature 16 to 22 C (61 to 72 F)

Frost-Free Days 185 to 265

Area and Percent of State 736,550 ha (1,820,000 ac), 2.5

Land Uses rangeland, wildlife habitat, irrigated cropland and some urban

This association consists of well-drained soils on valley plains and wide floodplains in the Santa Cruz, Sulphur Springs and San Simon valleys. The soils formed in mixed old and recent alluvium derived mostly from igneous rocks.

Tubac and the similar Continental soils make up about 50 percent of the association, Sonoita soils 20 percent, Grabe soils 20 percent and minor soils about 10 percent. The minor soils include Comoro, Pima, Guest, Anthony, Pinaleno, Eba, Forrest, McAllister, Anway, Tres Hermanos and Nickel.

Good yields of cotton, grain sorghum, alfalfa, small grain and vegetables are produced when the soils of this association are irrigated. The native vegetation is mostly grass in the higher elevations and desert shrubs and cacti at the lower elevations. Principal grasses are gramas, plains lovegrass, tobosa and annuals. Shrubs are mesquite, whitethorn, catclaw, burroweed, wolfberry and cacti. Paloverde and ironwood occur at lower elevations.

Under good range management, these soils have fair to good potential for the production of livestock forage. Many areas are in poor condition from overgrazing due to their easy accessibility.

Factors limiting the potential of these areas for development of homesites and other community uses are slow permeability and clayey subsoils in the Tubac and Continental soils and the possibility of flooding of Grabe soils. Sonoita soils are well suited for community uses.

TS4 White House-Bernardino-Hathaway Association

Deep, fine-textured and gravelly, moderately coarse to moderately fine-textured, nearly level to moderately steep soils on alluvial fan surfaces and steep side slopes.

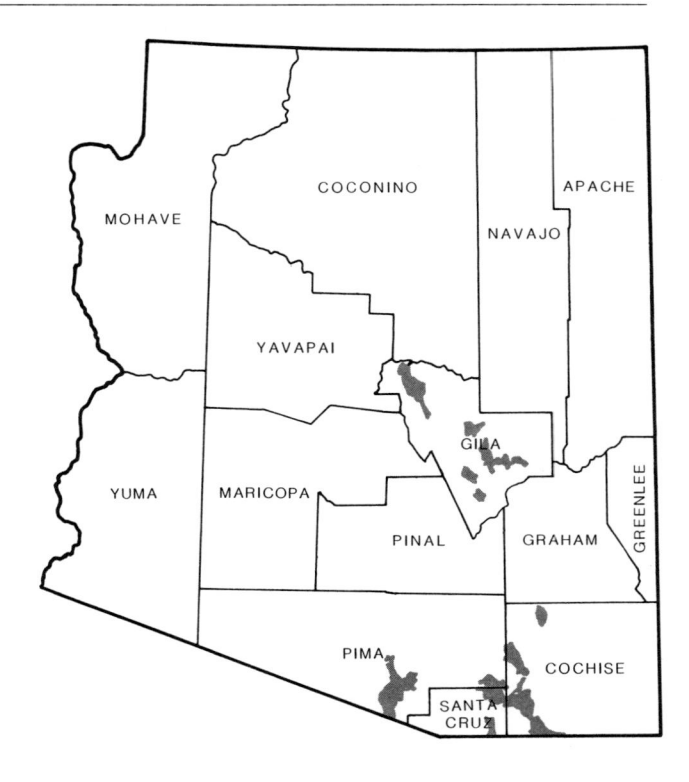

Soil Classification
  Ustollic Haplargids
  Aridic Calciustolls

Percent Slope mostly 1 to 20, but to 50 where Hathaway soils occur

Elevation 1,030 to 1,550 m (3,400 to 5,100 ft)

Mean Annual Precipitation 300 to 460 mm (12 to 18 in)

Winter Precipitation as the Percentage of Annual Precipitation 50 to 60

Mean January Air Temperature 2 to 7 C (35 to 45 F)

Mean July Air Temperature 24 to 29 C (75 to 85 F)

Mean Annual Soil Temperature 15 to 21 C (59 to 70 F)

Frost-Free Days 160 to 250

Area and Percent of State 326,920 ha (807,800 ac), 1.1

Land Uses rangeland, wildlife habitat and some homesites

This association consists of well-drained soils on nearly level to hilly valley plains and dissected old terraces. The soils formed in old alluvium derived from granitic, volcanic and sedimentary rocks.

White House soils make up about 40 percent of the association, Bernardino soils 20 percent, Hathaway soils 20 percent and minor soils 20 percent. The minor soils are mostly Caralampi and small areas of Bonita, Nickel, Pinaleno and Chiricahua and rock outcrop. Narrow bodies of Guest and Pima soil are along drainageways.

The dominant native vegetation is plains lovegrass, beargrass and blue, black, hairy, sideoats and slender grama. Scattered mesquite and cacti, with some oak trees, grow at higher elevations.

This association has some of the best rangeland in the state, and the potential for range improvement is high. It has fair potential for rangeland wildlife habitat.

Factors limiting the potential of the soils for community uses are slow permeability and high shrink-swell in the White House and Bernardino soils and excessive slope and high lime in Hathaway and some Bernardino soils.

TS5 Caralampi-Hathaway Association

Deep, gravelly, moderately coarse to moderately fine-textured, moderately steep to very steep soils on highly dissected old fan surfaces.

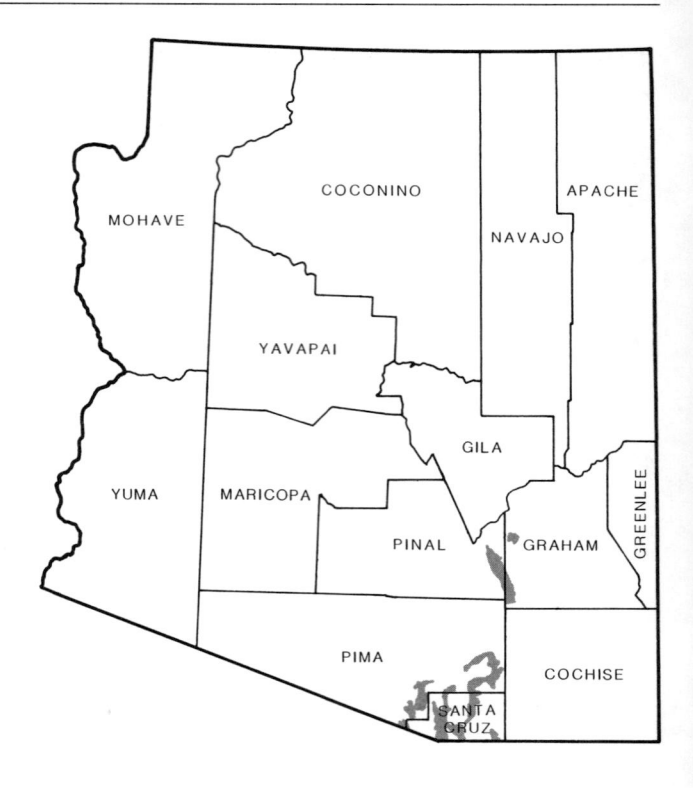

Soil Classification
Ustollic Haplargids
Aridic Calciustolls

Percent Slope mostly 20 to 60; range is 10 to 70

Elevation 910 to 1,520 m (3,000 to 5,000 ft)

Mean Annual Precipitation 300 to 410 mm (12 to 16 in)

Winter Precipitation as the Percentage of Annual Precipitation 35 to 45

Mean January Air Temperature 4 to 10 C (40 to 45 F)

Mean July Air Temperature 21 to 27 C (70 to 80 F)

Mean Annual Soil Temperature 16 to 21 C (60 to 70 F)

Frost-Free Days 170 to 250

Area and Percent of State 203,360 ha (502,500 ac), 0.7

Land Uses rangeland and wildlife habitat; some homesites near Nogales

This association consists of well-drained, very gravelly soils on long narrow ridges formed by deep dissection of old fan surfaces. The soils formed in old alluvium derived from igneous and sedimentary rocks.

Caralampi soils make up about 45 percent of the association, Hathaway soils 35 percent and minor soils about 20 percent. The minor soils include White House, Bernardino, Nolam and Signal on uplands and narrow bands of Torrifluvents and Torriorthents along drainageways. The minor soils may be extensive in some areas.

The dominant native vegetation is plains lovegrass, cane and Texas bluestem, calliandra, beargrass, sideoats, blue, black, hairy and slender grama, and scattered mesquite and cacti. Some oak and juniper grow at higher elevations.

Slopes in excess of 30 percent are somewhat steep for grazing by cattle and horses. The good wildlife habitat is provided in drainageways.

The major factors limiting the potential of these soils for community uses are slopes in excess of 15 percent and high gravel content. Hathaway soils are high in lime.

TS6 Lithic Torriorthents-Lithic Haplustolls-Rock Outcrop Association

Shallow, cobbly and gravelly, strongly sloping to very steep soils and rock outcrop on hills and mountains.

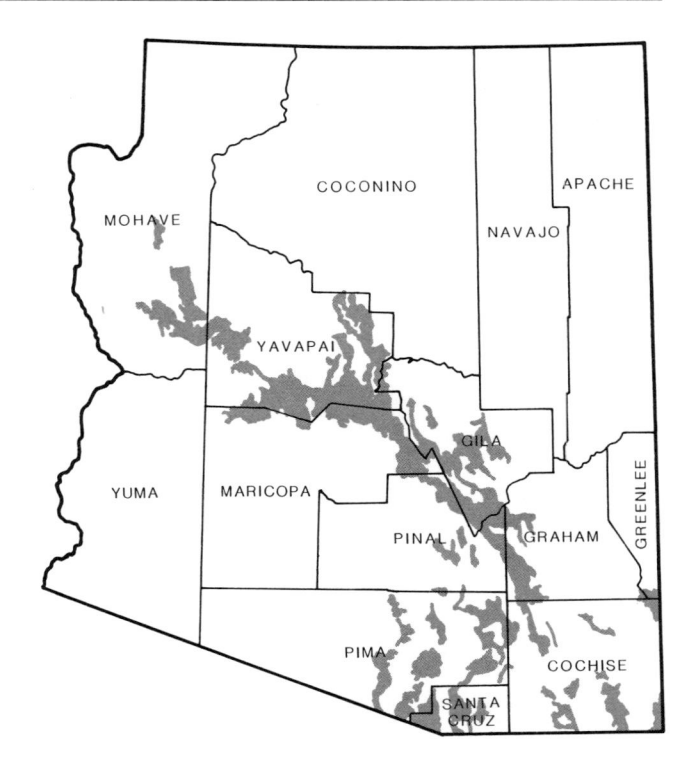

Soil Classification
Lithic Torriorthents
Lithic Haplustolls

Percent Slope mostly 20 to 50; range is 10 to 70

Elevation 760 to 1,760 m (2,500 to 5,800 ft)

Mean Annual Precipitation 250 to 510 mm (10 to 20 in)

Winter Precipitation as the Percentage of Annual Precipitation 35 to 60

Mean January Air Temperature 2 to 10 C (35 to 50 F)

Mean July Air Temperature 21 to 29 C (70 to 85 F)

Mean Annual Soil Temperature 15 to 22 C (59 to 72 F)

Frost-Free Days 160 to 260

Area and Percent of State 2,040,090 ha (5,041,000 ac), 6.9

Land Uses wildlife habitat, rangeland, recreation and mining of copper, lead, zinc, silver and gold

This association consists of well-drained, shallow soils and rock outcrop on semiarid, mid-elevation hills and mountains. The soils formed in residuum weathered from many rocks including granite, gneiss, rhyolite, andesite, tuffs, limestone, sandstone and basalt.

Lithic Torriorthents make up about 30 percent of the association, Lithic Haplustolls 25 percent, rock outcrop about 25 percent and Haplargids, other minor soils and Torrifluvents along drainageways, about 20 percent.

The Lithic Torriorthents include House Mountain, Cellar, Retriever, Courthouse, St. Thomas and Schrap series. The Lithic Haplustolls include Lampshire and Mabray soils and similar soils such as Atascosa and Romero. The dominant included Haplargids are Anklam, Chiminea, Chiricahua, Lehmans, Graham, Deloro and Oracle. Other minor soils include Pantano, Pinaleno, Caralampi and Nickel.

Parker Canyon, Arivaipa, Pena Blanca and Patagonia lakes are within areas bounded by this association. The native vegetation varies with elevation and precipitation. At lower elevations the dominant vegetation is paloverde, mesquite, whitethorn, catclaw, jojoba, calliandra, saguaro and other cacti and some grasses. At higher elevations grow many more perennial grasses including Arizona cottontop, cane and Texas bluestem, plains lovegrass, green sprangletop, wolftail and blue, hairy, black, sideoats and slender grama and many others. Oak and juniper may grow above 1,220 m (4,000 ft), but paloverde and saguaro do not grow at this elevation.

Smoother areas of this association have good potential for livestock grazing. The steeper, stonier areas are little used by domestic livestock, but they are good habitat for deer and, in some areas, bighorn sheep.

Factors limiting the potential of these areas for community uses are steep slopes, shallow depth to bedrock and rock fragments on the surface. Selected areas are suitable for campgrounds, picnic areas and pack trails.

TS7 White House-Caralampi Association

Deep, fine-textured and gravelly, moderately fine-textured, gently to moderately sloping soils on old alluvial fan surfaces.

Soil Classification Ustollic Haplargids

Percent Slope mostly 2 to 8; some to 15

Elevation 760 to 1,340 m (2,500 to 4,400 ft)

Mean Annual Precipitation 300 to 410 mm (12 to 16 in)

Winter Precipitation as the Percentage of Annual Precipitation 55 to 60

Mean January Air Temperature 7 to 10 C (45 to 50 F)

Mean July Air Temperature 29 to 32 C (85 to 90 F)

Mean Annual Soil Temperature 18 to 21 C (64 to 70 F)

Frost-Free Days 240 to 270

Area and Percent of State 67,180 ha (166,000 ac), 0.2

Land Uses rangeland and wildlife habitat; some homesites

This association consists of well-drained soils on old alluvial fan surfaces. They formed in old mixed alluvium weathered mostly from granitic and sedimentary rocks.

White House soils make up about 50 percent of the association, Caralampi soils 35 percent and minor soils 15 percent. The minor soils include small areas of Hathaway, Nolam, Bernardino and Palos Verdes and narrow bands of Torrifluvents along drainageways.

The dominant native vegetation is grama grasses, plains lovegrass, bush muhly, threeawn, calliandra, mesquite, cacti, burroweed and annual grasses and forbs.

Under good management, the soils have fair to good potential for the production of range forage. Good wildlife habitat is along drainageways.

Factors limiting the potential of these areas for homesite development are high shrink-swell and slow permeability in White House soils and high gravel content in Caralampi soils.

TS8 Caralampi-White House Association

Deep, gravelly, moderately fine and fine-textured, moderately sloping to steep soils on dissected old alluvial fan surfaces.

Soil Classification Ustollic Haplargids

Percent Slope 5 to 30

Elevation 700 to 1,220 m (2,300 to 4,000 ft)

Mean Annual Precipitation 300 to 410 mm (12 to 16 in)

Winter Precipitation as the Percentage of Annual Precipitation 55 to 60

Mean January Air Temperature 7 to 10 C (45 to 50 F)

Mean July Air Temperature 29 to 32 C (85 to 90 F)

Mean Annual Soil Temperature 18 to 21 C (65 to 70 F)

Frost-Free Days 200 to 270

Area and Percent of State 102,390 ha (253,000 ac), 0.3

Land Uses rangeland and wildlife habitat; some building sites

This association consists of well-drained soils on dissected old alluvial fans adjacent to the San Pedro River. The soils formed in old alluvium derived from granite and sedimentary rocks.

Caralampi soils make up about 65 percent of the association, White House soils 20 percent and other soils about 15 percent. The included soils consist mostly of small areas of Oracle and Hathaway on uplands and Comoro and other alluvial soils along narrow drainageways.

The native vegetation is mostly grama grasses, plains lovegrass, Arizona cottontop, bush muhly, sand dropseed, calliandra, catclaw, whitethorn, pricklypear and cholla cacti, and annual grasses and forbs. A few oak and juniper may grow on north-facing slopes at higher elevations.

Under good management the soils have fair to good potential for the production of range forage. Good wildlife habitat is along drainageways.

Factors limiting the potential of these areas for homesite development are high shrink-swell and slow permeability in the White House soils and slope and high gravel content in the Caralampi soils.

TS9 Latene-Nickel-Pinaleno Association

Deep, gravelly, limy, moderately coarse to moderately fine-textured, nearly level to very steep soils on dissected alluvial fan surfaces.

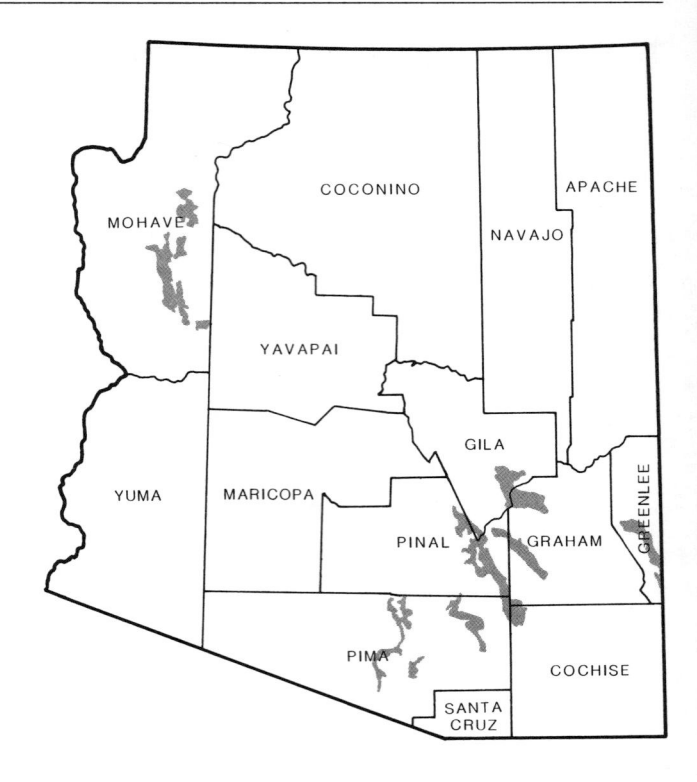

Soil Classification
　Typic Calciorthids
　Typic Haplargids

Percent Slope 1 to 60

Elevation 700 to 1,640 m (2,300 to 5,400 ft)

Mean Annual Precipitation 250 to 410 mm (10 to 16 in)

Winter Precipitation as the Percentage of Annual Precipitation 40 to 65

Mean January Air Temperature 4 to 10 C (40 to 50 F)

Mean July Air Temperature 27 to 29 C (80 to 85 F)

Mean Annual Soil Temperature 16 to 22 C (61 to 72 F)

Frost-Free Days 180 to 270

Area and Percent of State 534,200 ha (1,320,000 ac), 1.8

Land Uses rangeland and wildlife habitat; some building sites on Pinaleno, Nickel and Palos Verdes soils near Tucson

This association consists of well-drained soils on deeply dissected old alluvial fans and terraces. The soils formed in old alluvium derived from granite, gneiss, limestone and other sedimentary and igneous rocks.

Latene soils make up about 30 percent of this association, Nickel soils 20 percent, Pinaleno soils 20 percent and numerous minor soils 30 percent. The minor included soils, which may be extensive at some sites, are Tres Hermanos, Palos Verdes, Whitlock, Rillino and Dona Ana. Glendale, Comoro, Anthony and other Torrifluvents are along drainageways.

The native vegetation consists of creosotebush, whitethorn, paloverde, bursage, scattered Mohave yucca and mesquite, saguaro and other cacti, black grama, bush muhly, threeawn and annual grasses and forbs.

This association has low potential for producing livestock forage. The soils have low to fair potential for wildlife habitat.

Factors limiting the potential of these areas for urban development are the excessive slope and high gravel content of the Nickel and Pinaleno soils and high lime content in the lower horizons of all major soils. These limitations have been partially overcome or compensated for in the foothills near Tucson by proper engineering design and construction techniques.

TS10 Chiricahua-Cellar Association*

Shallow, gravelly, moderately coarse to fine-textured, gently rolling to hilly soils on low granitic foothills.

Soil Classification
   Ustollic Haplargids
   Lithic Torriorthents

Percent Slope 5 to 25

Elevation 610 to 1,460 m (2,000 to 4,800 ft)

Mean Annual Precipitation 280 to 480 mm (11 to 19 in)

Winter Precipitation as the Percentage of Annual Precipitation 55 to 60

Mean January Air Temperature 7 to 10 C (45 to 50 F)

Mean July Air Temperature 27 to 29 C (80 to 85 F)

Mean Annual Soil Temperature 15 to 22 C (60 to 72 F)

Frost-Free Days 220 to 260

Area and Percent of State 58,070 ha (143,500 ac), 0.2

Land Uses rangeland and wildlife habitat; some recreation and homesites

*This association has been correlated Oracle-Romero Association since the field work for the state general map was completed.

This association consists of well-drained soils formed on low granitic mountains and pediments. The soils formed in the residuum of strongly weathered coarse-grained granite.

Oracle (Chiricahua) soils make up about 50 percent of this association, Romero (Cellar) soils 30 percent and included minor soils and rock outcrop about 20 percent. The included soils are mostly small areas of Caralampi and White House and narrow bodies of Comoro and other Torrifluvents along drainageways.

The dominant native vegetation is cacti and shrubs and grasses including shrub live oak, manzanita, calliandra, turpentinebush, range ratany, bullgrass, grama grasses, plains lovegrass, wolftail, Arizona cottontop, and numerous other grasses and shrubs. Emory oak and juniper grow at higher elevations.

Under good management these soils have good potential for the production of livestock and wildlife forage. Areas of this association receive somewhat higher precipitation than is normal for the Thermic Semiarid associations, 250 to 310 mm (10 to 16 in). These areas receive up to 480 mm (19 in) mean annual precipitation.

Factors limiting the potential of these areas for development of homesites are slope, depth to bedrock and water supply.

TS11 Gothard-Crot-Stewart Association

Shallow to deep, moderately well and somewhat poorly drained, saline-sodic, nearly level soils on playas.

Soil Classification
 Typic Natrargids
 Aquic Natrustalfs
 Typic Durorthids

Percent Slope mostly less than 1; some short, steep slopes

Elevation 1,180 to 1,300 m (3,900 to 4,300 ft)

Mean Annual Precipitation 280 mm (11 in)

Winter Precipitation as the Percentage of Annual Precipitation 30 to 35

Mean January Air Temperature 4 to 7 C (40 to 45 F)

Mean July Air Temperature 24 to 27 C (75 to 80 F)

Mean Annual Soil Temperature 17 C (62 F)

Frost-Free Days 180 to 220

Area and Percent of State 58,280 ha (144,000 ac), 0.2

Land Uses rangeland and wildlife habitat; some homesites and, on Gothard soils, irrigated cropland

This association consists of imperfectly drained, saline-sodic soils surrounding the Willcox Playa. The soils formed in old lacustrine sediments derived from igneous and sedimentary rocks.

Gothard soils make up about 35 percent of the association, Crot soils 25 percent, Stewart soils 25 percent and several minor soils and the Willcox Playa about 15 percent. The minor soils include areas of Duncan, Dry Lake, Karro, Elfrida, Cogswell, Guest and Comoro.

The native vegetation is alkali sacaton, saltgrass, tobosa, scattered mesquite and annual grasses. Many small barren playas are common.

This association has limited potential for the production of livestock forage. Cattle graze these coarse grasses when they are green and tender, but the grasses have low palatability when dry and mature. Excess salt and alkali severely limit these soils for crop production.

The major limiting factors for most community uses are excess salt and alkali, very slow permeability, poor drainage, high water table in Crot soils and depth to pan in Stewart soils.

TS12 Continental-Latene-Pinaleno-Association

Deep, gravelly, medium to fine-textured, nearly level to steep soils on dissected alluvial fan surfaces.

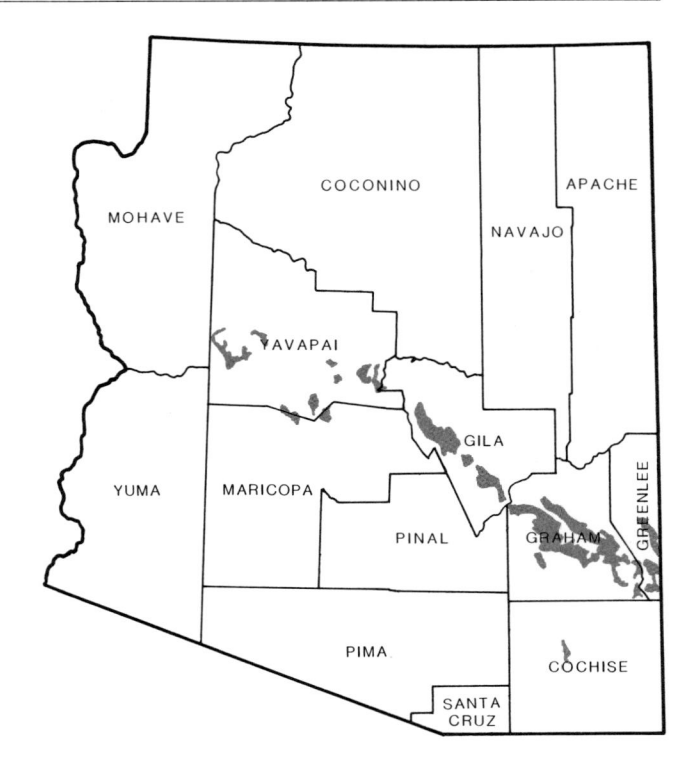

Soil Classification
  Typic Haplargids
  Typic Calciorthids

Percent Slope mostly 1 to 35; to 60 on some short escarpments

Elevation 640 to 1,400 m (2,100 to 4,600 ft)

Mean Annual Precipitation 250 to 360 mm (10 to 14 in)

Winter Precipitation as the Percentage of Annual Precipitation 35 to 60

Mean January Air Temperature 4 to 7 C (40 to 45 F)

Mean July Air Temperature 24 to 29 C (75 to 85 F)

Mean Annual Soil Temperature 17 to 22 C (62 to 72 F)

Frost-Free Days 205 to 275

Area and Percent of State 458,530 ha (1,133,000 ac), 1.6

Land Uses rangeland and wildlife habitat; some homesites and, around Roosevelt Lake, recreation

This association consists of well-drained soils on dissected old terraces and alluvial fans.

Continental soils make up about 30 percent of the association, Latene soils 25 percent, Pinaleno soils 25 percent and several minor soils 20 percent. The minor soils are mostly small areas of Nickel, Eba, Tres Hermanos, Cave, Mohave, Dona Ana and Whitlock and narrow bands of Anthony, Gila and other Torrifluvents along drainageways.

The dominant native vegetation is mesquite, whitethorn, catclaw, cholla and pricklypear cacti, wolfberry, creosotebush, burroweed, grama grasses, bush muhly, tobosa, threeawn and annuals.

The potential for the production of livestock and wildlife forage is only fair due to low precipitation and the predominance of brushy plant species. Good wildlife habitat is along drainageways.

Factors limiting the potential of these areas for the development of homesites and recreation are slow permeability and high shrink-swell in Continental soils, slope and high gravel content in Pinaleno soils and high lime content in the substrata of Latene soils.

TS13 Karro-Gothard Association

Deep, well and moderately well-drained, medium and moderately fine-textured, nearly level soils on lower valley slopes and playas.

Soil Classification
  Ustollic Calciorthids
  Typic Natrargids

Percent Slope mostly none to 2; some to 15

Elevation 1,030 to 1,400 m (3,400 to 4,600 ft)

Mean Annual Precipitation 250 to 300 mm (10 to 12 in)

Winter Precipitation as the Percentage of Annual Precipitation 30 to 45

Mean January Air Temperature 4 to 7 C (40 to 45 F)

Mean July Air Temperature 24 to 27 C (75 to 80 F)

Mean Annual Soil Temperature 16 to 19 C (60 to 66 F)

Frost-Free Days 170 to 230

Area and Percent of State 71,230 ha (176,000 ac), 0.2

Land Uses rangeland, wildlife habitat and irrigated cropland

This association consists of calcareous and sodic soils. The soils formed in old lacustrine sediments derived from igneous and sedimentary rocks.

Karro and Hondale* soils make up about 40 percent of the association, Gothard soils 35 percent and several minor soils about 25 percent. These minor soils consist of scattered areas of Elfrida, McAllister, Pima and Grabe in the Sulphur Springs Valley in Cochise County and areas of Hantz, Dona Ana, Glendale and Artesia in the San Simon Valley in Cochise and Graham counties.

The dominant native vegetation is alkali sacaton, tobosa, fluffgrass, burroweed, creosotebush, fourwing saltbush, blackbrush, mesquite and annuals.

The association has low potential for producing livestock forage. It has fair potential for growing irrigated crops where the sodium content is low or can be leached by adding amendments. The principal crops grown under irrigation are alfalfa, cotton and small grains. The high lime content may cause chlorosis in small grains. The potential for the improvement of wildlife habitat is good where the soils are irrigated and poor where nonirrigated.

Factors limiting the potential of these soils for development of homesites are the moderately slow permeability of the Karro soils and the very slow permeability of the Gothard soils, which are poor for use as septic tank absorption fields. Also, the Karro soils are high in lime content and the Gothard soils are saline-sodic.

*Soils originally called Karro in the San Simon area were correlated Hondale.

TS14 Nickel-Latene-Cave Association

Deep and shallow, limy and gravelly, medium and moderately coarse-textured, nearly level to very steep soils on dissected alluvial fan surfaces.

Soil Classification
Typic Calciorthids
Typic Paleorthids

Percent Slope mostly 5 to 30; range is none to 60

Elevation 730 to 1,520 m (2,400 to 5,000 ft)

Mean Annual Precipitation 250 to 360 m (10 to 14 in)

Winter Precipitation as the Percentage of Annual Precipitation 35 to 50

Mean January Air Temperature 7 to 10 C (45 to 50 F)

Mean July Air Temperature 24 to 29 C (75 to 85 F)

Mean Annual Soil Temperature 18 to 22 C (64 to 72 F)

Frost-Free Days 180 to 265

Area and Percent of State 414,820 ha (1,025,000 ac), 1.4

Land Uses rangeland and wildlife habitat; most of Tucson is in this association

This association consists of well-drained soils on dissected old alluvial fans and terrace escarpments that are mostly along the San Pedro River and the San Simon Creek in Cochise County and in the Tucson Basin. The soils formed in calcareous old alluvium derived from igneous and sedimentary rocks.

Nickel soils make up about 25 percent of the association, Latene soils 25 percent, Cave soils 15 percent, Rillino soils 20 percent and minor soils 15 percent. These minor soils, which may be extensive in some areas, include Tres Hermanos, Pinaleno, Dona Ana, Continental, Kimbrough, Hathaway and Forrest. Narrow bands of Pima, Guest, Stellar and Grabe soils are in and along drainageways.

The dominant native vegetation is creosotebush, whitethorn, cacti, paloverde (Tucson area), range ratany, scattered Mohave yucca and mesquite, fluffgrass, black grama and annual grasses and forbs.

The soils have limitations for development of homesites and commercial sites due to the high lime content of all soils, shallow depth to hardpan in Cave soils, high gravel content in Nickel and Rillino soils and excessive slope in Nickel and some Rillino soils. In the metropolitan Tucson area these limitations have been partially overcome or compensated for by proper engineering design and construction techniques.

TS15 Bonita-Graham-Rimrock Association

Shallow to deep, fine-textured, nearly level to steep soils on plains, hills and mountains.

Soil Classification
Typic Chromusterts
Typic Argiustolls

Percent Slope none to 15 on basalt and ash-flow tuffs; 50 or more on large cinder cones and andesitic mountains

Elevation 1,030 to 1,670 m (3,400 to 5,500 ft)

Mean Annual Precipitation mostly 300 to 410 mm (12 to 16 in); up to 500 mm (20 in) in some areas

Winter Precipitation as the Percentage of Annual Precipitation 30 to 65

Mean January Air Temperature 4 to 7 C (40 to 45 F)

Mean July Air Temperature 24 to 29 C (75 to 85 F)

Mean Annual Soil Temperature 15 to 20 C (59 to 68 F)

Frost-Free Days 160 to 220

Area and Percent of State 330,440 ha (816,500 ac), 1.1

Land Uses rangeland and wildlife habitat; some building sites

This association consists of well-drained soils formed in residuum and alluvium weathered from basalt, andesite, ash-flow tuffs, cinders and related volcanic rocks.

Bonita soils make up about 30 percent of the association, Graham soils 20 percent, Rimrock soils 20 percent and minor associated soils 30 percent. These minor soils may be quite extensive in some areas and are mostly Bernardino, White House, Sontag, Krentz, Guest and Pima and small areas of rock outcrop.

The native vegetation is mostly grass. Grasses include tobosa, vine-mesquite, curlymesquite, plains bristlegrass, grama grasses and cane bluestem. Scattered mesquite, catclaw, agave, pricklypear and annual grasses and forbs also occur. Ocotillo is present on Graham soils and oak and juniper occur at higher elevations.

This association has low to fair potential for livestock grazing. Livestock avoid areas with cobbly, stony surfaces and areas that have wide cracks. In addition, the dominant tobosa has low palatability when dry and mature. Wildlife use on these soils is mostly transient due to lack of cover.

Factors limiting the potential of these areas for building sites are high shrink-swell, depth to bedrock on Graham and Rimrock soils and slope on Graham soils.

TS16 Penthouse-Latene-Cornville Association

Deep, medium to fine-textured, nearly level to moderately steep soils on dissected fan surfaces and valley slopes.

Soil Classification
Ustollic Haplargids
Typic Calciorthids
Typic Haplargids

Percent Slope mostly 1 to 10; 30 or more on some escarpments

Elevation 910 to 1,400 m (3,000 to 4,600 ft)

Mean Annual Precipitation 280 to 410 mm (11 to 16 in)

Winter Precipitation as the Percentage of Annual Precipitation 50 to 55

Mean January Air Temperature 2 to 4 C (35 to 40 F)

Mean July Air Temperature 24 to 27 C (75 to 80 F)

Mean Annual Soil Temperature 15 to 19 C (59 to 66 F)

Frost-Free Days 180 to 235

Area and Percent of State 31,160 ha (77,000 ac), 0.1

Land Uses rangeland and wildlife habitat; some homesites and irrigated cropland and pasture on Cornville soils

This association consists of well-drained soils on dissected fan surfaces along the Verde River and its tributaries. They formed in old alluvial and lacustrine sediments derived from calcareous sedimentary and volcanic rocks.

Penthouse soils make up about 30 percent of the association, Latene soils 30 percent, Cornville 25 percent and minor included soils about 15 percent. These minor soils are mostly small areas of Continental, Nickel and Rillino and narrow bands of Torrifluvents along drainageways.

The native vegetation is mostly mesquite, snakeweed, creosotebush, fourwing saltbush, cacti, tobosa, grama grasses, wolftail and annual grasses and forbs. A few juniper grow at higher elevations.

These soils have fair to good potential for forage production but are limited by low precipitation. They commonly are used as winter range for cattle and sheep due to their proximity to the high plateaus to the east. Crops grown on Cornville and possibly Latene soils are alfalfa, corn, small grains and pasture grasses.

The dominant limiting factors for community uses on these soils are the cobbly surface and clayey, slow permeability of subsoils of Penthouse soils and the high lime content in the substratum of all major soils.

TS17 Signal-Grabe Association

Deep, medium and fine-textured, nearly level to gently sloping soils on high valley plains.

Soil Classification
  Aridic Paleustolls
  Typic Torrifluvents

Percent Slope none to 5

Elevation 1,460 to 1,670 m (4,800 to 5,500 ft)

Mean Annual Precipitation 350 to 430 mm (14 to 17 in)

Winter Precipitation as the Percentage of Annual Precipitation 45 to 55

Mean January Air Temperature 4 to 7 C (40 to 45 F)

Mean July Air Temperature 21 to 29 C (70 to 85 F)

Mean Annual Soil Temperature 15 to 19 C (60 to 66 F)

Frost-Free Days 160 to 210

Area and Percent of State 43,910 ha (108,500 ac), 0.1

Land Uses rangeland and wildlife habitat

This association consists of well-drained soils on gently sloping uplands, broad floodplains and low fans. The soils formed in alluvium derived mostly from volcanic rocks, but also minor sedimentary rocks.

Signal soils make up about 50 percent of the association, Grabe soils 30 percent and the other included soils about 20 percent. The included soils are mostly areas of Bonita, Rimrock and Graham on the uplands and Pima and Guest in floodplains.

The dominant vegetation is grama grasses, tobosa, curlymesquite, calliandra, catclaw and cacti on the Signal soils and mesquite, gramas, threeawn and bush muhly on the Grabe soils. Under good range management they have fair to good potential for the production of livestock forage.

One factor limiting the potential of these soils for homesite development is the slow permeability of the Signal soils, which are poor for use as septic tank absorption fields. Also, the Signal soils have high shrink-swell potential and the Grabe soils are subject to flooding.

TS18 Graham-Lampshire-House Mountain Association

Shallow, gravelly and cobbly, medium to fine-textured, rolling to very steep soils on hills and mountains.

Soil Classification
Lithic Argiustolls
Lithic Haplustolls
Lithic Torriorthents

Percent Slope 5 to 60

Elevation 1,060 to 1,670 m (3,500 to 5,500 ft)

Mean Annual Precipitation 300 to 410 mm (12 to 16 in)

Winter Precipitation as the Percentage of Annual Precipitation 40 to 60

Mean January Air Temperature 4 to 7 C (40 to 45 F)

Mean July Air Temperature 24 to 29 C (75 to 85 F)

Mean Annual Soil Temperature 15 to 20 C (59 to 68 F)

Frost-Free Days 160 to 220

Area and Percent of State 242,420 ha (599,000 ac), 0.8

Land Uses rangeland and wildlife habitat

This association consists of shallow and very shallow, well-drained soils and rock outcrop on volcanic hills and mountains. The soils formed in residuum weathered from basalt, ash-tuff, andesite, dacite and other related volcanic rocks.

Graham soils make up about 40 percent of the association, Lampshire soils 30 percent, House Mountain soils 15 percent and rock outcrop and included closely associated soils about 15 percent. The included soils are mostly small areas of Limpia, Rimrock, Bonita, Guest and Pima.

The native vegetation includes mesquite, catclaw, calliandra, ocotillo, shrub live oak, tobosa, grama grasses, curlymesquite, plains lovegrass, cane bluestem, agave, sotol, cacti and forbs.

These soils have fair to good potential to produce livestock and wildlife forage; however, the cobbly, stony surface and steep slopes preclude heavy use by livestock.

Factors limiting the potential for homesite development on these soils are shallow depth to rock, high content of gravels, cobbles or stones and steep slopes.

TS19 Anthony-Sonoita Association

Deep, moderately coarse-textured, nearly level to gently sloping soils on alluvial fans.

Soil Classification
  Typic Torrifluvents
  Typic Haplargids

Percent Slope none to 5

Elevation 670 to 1,150 m (2,200 to 3,800 ft)

Mean Annual Precipitation 250 to 350 mm (10 to 14 in)

Winter Precipitation as the Percentage of Annual Precipitation 35 to 50

Mean January Air Temperature 7 to 10 C (45 to 50 F)

Mean July Air Temperature 29 to 32 C (85 to 90 F)

Mean Annual Soil Temperature 20 to 22 C (68 to 72 F)

Frost-Free Days 250 to 270

Area and Percent of State 84,180 ha (208,000 ac), 0.3

Land Uses rangeland, wildlife habitat, recreation and urban

This association consists of well-drained soils on alluvial fans at the base of granitic mountains near Tucson. They formed in mixed alluvium weathered from granite, gneiss and some volcanic rocks.

Anthony soils make up about 35 percent of the association, Sonoita soils 30 percent and minor included soils about 35 percent. These minor soils, which may be extensive in some places, are mostly small areas of Bucklebar, Valencia, Palos Verdes, Pinaleno, Tres Hermanos and Tubac. Narrow areas of sandy Torrifluvents are along drainageways.

The native vegetation consists of paloverde, mesquite, whitethorn, bursage, some ironwood, cholla, saguaro and pricklypear cacti, snakeweed, bush muhly, threeawn and annual grasses and forbs.

Under good management these soils have fair to good potential for the production of livestock and wildlife forage. Many areas are in poor condition from overgrazing due to their easy accessibility.

The association has good potential for homesites and other community development if protected from flooding.

Mesic Arid Soils

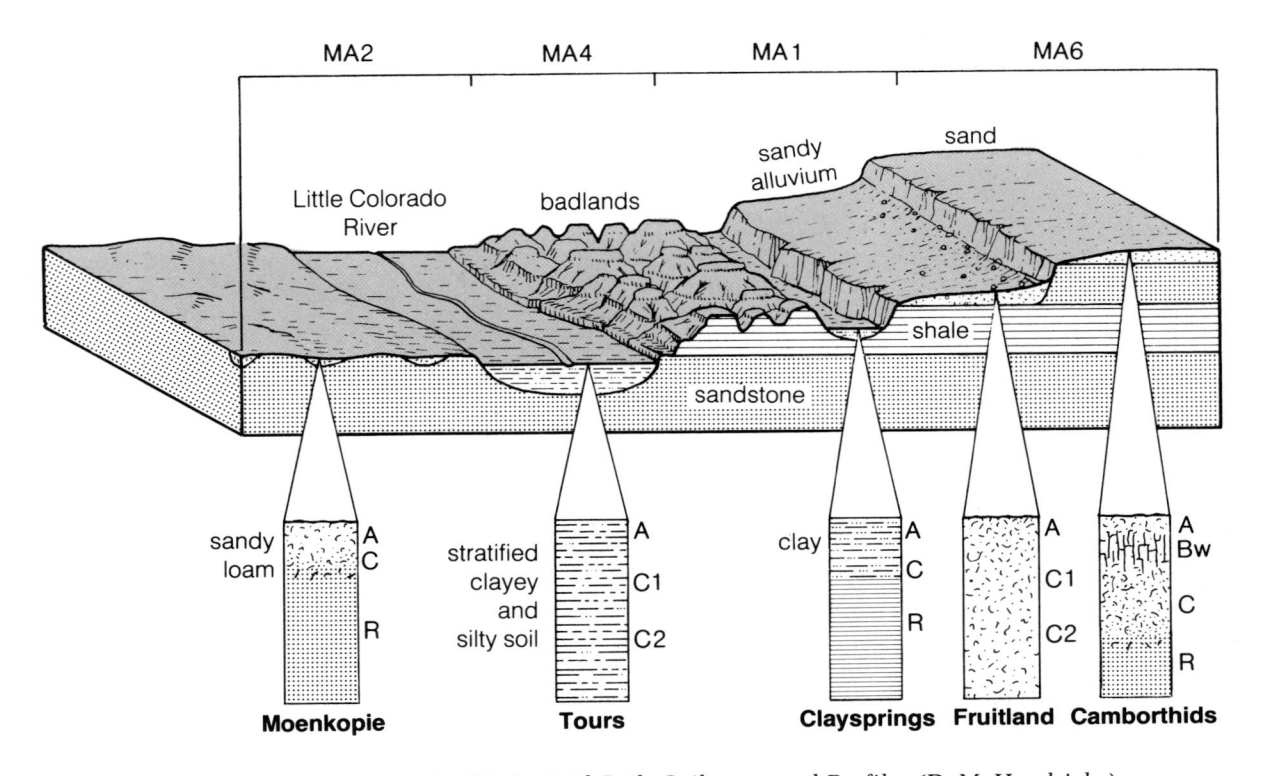

FIGURE 48. Representative Mesic Arid Soils Soilscape and Profiles (D. M. Hendricks)

MA Mesic Arid Soils

FIGURE 49. *Geographic Distribution of Mesic Arid (MA) Soils in Arizona (D. M. Hendricks)*

- MA1 Badland-Torriorthents-Torrifluvents Association
- MA2 Moenkopie-Shalet-Tours Association
- MA3 Sheppard-Fruitland-Rock Outcrop Association
- MA4 Tours-Navajo Association
- MA5 Torriorthents-Torrifluvents Association
- MA6 Fruitland-Camborthids-Torrifluvents Association

MA Mesic Arid Soils

Mesic Arid Soils have a mean annual soil temperature of 8 C (47 F) or more, but less than 15 C (59 F). The difference between mean summer and mean winter temperatures is greater than 5 C (9 F) measured at a depth of 50 cm (20 in) or at the soil-bedrock interface in shallow soils. These soils receive 150 to 250 mm (6 to 10 in) mean annual precipitation. Mesic Arid Soils are extensive on the Colorado Plateau. The elevation of Mesic Arid Soils is intermediate. They cover about 3,546,990 ha (8,764,500 ac) or 12 percent of Arizona.

MA1 Badland-Torriorthents-Torrifluvents Association

Shallow and deep, moderately fine and fine-textured, moderately sloping to very steep soils on eroded uplands and nearly level floodplains.

Soil Classification
 Typic Torriorthents
 Typic Torrifluvents

Percent Slope Badland, none to 60 or more; Torriorthents-Torrifluvents, none to 8

Elevation 1,400 to 1,820 m (4,600 to 6,000 ft)

Mean Annual Precipitation 150 to 300 mm (6 to 12 in)

Winter Precipitation as the Percentage of Annual Precipitation 55 to 60

Mean January Air Temperature -1 to 2 C (30 to 35 F)

Mean July Air Temperature 21 to 27 C (70 to 80 F)

Mean Annual Soil Temperature 11 to 15 C (51 to 59 F)

Frost-Free Days 130 to 180

Area and Percent of State 489,280 ha (1,209,000 ac), 1.7

Land Uses rangeland, wildlife habitat and recreation

This association consists of Badland and shallow, well-drained soils formed on shale and sandstone and deep, well-drained soils formed in recent alluvium derived from sedimentary rocks.

Badland makes up about 40 percent of the association, Torriorthents 25 percent, Torrifluvents 25 percent and minor areas of associated soils and rock outcrop, about 10 percent.

Claysprings is the dominant Torriorthent and Tours and Navajo are the major Torrifluvents in the association. The minor soils are small areas of Jocity, Sheppard, Clovis, Shalet, Moenkopie and Fruitland.

The native vegetation is very sparse and consists mostly of fourwing saltbush, shadscale, Mormon-tea, alkali sacaton, galleta, blue grama, Indian ricegrass and annuals.

The Badland and Torriorthents have low potential for forage production. The Torrifluvents that are not entrenched or eroded have fair to good potential under good management to produce forage. The major limitation is low rainfall. Small areas of the Tours and Navajo soils are used as irrigated cropland.

Limitations of the potential for construction of structures on these soils are shallow depths, steep slopes in the badlands, high shrink-swell and slow permeability of the Torriorthents. The Torrifluvents may be subject to flooding and generally have low strength.

Areas of the Badlands are unique scenic attractions. They include the Painted Desert, Monument Valley and the Petrified Forest National Monument.

MA2 Moenkopie-Shalet-Tours Association

Shallow and deep, moderately coarse to moderately fine-textured, nearly level to rolling soils on sandstone and shale plateaus.

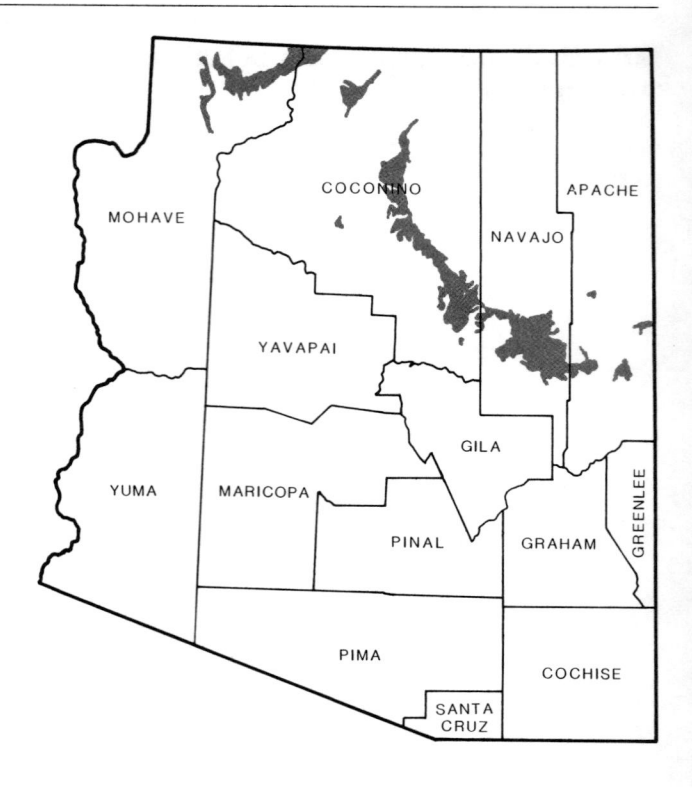

Soil Classification
Lithic Torriorthents
Typic Torriorthents
Typic Torrifluvents

Percent Slope mostly 1 to 5; range is none to 20

Elevation 1,340 to 1,970 m (4,400 to 6,500 ft)

Mean Annual Precipitation 150 to 250 mm (6 to 10 in)

Winter Precipitation as the Percentage of Annual Precipitation 55 to 65

Mean January Air Temperature -1 to 2 C (30 to 35 F)

Mean July Air Temperature 21 to 29 C (70 to 85 F)

Mean Annual Soil Temperature 9 to 15 C (49 to 59 F)

Frost-Free Days 130 to 185

Area and Percent of State 563,750 ha (1,393,000 ac), 1.9

Land Uses rangeland and wildlife habitat

This association consists of well-drained soils on plateaus and floodplains. The soils formed in residuum and alluvium weathered from sandstone, shale and conglomerate rocks.

Moenkopie soils make up about 60 percent of the association, Shalet soils 15 percent, Tours soils 15 percent, and minor areas of associated soils, 10 percent. The minor soils are mostly small areas of Ives, Jocity, Trail, Clovis, Palma, Claysprings and Purgatory series. Also included are small areas of sandstone rock outcrop.

The dominant native vegetation includes alkali sacaton, galleta, fourwing saltbush, blue and black grama, sand dropseed and Mormon-tea, and some sagebrush and scattered juniper.

Moenkopie and Shalet soils have low potential for forage production. Tours soils that receive extra water from runoff have fair to good potential under good management to produce forage.

Factors limiting the potential of these soils for homesite development are the shallow depths to rock in the Moenkopie and Shalet soils. Also, the Tours soils may be subject to flooding and have moderately slow permeability, which is poor for use as septic tank absorption fields.

MA3 Sheppard-Fruitland-Rock Outcrop Association

Deep, coarse and moderately coarse-textured, nearly level to rolling soils and rock outcrop on plains.

Soil Classification
 Typic Torripsamments
 Typic Torriorthents

Percent Slope 2 to 15; some to 30

Elevation 1,520 to 2,120 m (5,000 to 7,000 ft)

Mean Annual Precipitation 150 to 250 mm (6 to 10 in)

Winter Precipitation as the Percentage of Annual Precipitation 50 to 65

Mean January Air Temperature -1 to 2 C (30 to 35 F)

Mean July Air Temperature 21 to 29 C (70 to 85 F)

Mean Annual Soil Temperature 10 to 14 C (50 to 58 F)

Frost-Free Days 130 to 170

Area and Percent of State 865,250 ha (2,138,000 ac), 2.9

Land Uses rangeland, wildlife habitat and recreation

This association consists of somewhat excessively drained and well-drained soils and rock outcrop on plains and plateaus. The plains are broken by prominent mesas, buttes and escarpments. Steep, rock-walled canyons form the sides of the drainages that traverse the areas. The soils formed in aeolian sandy material weathered from sandstone and shale.

Sheppard soils make up about 35 percent of the association, Fruitland soils 35 percent, rock outcrop about 15 percent and minor areas of associated soils and dune land and Badland about 15 percent. The minor soils are mostly small areas of Moenkopie, Shalet and Palma. The dune land occurs as scattered areas of low, poorly stabilized dunes 0.3 to 3 m (1 to 10 ft) high. The Badland consists of small areas of eroded shaly materials.

The native vegetation is dominantly sand sagebrush and sparse areas of Mormon-tea, blackbrush, rabbitbrush, Indian ricegrass, sand dropseed, galleta, grama grasses, threeawn, Russian thistle and other annual weeds and grasses. A few scattered stands of juniper and pinyon pine grow in rocky areas. The potential for livestock and wildlife forage production is low due to low precipitation and low water-holding capacity of the major soils.

Limitations for potential homesite development on these soils are few. The sandy texture of the Sheppard soils, however, is a severe limitation to shallow excavations.

Some areas, such as Monument Valley and Nakai and Piute canyons, are noted scenic attractions because of their vivid colors and unique rock formations. The Betatakin and Keet Seel Ruins and other cliff dwellings in the Navajo National Monument are in this mapping unit. However, until more facilities are built, much of the area can be reached only with great difficulty.

MA4 Tours-Navajo Association

Deep, moderately fine and fine-textured, nearly level to gently sloping soils on floodplains.

Soil Classification Typic Torrifluvents

Percent Slope none to 3

Elevation 1,400 to 1,730 m (4,600 to 5,700 ft)

Mean Annual Precipitation 150 to 250 mm (6 to 10 in)

Winter Precipitation as the Percentage of Annual Precipitation 45 to 60

Mean January Air Temperature -1 C (30 F)

Mean July Air Temperature 21 to 24 C (70 to 75 F)

Mean Annual Soil Temperature 10 to 15 C (50 to 59 F)

Frost-Free Days 155 to 175

Area and Percent of State 224,000 ha (553,500 ac), 0.8

Land Uses rangeland, wildlife habitat and irrigated cropland; some homesites and community uses

This association consists of well-drained soils on the floodplains and adjacent low alluvial fans of the Little Colorado River and its major tributaries. The soils formed in recent alluvium derived from sedimentary and volcanic rocks.

Tours soils make up about 35 percent of the association, Navajo soils 35 percent and minor areas of associated soils and riverwash about 30 percent. The minor soils are mostly Ives, Trail and Jocity.

The native vegetation is mostly fourwing saltbush, shadscale, rabbitbrush, sand sagebrush, galleta, alkali sacaton, blue and black grama, threeawn, saltgrass and annuals. Tamarisk, desertwillow and cottonwood trees grow along the streams in some areas.

The soils have fair potential under good management for producing livestock forage. Limited precipitation is the major factor, but these soils receive runoff from adjacent areas during wet periods. Riparian vegetation and adjacent irrigated cropland provide elements for good wildlife habitat in this association. The short growing season limits crops grown in irrigated areas to alfalfa, corn, small grain and pasture grasses. Saline areas require careful management and reclamation practices.

Factors limiting the potential of these areas for homesites and community uses are flooding hazard, moderately slow to very slow permeability, low bearing strength, salinity, erosion hazard and potential frost action.

MA5 Torriorthents-Torrifluvents Association

Shallow and deep, moderately fine and fine-textured, nearly level to gently sloping soils on valley slopes and floodplains.

Soil Classification
  Typic Torriorthents
  Typic Torrifluvents

Percent Slope none to 5

Elevation 1,520 to 1,760 m (5,000 to 5,800 ft)

Mean Annual Precipitation 200 to 250 mm (8 to 10 in)

Winter Precipitation as the Percentage of Annual Precipitation 55 to 60

Mean January Air Temperature -1 C (30 F)

Mean July Air Temperature 21 to 24 C (70 to 75 F)

Mean Annual Soil Temperature 11 to 13 C (52 to 56 F)

Frost-Free Days 130 to 170

Area and Percent of State 38,450 ha (95,000 ac), 0.1

Land Uses rangeland and wildlife habitat; some irrigated cropland and pasture, and some villages

This association consists of well-drained soils on the floodplains and adjacent fans and low uplands along Chinle Wash and Lukachukai Creek in the Navajo Indian Reservation. The soils formed in residuum and alluvium derived from clay, shale, sandstone and conglomerate.

The Torriorthents and small areas of Badland make up about 70 percent of the association and Torrifluvents 30 percent.

The shallow Claysprings and Shalet soils are the dominant Torriorthents. The deep Tours and Navajo soils are the major Torrifluvents. Included with the Torrifluvents are small areas of Ives, Jocity and Trail, and riverwash.

The native vegetation is mostly alkali sacaton, saltgrass, galleta, blue and black grama, threeawn, fourwing saltbush, sagebrush, rabbitbrush and annual weeds and grasses. Tamarisk, desertwillow and cottonwood trees grow along streams in some areas.

The soils have fair potential under good management for producing livestock forage. Limited precipitation is the major factor, but these soils receive runoff from adjacent areas during wet periods. Riparian vegetation along the streams and adjacent irrigated cropland provide good wildlife habitat. The short growing season limits crops grown in irrigated areas to alfalfa, corn and pasture grasses. Saline areas require careful management and reclamation practices.

Factors limiting the potential of these areas for homesites and community uses are flooding hazard, moderately slow to very slow permeability, low bearing strength, salinity, erosion hazard and potential frost action.

MA6 Fruitland-Camborthids-Torrifluvents Association

Shallow to deep, moderately coarse to moderately fine-textured, nearly level to hilly soils on upland plains.

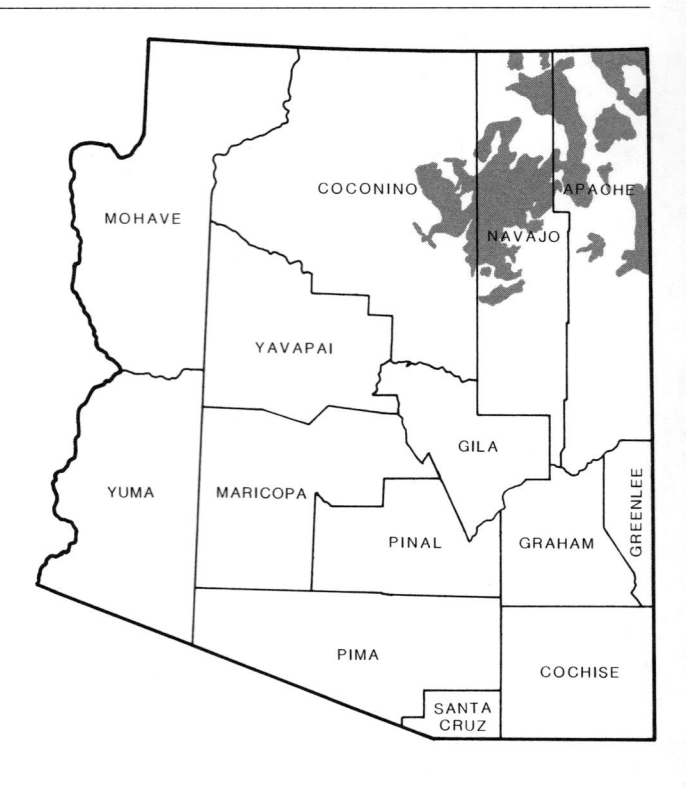

Soil Classification
 Typic Torriorthents
 Camborthids
 Typic Torrifluvents

Percent Slope 2 to 20

Elevation 1,520 to 2,060 m (5,000 to 6,800 ft)

Mean Annual Precipitation 150 to 300 mm (6 to 12 in)

Winter Precipitation as the Percentage of Annual Precipitation 55 to 60

Mean January Air Temperature -1 to 2 C (30 to 35 F)

Mean July Air Temperature 21 to 27 C (70 to 80 F)

Mean Annual Soil Temperature 10 to 14 C (50 to 58 F)

Frost-Free Days 130 to 170

Area and Percent of State 1,366,270 ha (3,376,000 ac), 4.6

Land Uses rangeland and wildlife habitat; some cropland

This association consists of well-drained soils on the high plains. The plains are broken by occasional steep-sided drainageways and scattered buttes. The soils formed in a thick to thin, wind- and water-laid mantle of alluvium weathered from sandstone and shale.

Fruitland soils and closely associated unnamed shallow and moderately deep Torriorthents make up about 50 percent of the association, Camborthids 30 percent, Torrifluvents 15 percent and small areas of rock outcrop and minor included soils 5 percent. The Torrifluvents are mostly Ives and Tours series.

The dominant native vegetation is sand sagebrush, galleta, blue and black grama, Indian ricegrass, sand dropseed, threeawn and rabbitbrush. Juniper and a few pinyon pine grow at higher elevations.

This association has low to fair potential under good management for livestock forage production. Limited precipitation is the major factor. The best potential is in those areas that receive runoff from adjacent areas. Crops grown are mostly small patches of corn, alfalfa and pasture grasses.

The Fruitland soils and Camborthids have good potential for homesites and community uses. The Torrifluvents are subject to seasonal flooding.

Mesic Semiarid Soils

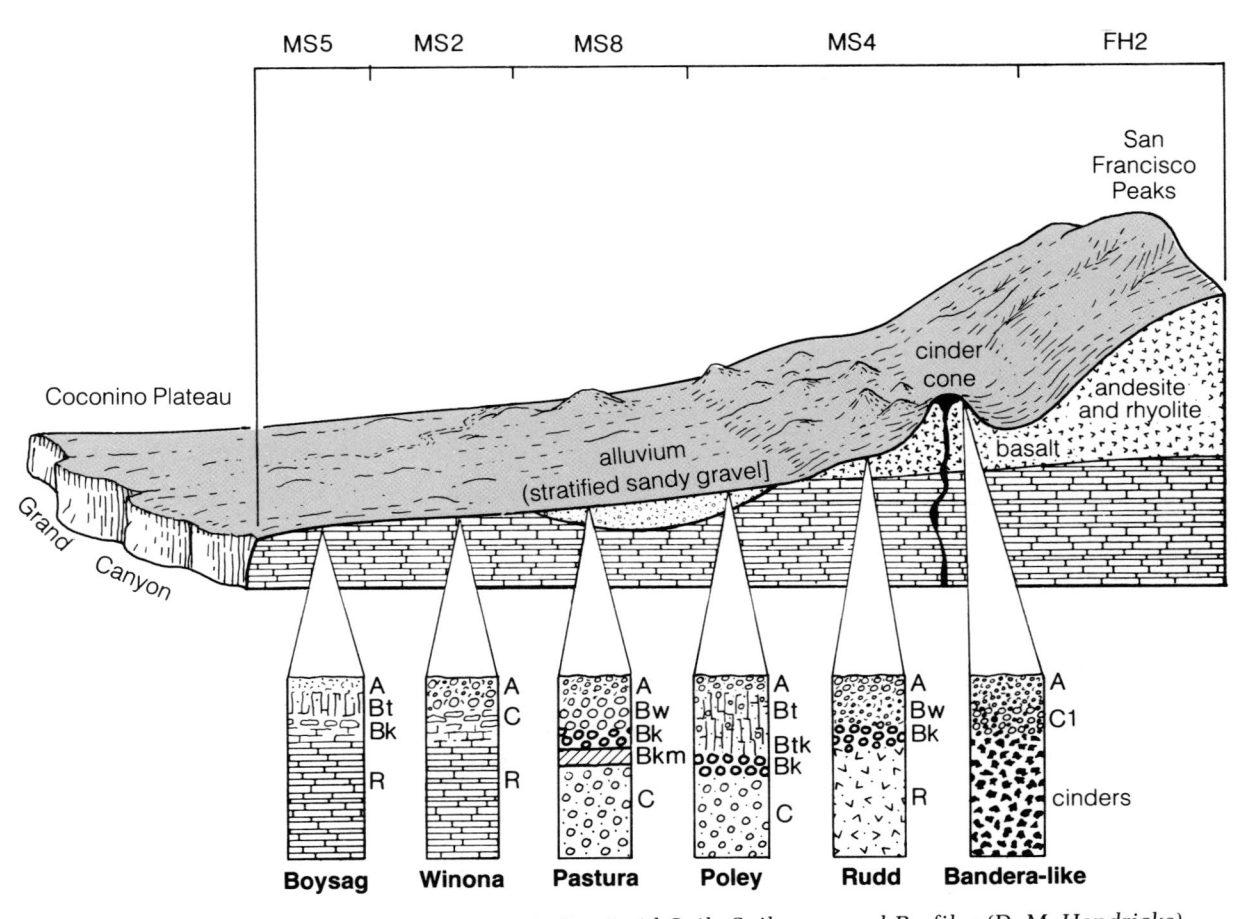

FIGURE 50. Representative Mesic Semiarid Soils Soilscape and Profiles (D. M. Hendricks)

FIGURE 51. *Geographic Distribution of Mesic Semiarid (MS) Soils in Arizona (D. M. Hendricks)*

MS1 Tortugas-Purner-Jacks Association
MS2 Winona-Boysag-Rock Outcrop Association
MS3 Palma-Clovis-Trail Association
MS4 Rudd-Bandera-Cabezon Association
MS5 Roundtop-Boysag Association
MS6 Lithic Torriorthents-Lithic Haplargids-Rock Outcrop Association
MS7 Cabezon-Thunderbird-Springerville Association
MS8 Pastura-Poley-Partri Association
MS9 Lonti-Balon-Lynx Association
MS10 Pastura-Abra-Lynx Association

MS Mesic Semiarid Soils

Mesic Semiarid Soils have a mean annual soil temperature of 8 C (47 F) or more, but less than 15 C (59 F). The difference between mean summer and mean winter temperatures is greater than 5 C (9 F) measured at a depth of 50 cm (20 in) or at the soil-bedrock interface in shallow soils. The mean annual precipitation on these soils ranges from 250 to 460 mm (10 to 18 in). Their elevation is intermediate. Mesic Semiarid Soils cover about 5,328,080 ha (13,165,500 ac), or 18 percent of Arizona.

MS1 Tortugas-Purner-Jacks Association

Shallow to moderately deep, gravelly and cobbly, medium to fine-textured, undulating to steep soils on hills and mountains.

Soil Classification
   Lithic Haplustolls
   Udic Haplustalfs

Percent Slope 2 to 45

Elevation 1,370 to 2,060 m (4,500 to 6,800 ft)

Mean Annual Precipitation mostly 300 to 410 mm (12 to 16 in); to 460 mm (18 in) in a few areas

Winter Precipitation as the Percentage of Annual Precipitation 50 to 70

Mean January Air Temperature -1 to 7 C (30 to 45 F)

Mean July Air Temperature 21 to 27 C (70 to 80 F)

Mean Annual Soil Temperature 9 to 15 C (48 to 59 F)

Frost-Free Days 140 to 160

Area and Percent of State 180,500 ha (446,000 ac), 0.6

Land Uses rangeland, wildlife habitat and recreation; a few rock quarries

This association consists of well-drained soils formed in residuum on limestone and sandstone ridges, hills and mountains south of the Grand Canyon.

Tortugas soils make up about 45 percent of the association, Purner soils 20 percent, Jacks soils 20 percent and other associated soils and rock outcrop 15 percent. The included soils are mostly small areas of Boysag, Dye, Thunderbird, Springerville, Pastura, Moenkopie and Lynx. These included soils may be fairly extensive in some areas.

The dominant native vegetation is pinyon pine, juniper, ceanothus, cliffrose, shrub live oak, mountainmahogany, wolftail, spike muhly and blue, black and sideoats grama. Scattered ponderosa pine grow on Jacks soils in some areas.

The potential for livestock forage production is fair to good. Areas that have steep slopes and areas that lack water facilities limit use by domestic livestock. The potential for use as woodland and rangeland wildlife habitat is good.

The major factors limiting the potential of these soils for building sites are depth to bedrock, slope and low bearing strength, and high shrink-swell of the Jacks soils.

MS2 Winona-Boysag-Rock Outcrop Association

Shallow, medium and fine-textured, undulating to rolling soils and rock outcrop on plateaus and plains.

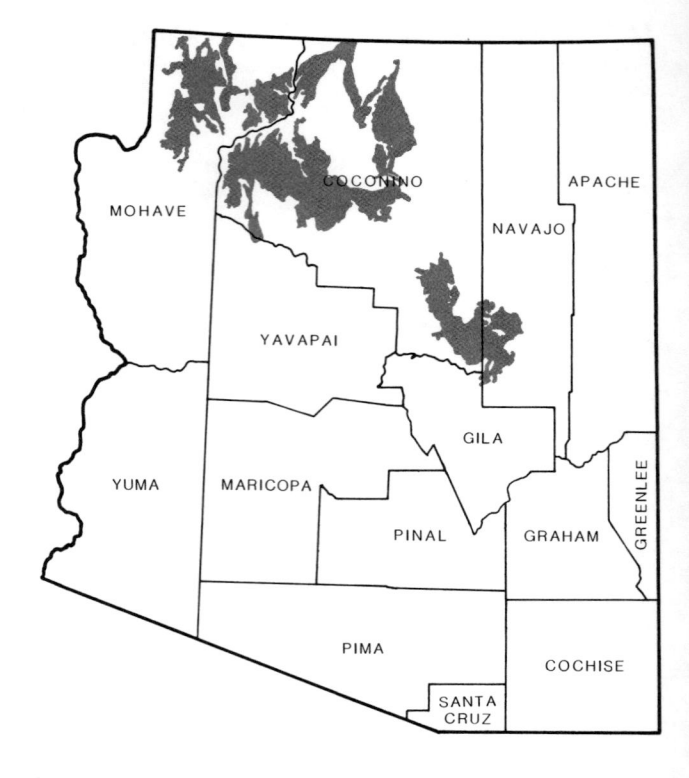

Soil Classification
 Lithic Torriorthents
 Lithic Ustollic Haplargids

Percent Slope mostly 2 to 15; 50 or more on breaks and escarpments

Elevation 1,400 to 2,000 m (4,600 to 6,600 ft)

Mean Annual Precipitation mostly 250 to 410 mm (10 to 16 in); to 500 mm (20 in) in a few higher places

Winter Precipitation as the Percentage of Annual Precipitation 45 to 70

Mean January Air Temperature -1 to 7 C (30 to 45 F)

Mean July Air Temperature 21 to 27 C (70 to 80 F)

Mean Annual Soil Temperature 10 to 14 C (50 to 58 F)

Frost-Free Days 130 to 180

Area and Percent of State 998,390 ha (2,467,000 ac), 3.4

Land Uses rangeland, wildlife habitat and recreation

This association consists of shallow, well-drained soils and rock outcrop on broad limestone and sandstone plateaus and plains. The soils formed in residuum on limestone and calcareous sandstone.

Winona soils make up about 60 percent of the association, Boysag soils 15 percent, limestone and sandstone rock outcrop 15 percent and included associated soils 10 percent. These included soils consist mostly of small areas of Tortugas, Welring, Moenkopie, Poley and Tusayan on uplands, and Tours and Lynx along drainageways.

The dominant native vegetation is juniper, pinyon pine, cliffrose, ceanothus, big sagebrush, blue and black grama, Indian ricegrass, pine dropseed, spike muhly and galleta.

This association has good potential to produce livestock forage but many areas have reduced gross yields due to juniper and pinyon encroachment. The potential use for rangeland wildlife habitat is fair. The best habitat is in the higher wooded areas.

Walnut Canyon and part of the Wupatki National Monument and Meteor Crater are within this association. The association borders the Grand Canyon in places and part lies within the Grand Canyon National Park.

Factors limiting the potential of these areas for building sites, sanitary facilities and recreational areas are shallow depth to bedrock, rock fragments on the surface and excessive slope on the steeper areas.

MS3 Palma-Clovis-Trail Association

Deep, coarse to moderately fine-textured, nearly level to rolling soils on plains.

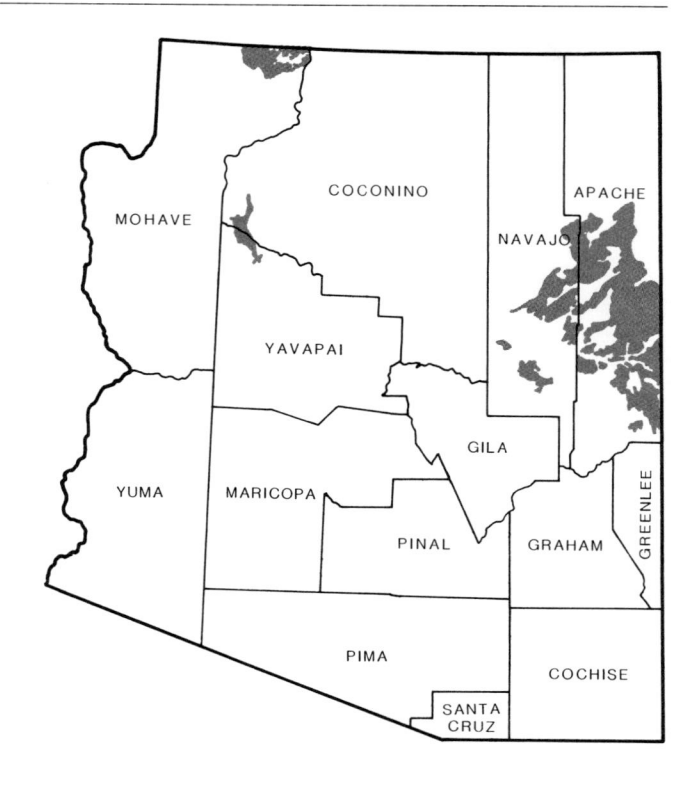

Soil Classification
  Ustollic Haplargids
  Typic Torrifluvents

Percent Slope 1 to 15

Elevation 1,460 to 2,120 m (4,800 to 7,000 ft)

Mean Annual Precipitation mostly 250 to 360 mm (10 to 14 in); range is 230 to 410 mm (9 to 16 in)

Winter Precipitation as the Percentage of Annual Precipitation 45 to 65

Mean January Air Temperature -1 to 2 C (30 to 35 F)

Mean July Air Temperature 21 to 24 C (70 to 75 F)

Mean Annual Soil Temperature 9 to 14 C (48 to 57 F)

Frost-Free Days 125 to 175

Area and Percent of State 1,743,040 ha (4,307,000 ac), 5.9

Land Uses rangeland and wildlife habitat; some cropland

This association consists of extensive areas of well-drained soils on the high plains. The soils formed in alluvium derived from sandstone, limestone and shale rocks.

Palma soils make up about 35 percent of the association, Clovis soils 30 percent, Trail soils 20 percent and included associated soils 15 percent. The included soils are mostly small areas of Moenkopie, Sheppard, Hubert and Millet on uplands, and narrow areas of Tours, Ives and Jocity along drainageways.

The dominant native vegetation is black and blue grama, Indian ricegrass, needleandthread, galleta, winterfat, rabbitbrush, sagebrush in some areas, and widely scattered juniper and pinyon pine in high-precipitation areas.

This association has fair potential under good management for the production of livestock forage but is somewhat limited by low amounts of precipitation. The potential for wildlife habitat is fair. Crops grown under irrigation near St. John are grain sorghum, alfalfa and corn.

Factors limiting the potential of these areas for home and recreation sites are flooding on Trail soils, moderate to low available water capacity on all soils and loose sandy surface soils subject to wind erosion.

MS4 Rudd-Bandera*-Cabezon Association

Shallow, gravelly, cobbly and stony, medium and fine-textured, undulating soils on plains and mesa tops and gently rolling to steep soils on cinder cones.

Soil Classification
Lithic Calciustolls
Torriorthentic Haplustolls
Lithic Argiustolls

Percent Slope 2 to 45

Elevation 1,520 to 2,250 m (5,000 to 7,400 ft)

Mean Annual Precipitation mostly 250 to 410 mm (10 to 16 in); to 500 mm (20 in) in a few areas

Winter Precipitation as the Percentage of Annual Precipitation 45 to 70

Mean January Air Temperature -1 to 7 C (30 to 45 F)

Mean July Air Temperature 16 to 27 C (60 to 80 F)

Mean Annual Soil Temperature 8 to 15 C (47 to 59 F)

Frost-Free Days 120 to 170

Area and Percent of State 495,760 ha (1,225,000 ac), 1.7

Land Uses rangeland, wildlife habitat and recreation

*Bandera soils are in a frigid temperature class.

This association consists of well-drained and somewhat excessively drained soils formed in residuum and alluvium weathered from basalt, andesite, ash-flow tuffs, cinders and related volcanic rock. They occur on plains, mesa tops and cinder cones.

Rudd soils make up about 40 percent of the association, Bandera-like soils 25 percent, Cabezon soils 20 percent and included associated soils, rock outcrop and cinder land 15 percent. The included soils are mostly small areas of Thunderbird, Springerville, Ziegler, Cross, Pastura and Apache.

The dominant native vegetation is blue, black and sideoats grama, galleta, western wheatgrass, juniper, shrub live oak, big sagebrush, cliffrose, some Mormon-tea, and some pinyon and ponderosa pine at higher elevations.

The association has fair potential for livestock and wildlife forage production. Excessively stony areas limit use by cattle and horses.

Factors that limit the potential of these areas for homesites and recreation are shallow depth to bedrock, high shrink-swell and stones on Cabezon soils; and slope on Bandera-like soils.

MS5 Roundtop-Boysag Association

Shallow and moderately deep, gravelly, fine-textured, undulating to moderately steep soils on plains and mountains.

Soil Classification
  Aridic Argiustolls
  Lithic Ustollic Haplargids

Percent Slope 2 to 30

Elevation 1,880 to 2,180 m (6,200 to 7,200 ft)

Mean Annual Precipitation 300 to 460 mm (12 to 18 in)

Winter Precipitation as the Percentage of Annual Precipitation 45 to 60

Mean January Air Temperature -1 to 2 C (30 to 35 F)

Mean July Air Temperature 18 to 24 C (65 to 75 F)

Mean Annual Soil Temperature 8 to 12 C (47 to 54 F)

Frost-Free Days 130 to 160

Area and Percent of State 157,020 ha (388,000 ac), 0.5

Land Uses rangeland, wildlife habitat and recreation; Grand Canyon National Monument is in part of this unit

This association consists of well-drained soils formed in residuum weathered from limestone and calcareous sandstone on plains and hills. They occur in Coconino County south of the Grand Canyon, an area just southeast of Flagstaff and a small area near Seligman.

Roundtop soils make up about 60 percent of the association, Boysag soils 25 percent and minor areas of associated soils and rock outcrop 15 percent. The minor soils are mostly small areas of Jacks, Winona and Rond on uplands and narrow areas of Lynx and other Torrifluvents along drainageways.

The dominant native vegetation is pinyon and some ponderosa pine, juniper, Gambel oak, mountainmahogany, cliffrose, beargrass, ring muhly, squirreltail, pine dropseed, Indian ricegrass, rabbitbrush, big sagebrush and blue, black and sideoats grama.

These soils have fair to good potential for livestock and wildlife forage production. Pinyon and juniper have invaded some areas and limited growth of better forage species.

The principal factors limiting the potential of these soils for homesites, community uses and recreation are depth to bedrock, slow permeability, high shrink-swell and low bearing strength. Some areas have excessive slope.

MS6 Lithic Torriorthents-Lithic Haplargids-Rock Outcrop Association

Shallow, gravelly and cobbly, moderately coarse to fine-textured, moderately sloping to very steep soils and rock outcrop on hills and mountains.

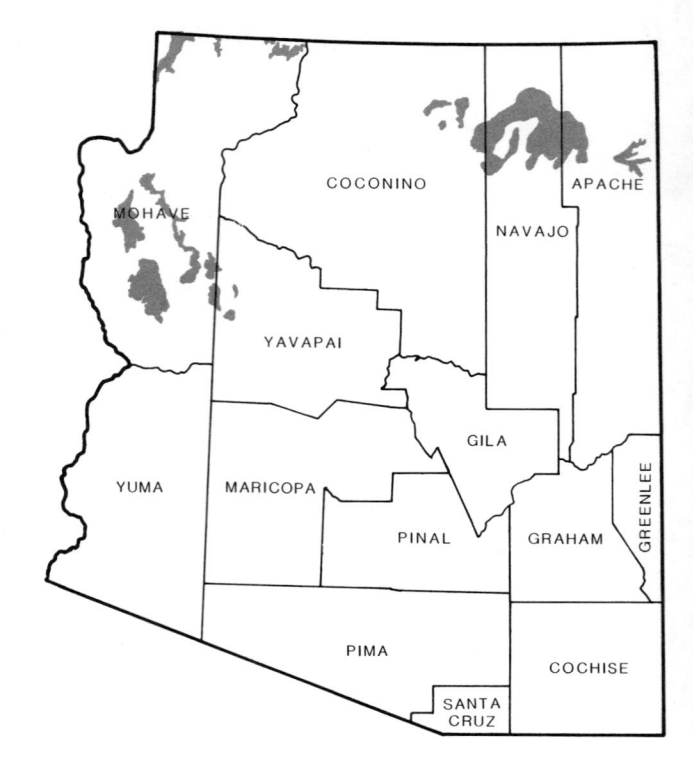

Soil Classification
Lithic Torriorthents
Lithic Haplargids

Percent Slope 10 to 60 or more

Elevation 1,520 to 2,120 m (5,000 to 7,000 ft)

Mean Annual Precipitation 250 to 410 mm (10 to 16 in)

Winter Precipitation as the Percentage of Annual Precipitation 50 to 70

Mean January Air Temperature -1 to 7 C (30 to 45 F)

Mean July Air Temperature 21 to 29 C (70 to 85 F)

Mean Annual Soil Temperature 8 to 15 C (47 to 59 F)

Frost-Free Days 130 to 190

Area and Percent of State 596,330 ha (1,473,500 ac), 2.0

Land Uses rangeland, wildlife habitat, recreation and mining

This association consists of rock outcrop and well-drained, very shallow and shallow soils on hills, mountains, mesa escarpments and canyon walls throughout the Mesic Semiarid zone. The soils formed in residuum and colluvium weathered from volcanic, granitic and sedimentary rocks.

Lithic Torriorthents make up about 40 percent of the association, Lithic Haplargids 25 percent, rock outcrop 15 percent and minor areas of included moderately deep and deep soils formed in alluvium on fans and along drainageways 20 percent. Representative deep and moderately deep soils included in the association are Pastura, Lonti, Poley, Balon and Jacks on uplands, and Lynx, Tours, Cordes, Redbank and Rune along drainageways.

The dominant native vegetation is juniper, pinyon pine, shrub live oak, manzanita, ceanothus, mountainmahogany, grama grass, bluestem, western wheatgrass, Indian ricegrass, big sagebrush in some areas, needlegrass and threeawn.

Smoother areas of this association have good potential for grazing livestock. The steeper, stonier areas receive limited use by domestic livestock but are good habitat for deer and, in some areas, bighorn sheep.

Factors limiting the potential of these areas for community uses are steep slopes, shallow depth to bedrock and rock fragments on the surface. Selected areas are suitable for campgrounds, picnic areas and pack trails. Canyon de Chelly National Monument, a noted scenic attraction, is included in this association. A portion of the Black Mesa is used for surface-mined coal production. Both of these areas are on the Navajo and Hopi Indian reservations.

MS7 Cabezon-Thunderbird-Springerville Association

Shallow to deep, gravelly, cobbly and stony, fine-textured, nearly level to very steep soils on basaltic plains, mesas and hills.

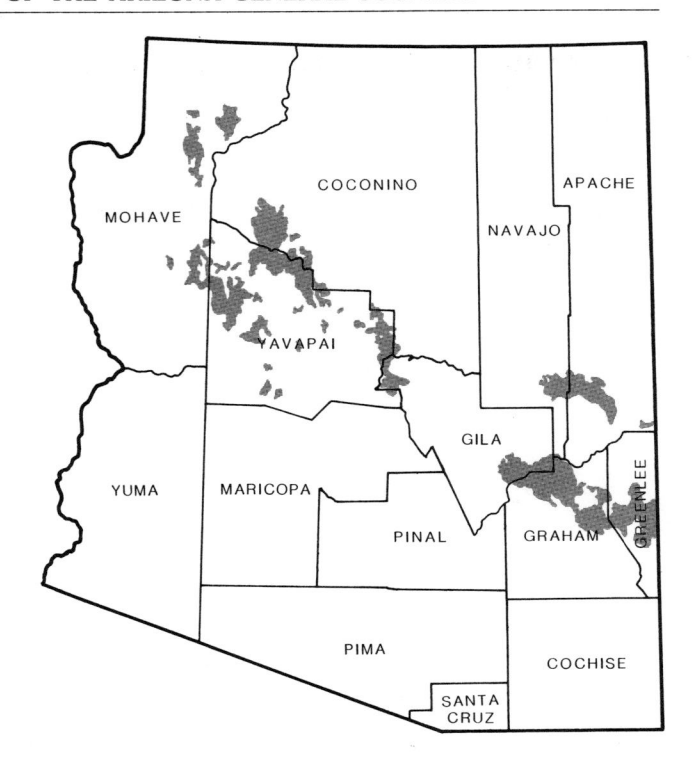

Soil Classification
  Lithic Argiustolls
  Aridic Argiustolls
  Typic Chromusterts

Percent Slope mostly none to 30; to 75 or more on escarpment-like areas and in canyons

Elevation 1,370 to 2,120 m (4,500 to 7,000 ft); some peaks to 2,430 m (8,000 ft)

Mean Annual Precipitation 300 to 510 mm (12 to 20 in)

Winter Precipitation as the Percentage of Annual Precipitation 45 to 65

Mean January Air Temperature -1 to 7 C (30 to 45 F)

Mean July Air Temperature 21 to 29 C (70 to 85 F)

Mean Annual Soil Temperature 8 to 15 C (47 to 59 F)

Frost-Free Days 120 to 180

Area and Percent of State 597,340 ha (1,476,000 ac), 2.0

Land Uses rangeland, wildlife habitat and recreation

This association consists of well-drained soils on plains, mesas, hills and very steep escarpments in scattered areas throughout the Mesic Semiarid zone. The soils formed in residuum and alluvium weathered from basalt and ash-flow tuffs, cinders and related volcanic materials.

Cabezon soils make up about 35 percent of the association, Thunderbird soils 30 percent, Springerville soils 15 percent and minor areas of associated soils and rock outcrop 20 percent. These included soils are mostly small areas of Venezia, Apache, Cross, Rudd, Waldroup and Ziegler on uplands, and narrow areas of Lynx and Jacques along drainageways.

The dominant native vegetation is blue, hairy and sideoats grama, tobosa, wolftail, algerita, ring muhly, squirreltail and juniper. Some pinyon and ponderosa pine grow in the areas that have high precipitation.

These soils have fair potential for production of livestock and wildlife forage. Domestic livestock avoid areas with broad fissures and areas with cobbly, stony surfaces.

The principal factors limiting the potential of these soils for development of homesites and recreation sites are high shrink-swell, clay textures, slow to very slow permeability and excessive rock fragments on the surface.

MS8 Pastura-Poley-Partri Association

Shallow, gravelly, medium-textured and deep fine-textured, nearly level to rolling soils on plains and hills.

Soil Classification
 Ustollic Paleorthids
 Ustollic Haplargids
 Aridic Argiustolls

Percent Slope 1 to 15

Elevation 1,460 to 1,940 m (4,800 to 6,400 ft)

Mean Annual Precipitation 300 to 410 mm (12 to 16 in)

Winter Precipitation as the Percentage of Annual Precipitation 50 to 65

Mean January Air Temperature -1 to 4 C (30 to 40 F)

Mean July Air Temperature 21 to 27 C (70 to 80 F)

Mean Annual Soil Temperature 10 to 14 C (50 to 57 F)

Frost-Free Days 135 to 175

Area and Percent of State 335,290 ha (828,500 ac), 1.1

Land Uses rangeland and wildlife habitat; some irrigated cropland on Poley soils

This association consists of well-drained soils on plains and hills. The soils formed on old alluvium weathered from limestone, sandstone and basalt.

Pastura soils make up about 35 percent of the association, Poley soils 35 percent, Partri soils 25 percent and minor areas of associated soils 5 percent. The included soils are mostly small areas of Showlow and Abra on uplands, and Rune, Cordes and Lynx along drainageways.

Small areas of Poley soils are used for irrigated cropland. The dominant native vegetation is blue, hairy, black and sideoats grama, wolftail, ring muhly, cane bluestem, squirreltail, winterfat and scattered juniper in some areas. Crops grown under irrigation are mostly feed grains, corn, alfalfa and pasture grasses.

These soils have good potential for producing livestock forage. Wildlife use, except by antelope, is transitory. The best cover is along drainageways.

The major factors limiting the potential of these soils for homesites and other community uses are shallow depth to hardpan and high lime in Pastura soils, and slow permeability, clay subsoils, high shrink-swell and low bearing strength in Poley and Partri soils.

MS9 Lonti-Balon-Lynx Association

Deep, moderately fine and gravelly, moderately fine and fine-textured, nearly level soils on floodplains and undulating to steep valley slopes and plains.

Soil Classification
Ustollic Haplargids
Cumulic Haplustolls

Percent Slope uplands, 1 to 45; floodplains, swales and lower fans, none to 5

Elevation 1,270 to 1,670 m (4,200 to 5,500 ft)

Mean Annual Precipitation 300 to 460 mm (12 to 18 in)

Winter Precipitation as the Percentage of Annual Precipitation 50 to 60

Mean January Air Temperature 2 to 7 C (35 to 45 F)

Mean July Air Temperature 21 to 29 C (70 to 85 F)

Mean Annual Soil Temperature 10 to 15 C (50 to 59 F)

Frost-Free Days 140 to 200

Area and Percent of State 128,290 ha (317,000 ac), 0.4

Land Uses rangeland and wildlife habitat; cropland, urban, recreation and placer mining on small areas of Lynx soils

This association consists of well-drained soils on dissected old alluvial fan surfaces and in swales and on floodplains. Lonti and Balon soils formed in mixed old alluvium, and Lynx soils in mixed recent alluvium, weathered mainly from granite, schist, sandstone, shale, limestone and volcanic rocks.

Lonti soils make up about 60 percent of the association, Balon soils 15 percent, Lynx soils 15 percent and minor areas of associated soils 10 percent. These minor soil areas are mostly Abra, Wineg and Springerville on the uplands and intermixed areas of Cordes on floodplains.

The dominant native vegetation is blue, hairy, black and sideoats grama, squirreltail, wolftail, muhly grasses, threeawn, shrub live oak, ceanothus, snakeweed and pubescent squawbush. Crops raised under irrigation are feed grains, corn, alfalfa and pasture grasses. There are a few small orchards.

These soils have good potential under good management for the production of range forage. Brush control and range seeding are practical on these soils. The best wildlife habitat is along drainageways.

The major factors limiting the potential of these areas for urban and recreational development are the flooding hazard on Lynx soils; moderately slow to slow permeability of all soils; high shrink-swell and low bearing strength of Lonti soils; and excessive slope and erosion hazard on some Balon and Lonti soils.

MS10 Pastura-Abra-Lynx Association

Shallow, gravelly, medium-textured and deep, medium and moderately fine-textured, nearly level to rolling soils on plains and hills.

Soil Classification
  Ustollic Paleorthids
  Ustollic Calciorthids
  Cumulic Haplustolls

Percent Slope uplands, 1 to 30; floodplains and swales, none to 5

Elevation 1,340 to 1,700 m (4,400 to 5,600 ft)

Mean Annual Precipitation 280 to 410 mm (11 to 16 in)

Winter Precipitation as the Percentage of Annual Precipitation 50 to 60

Mean January Air Temperature -1 to 4 C (30 to 40 F)

Mean July Air Temperature 21 to 27 C (70 to 80 F)

Mean Annual Soil Temperature 12 to 14 C (53 to 58 F)

Frost-Free Days 140 to 170

Area and Percent of State 96,120 ha (237,500 ac), 0.3

Land Uses rangeland and wildlife habitat; cropland, urban and recreation on small areas of Lynx soils

This association consists of well-drained, limy soils on nearly level to undulating plains, undulating to rolling ridges and hill toeslopes in the Chino Valley and adjacent area. Pastura and Abra soils formed in mixed old alluvium, and Lynx soils in recent alluvium, both weathered from limestone, sandstone, shale, granite, schist and volcanic rocks.

Pastura soils make up about 35 percent of the association, Abra soils 35 percent, Lynx soils 20 percent and minor areas of Lonti, Poley and Wineg soils 10 percent.

The dominant native vegetation is blue, hairy, black and sideoats grama, squirreltail, wolftail, muhly grasses, threeawn, shrub live oak, ceanothus, snakeweed and pubescent squawbush. Juniper and pinyon pine grow in some areas. Crops raised under irrigation are small grain, corn, alfalfa and pasture grasses.

This association has good potential for the production of range forage. Brush control and range seeding are possible on these soils. Juniper eradication is practical on the Pastura and Abra soils. These soils have good potential for rangeland wildlife habitat improvement, such as range seeding and water developments.

The major factors limiting the potential of these areas for development of homesites and recreation facilities are the possible flooding and the moderately slow permeability of Lynx soils; high lime content of Abra and Pastura soils and shallow depth to hardpan in Pastura soils; and excessive slope in some areas of Abra and Pastura soils.

Mesic Subhumid Soils

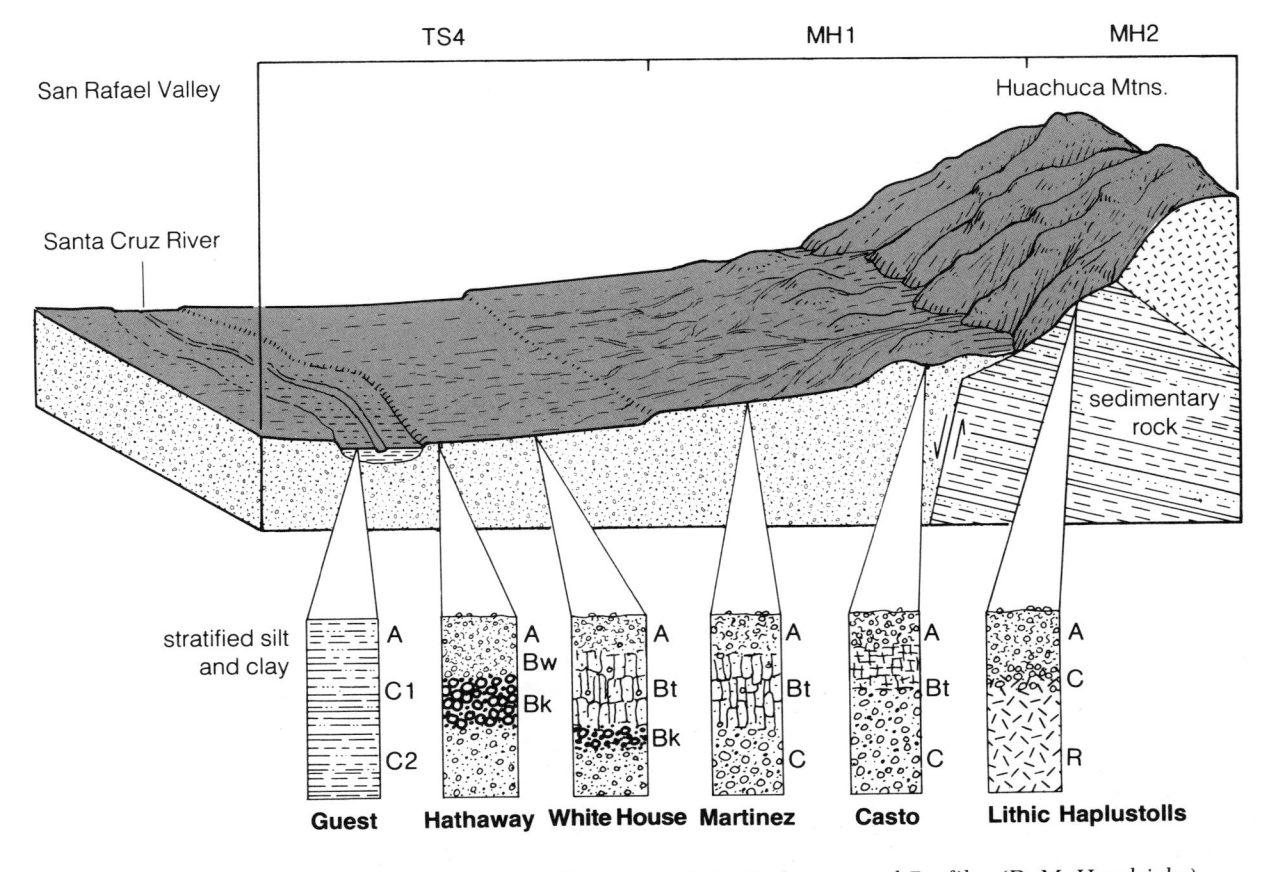

FIGURE 52. Representative Mesic Subhumid Soils Soilscape and Profiles (D. M. Hendricks)

MH
Mesic Subhumid Soils

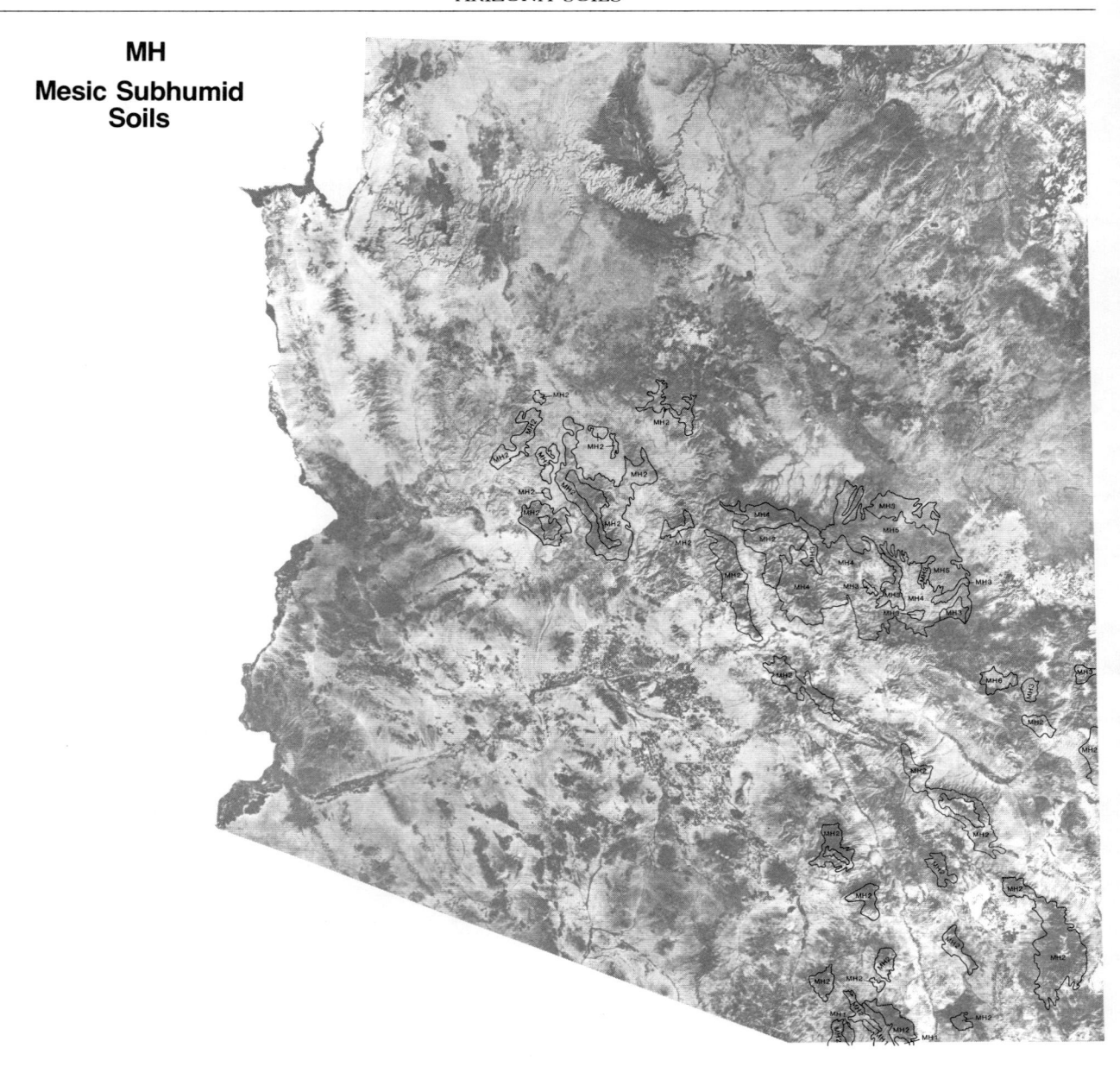

FIGURE 53. *Geographic Distribution of Mesic Subhumid (MH) Soils in Arizona (D. M. Hendricks)*

- MH1 Casto-Martinez-Canelo Association
- MH2 Lithic Haplustolls-Lithic Argiustolls-Rock Outcrop Association
- MH3 Showlow-Disterheff-Cibeque Association
- MH4 Roundtop-Tortugas-Jacks Association
- MH5 Overgaard-Elledge-Telephone Association
- MH6 Pachic Argiustolls-Lynx Association

MH Mesic Subhumid Soils

Mesic Subhumid Soils have a mean annual soil temperature of 8 C (47 F) or more, but less than 15 C (59 F). The difference between mean summer and mean winter temperatures is greater than 5 C (9 F). The mean annual precipitation is more than 410 mm (16 in). These soils are at intermediate elevations. They cover about 2,021,270 ha (4,994,500 ac), or 7 percent of Arizona.

MH1 Casto-Martinez-Canelo Association

Deep, very fine-textured and deep, gravelly, moderately fine and fine-textured, nearly level to steep soils on dissected fan surfaces and valley slopes.

Soil Classification
 Udic Haplustalfs
 Udic Paleustalfs
 Aquic Haplustalfs

Percent Slope 2 on mesas to 40 on sides of ridges and mesas; vertical relief 8 to 60 m (25 to 200 ft)

Elevation 1,670 to 1,880 m (5,500 to 6,200 ft)

Mean Annual Precipitation 410 to 510 mm (16 to 20 in)

Winter Precipitation as the Percentage of Annual Precipitation 30 to 35

Mean January Air Temperature 4 to 7 C (40 to 45 F)

Mean July Air Temperature 18 to 24 C (65 to 75 F)

Mean Annual Soil Temperature 12 to 15 C (54 to 59 F)

Frost-Free Days 150 to 180

Area and Percent of State 28,530 ha (70,500 ac), 0.1

Land Uses rangeland, wildlife habitat and recreation

This association consists of deep, well-drained and somewhat poorly drained soils on high, dissected old alluvial fans. The soils formed in old alluvium derived from sedimentary and igneous rocks. The fan remnants range from long, narrow, rounded ridges with moderately steep slopes to fairly broad-topped, gently sloping mesas with steep side slopes.

Casto soils make up about 60 percent of the association, Martinez soils 20 percent, Canelo soils 15 percent and minor areas of White House soils on uplands and Comoro and Grabe soils in drainageways about 5 percent.

The dominant native vegetation on Casto and Canelo soils is Emory oak, pinyon pine, juniper, manzanita, Arizona cottontop, Texas and cane bluestem, grama grasses, curlymesquite, wolftail and perennial forbs. Martinez soils mostly have the same grasses but not the woody species.

This association has good potential for the production of livestock and wildlife forage. Brush control and range seeding are possible, where needed, on these soils. The principal kinds of wildlife are deer, antelope, quail, javelina, rabbits and some coatimundi. Parker Canyon Lake, in this unit, is stocked with bass, catfish and trout.

The principal factors limiting the potential of these soils for homesites, community uses and recreation sites are slow or very slow permeability and clayey texture in all soils, and excessive rock fragment content and slope on the Casto and Canelo soils.

MH2 Lithic Haplustolls-Lithic Argiustolls-Rock Outcrop Association

Shallow, gravelly and cobbly, moderately coarse to moderately fine-textured, gently sloping to very steep soils and rock outcrop on hills and mountains.

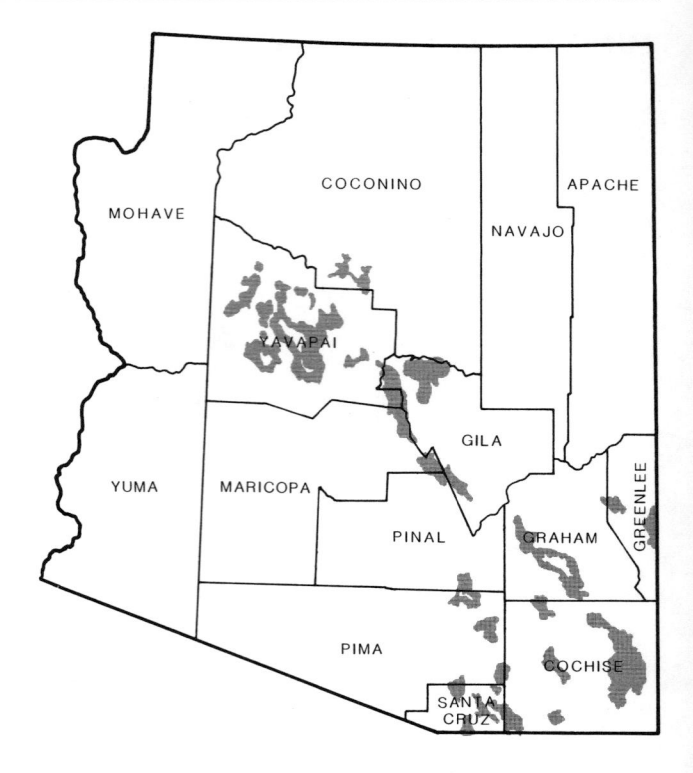

Soil Classification
 Lithic Haplustolls
 Lithic Argiustolls

Percent Slope mostly 5 to 60

Elevation central, 1,370 m (4,500 ft); southern, 2,135 m (7,000 ft)

Mean Annual Precipitation 410 to 630 mm (16 to 25 in)

Winter Precipitation as the Percentage of Annual Precipitation 30 to 60

Mean January Air Temperature 2 to 7 C (35 to 45 F)

Mean July Air Temperature 21 to 29 C (70 to 85 F)

Mean Annual Soil Temperature 8 to 15 C (47 to 59 F)

Frost-Free Days 140 to 230

Area and Percent of State 1,147,320 ha (2,835,000 ac), 3.9

Land Uses rangeland, wildlife habitat, water supply and recreation

This association consists of rock outcrop and dark-colored, well-drained, shallow and very shallow soils formed in the residuum on igneous and sedimentary hills and mountains.

Lithic Haplustolls make up about 50 percent of the association, Lithic Argiustolls 20 percent, rock outcrop 15 percent and minor areas of other shallow, moderately deep and deep soils 15 percent.

Faraway, Tortugas and the similar Barkerville soils are representative of Lithic Haplustolls. Luzena and Cabezon soils are representative Lithic Argiustolls. Cross, Gaddes, Mokiak and Moano soils on uplands and Cordes and Lynx soils along drainageways are some of the minor included soils.

The dominant native vegetation is oak, juniper, pinyon pine, manzanita, cliffrose, grama grasses, plains lovegrass, Texas and cane bluestem, tanglehead and galleta. Some ponderosa and other pine trees grow above 1,670 m (5,500 ft). These soils have fair to good potential for the production of livestock forage on much of the association. Wildlife commonly observed on the association include white- and blacktail deer, bighorn sheep, quail, dove, rabbits, squirrel, wild turkey and small rodents and birds.

The dominant factors limiting the potential of these areas for the development of homesites and other community and recreational uses are excessive slope, excessive rock fragments on the surface and shallow depth to bedrock. Small areas on the fans and drainageways may be suitable for campsite and picnic areas.

MH3 Showlow-Disterheff-Cibeque Association

Deep, gravelly, medium and fine-textured, nearly level to steep soils on dissected uplands.

Soil Classification
  Aridic Argiustolls
  Typic Haplustalfs
  Ustollic Calciorthids

Percent Slope plains, 2 to 20; dissected uplands, to 60

Elevation 1,370 to 2,030 m (4,500 to 6,700 ft)

Mean Annual Precipitation 410 to 560 mm (16 to 22 in)

Winter Precipitation as the Percentage of Annual Precipitation 45 to 55

Mean January Air Temperature -1 to 4 C (30 to 40 F)

Mean July Air Temperature 16 to 24 C (60 to 75 F)

Mean Annual Soil Temperature 8 to 15 C (47 to 59 F)

Frost-Free Days 115 to 170

Area and Percent of State 227,440 ha (562,000 ac), 0.8

Land Uses rangeland, woodland, wildlife habitat, water supply and recreation

This association consists of well-drained soils in narrow valleys and on plains and dissected uplands on the Mogollon Plateau and in other small areas. The soils formed in mixed gravelly old alluvium derived from sedimentary and some volcanic rocks.

Showlow soils make up about 35 percent of the association, Disterheff soils 35 percent, Cibeque soils 20 percent and minor areas of associated soils 10 percent. These minor soils are mostly small areas of Elledge, Chevelon and Millard on uplands and narrow areas of Lynx, Jaques, Tours and Zeniff along drainageways.

A few small areas of Showlow soils were dry farmed in the past. Corn, small grain and grain sorghum were produced but crop failure due to lack of moisture was common. The dominant native vegetation is juniper, pinyon pines, shrub live oak, cliffrose, mountainmahogany, grama grasses, western wheatgrass and muhly grasses. Ponderosa pine grows at high elevations and on the north and east aspects at lower elevations.

These soils have only fair potential for the production of livestock forage due to the competition of woody species. Grazing use is mainly during the summer months. They have good potential for woodland and rangeland wildlife habitat. Dominant wildlife species are deer, squirrel, Wild Turkey and elk at some high elevations, rabbits, quail, dove and small rodents and birds.

The major factors limiting the potential of these soils for homesites, community uses and recreation sites are the slow permeability and high shrink-swell potential of the Showlow and Disterheff soils, excessive slope of the Cibeque and some Disterheff soils and high lime in the Cibeque soils. Smooth areas of Showlow and Disterheff soils are suitable for campgrounds and picnic areas. Hunting, fishing and camping are good in many areas of this association.

MH4 Roundtop-Tortugas-Jacks Association

Shallow to moderately deep, gravelly and cobbly, medium and fine-textured, undulating to very steep soils on hills and mountains.

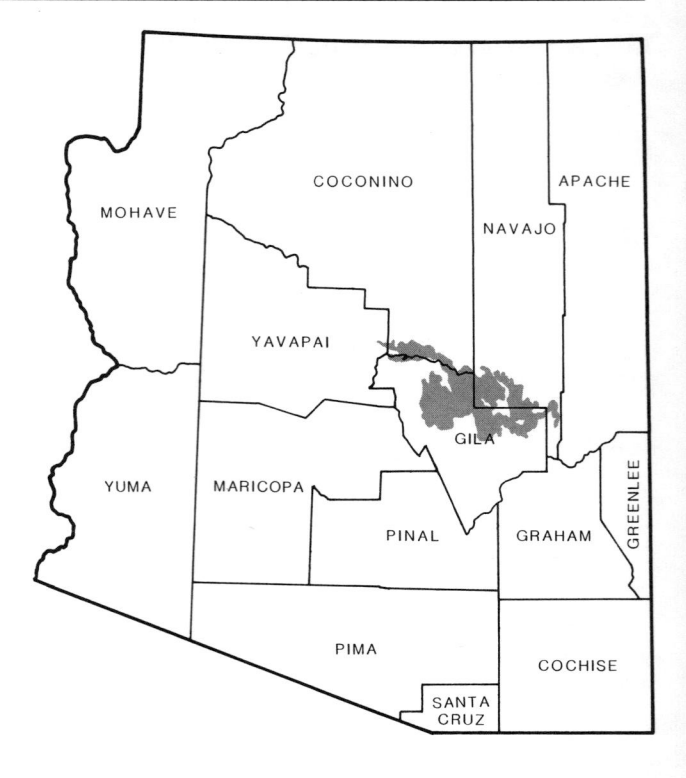

Soil Classification
Aridic Argiustolls
Lithic Haplustolls
Udic Haplustalfs

Percent Slope 2 to 60 or more

Elevation 1,370 to 2,060 m (4,500 to 6,800 ft)

Mean Annual Precipitation 410 to 630 mm (16 to 25 in)

Winter Precipitation as the Percentage of Annual Precipitation 50 to 60

Mean January Air Temperature -1 to 7 C (30 to 45 F)

Mean July Air Temperature 21 to 29 C (70 to 85 F)

Mean Annual Soil Temperature 8 to 15 C (47 to 59 F)

Frost-Free Days 120 to 170

Area and Percent of State 413,600 ha (1,022,000 ac), 1.4

Land Uses rangeland, woodland, wildlife habitat, water supply and recreation

This association consists of well-drained soils on limestone and sandstone mountains just south of the Mogollon Rim.

Roundtop soils make up about 35 percent of the association, Tortugas soils 35 percent, Jacks soils 15 percent and minor areas of rock outcrop and associated soils 15 percent. The minor soils are mostly small areas of Dye, Rond, Telephone and Barkerville on uplands, and narrow areas of Tours, Lynx and Jacques along drainageways.

The dominant native vegetation is pinyon pine, juniper, Gambel oak, manzanita, shrub live oak, ceanothus, cliffrose, mountainmahogany, grama and muhly grasses, western wheatgrass, wolftail, squirreltail and pine dropseed. Ponderosa pine grows at high elevations and on north and east aspects at lower elevations.

These soils have only fair potential for the production of livestock forage due to the competition from woody species. Brush control is practical on the smoother areas. The association has good potential for woodland and rangeland wildlife habitat. The dominant wildlife species are deer, squirrel, turkey and elk at some high elevations, rabbits, quail, dove and small rodents and birds.

The major factors limiting the potential of these soils for homesites, community uses and recreation sites are shallow and moderate depths to bedrock and slow permeability of Roundtop and Jacks soils and steep slopes in some areas of all soils. Smooth areas of Jacks soils and some of the included soils are suitable for campgrounds and picnic areas. Hunting, fishing and camping are good in many areas of this association. Several large Indian ruins are in the area.

MH5 Overgaard-Elledge-Telephone Association

Shallow to deep, gravelly and cobbly, moderately coarse and fine-textured, undulating to very steep soils on mountains and hills.

Soil Classification
Typic Haplustalfs
Udic Haplustalfs
Lithic Ustorthents

Percent Slope 2 to 60 or more

Elevation 1,670 to 2,340 m (5,500 to 7,700 ft)

Mean Annual Precipitation 460 to 630 mm (18 to 25 in)

Winter Precipitation as the Percentage of Annual Precipitation 45 to 50

Mean January Air Temperature -1 to 2 C (30 to 35 F)

Mean July Air Temperature 18 to 21 C (65 to 70 F)

Mean Annual Soil Temperature 7 to 13 C (45 to 56 F)

Frost-Free Days 110 to 150

Area and Percent of State 179,280 ha (443,000 ac), 0.6

Land Uses forestry, rangeland, water supply, wildlife habitat and recreation

This association consists of well-drained soils on the forested highlands along the Mogollon Rim and the deeply dissected breaks along the Rim's south edge. The soils formed in residuum and colluvium weathered from sandstone, cherty limestone and shale.

Overgaard soils make up about 35 percent of the association, Elledge soils 30 percent, Telephone soils 25 percent and minor areas of associated soils and rock outcrop 10 percent. The minor included soils are mostly Amos, Showlow, Jacks, Tortugas and Roundtop on uplands, and Lynx, Tours and Jocity along drainageways.

The dominant vegetation is ponderosa and pinyon pine, Gambel oak, juniper, mountainmahogany, manzanita, mountain muhly, grama grasses, Arizona fescue, little bluestem, junegrass, pine dropseed and western wheatgrass.

Timber is harvested in the more accessible areas. The association has low potential for domestic livestock forage production due to steep slopes and competition from woody species. Most grazing is during the summer months. The potential is good for woodland wildlife habitat. The dominant game species are deer, elk, black bear, wild turkey, mountain lion, squirrel and Band-tailed Pigeon. Camping, hunting and fishing are good in this association.

The major factors limiting the potential of these soils for homesites, community uses and recreation are excessive slope on much of the area, moderate and shallow depth to bedrock in the Elledge and Telephone soils, and slow permeability in Overgaard and Elledge soils. All of these soils are subject to moderate frost action. Smooth and less sloping areas of Overgaard and Elledge soils and some of the included minor soils are suitable for campgrounds and picnic areas.

MH6 Pachic Arguistolls-Lynx Association

Deep, moderately fine and fine-textured, nearly level to gently rolling soils on plains.

Soil Classification
Pachic Arguistolls
Cumulic Haplustolls

Percent Slope mostly 2 to 8; range is 2 to 15

Elevation 1,760 to 2,000 m (5,800 to 6,600 ft)

Mean Annual Precipitation 410 to 510 mm (16 to 20 in)

Winter Precipitation as the Percentage of Annual Precipitation 45 to 50

Mean January Air Temperature -1 to 2 C (30 to 35 F)

Mean July Air Temperature 21 to 24 C (70 to 75 F)

Mean Annual Soil Temperature 8 to 12 C (47 to 54 F)

Frost-Free Days 115 to 150

Area and Percent of State 25,090 ha (62,000 ac), less than 0.1

Land Uses rangeland, wildlife habitat and recreation

This association consists of well-drained soils of the Big Prairie area near Point of Pines on the San Carlos Apache Indian Reservation. The soils formed in mixed old and recent alluvium weathered from basalt, tuffs and agglomerates.

Pachic Arguistolls (Showlow-like soils) make up about 55 percent of the association, Lynx soils 30 percent and minor areas of associated soils and rock outcrop 15 percent. The included soils are mostly small areas of Thunderbird, Cabezon and Faraway on uplands and small areas of Cordes intermixed with Lynx along drainageways.

The dominant native vegetation is grass with scattered ponderosa and pinyon pine and juniper. Grasses are sideoats, blue and hairy grama, tobosa, Arizona fescue, muhly grasses and western wheatgrass.

This association has good potential for producing livestock forage. Some areas are overgrazed due to easy accessibility. The potential for rangeland wildlife habitat is good. Dominant species are antelope, elk (winter range), quail, dove, rabbits and small mammals and birds. Several small, trout-stocked lakes are in the area.

Factors limiting the potential of these soils for development of homesites and recreation sites are moderately slow and slow permeability and possible brief flooding on some areas of Lynx soils. In addition, the Arguistolls have high shrink-swell in the subsoil.

Frigid Subhumid Soils

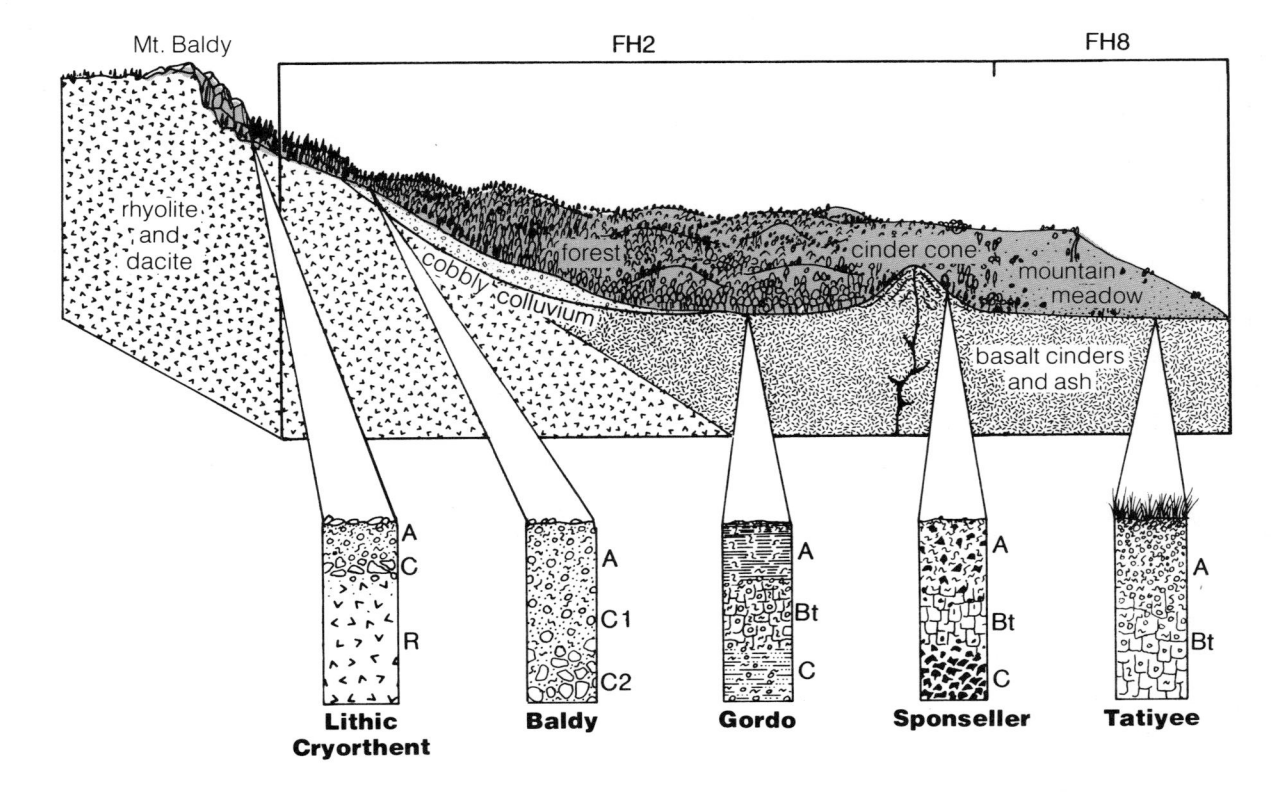

FIGURE 54. Representative Frigid Subhumid Soils Soilscape and Profiles (D. M. Hendricks)

FIGURE 55. Geographic Distribution of Frigid Subhumid (FH) Soils in Arizona (D. M. Hendricks)
- FH1   Mirabal-Dandrea-Brolliar Association
- FH2   Sponseller-Ess-Gordo Association
- FH3   Soldier-Hogg-McVickers Association
- FH4   Soldier-Lithic Cryoborolls Association
- FH5   Mirabal-Baldy-Rock Outcrop Association
- FH6   Eutroboralfs-Mirabal Association
- FH7   Cryorthents-Eutroboralfs Association
- FH8   Gordo-Tatiyee Association

FH Frigid Subhumid Soils

Frigid Subhumid Soils have mean annual soil temperatures lower than 8 C (47 F). The difference between mean winter and mean summer soil temperataure is more than 5 C (9 F) at a depth of 50 cm (20 in) or at a bedrock contact in shallow soils. The mean annual precipitation is more than 410 cm (16 in), with one-half or more usually falling during the winter and early spring months as snow, sleet or rain. These soils are at elevations mostly more than 1,970 m (6,500 ft) on the Colorado Plateau and in a few of the high mountains of the Basin and Range Province. They cover about 2,093,110 ha (5,172,000 ac), or 7 percent of Arizona.

FH1 Mirabal-Dandrea-Brolliar Association

Moderately deep and deep, gravelly and cobbly, moderately coarse and fine-textured, gently sloping to very steep mountain soils.

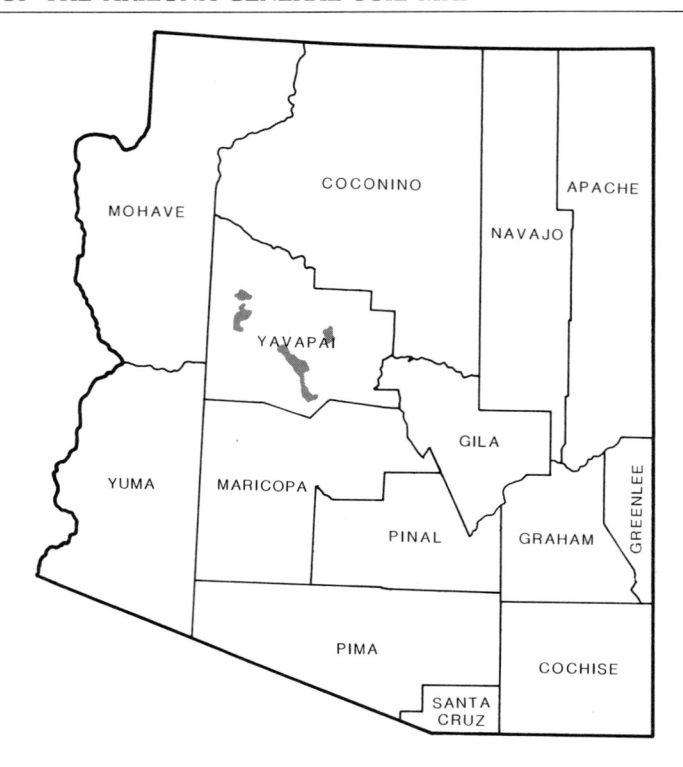

Soil Classification
  Typic Ustorthents
  Mollic Eutroboralfs
  Typic Argiborolls

Percent Slope mountain and mesa tops, 1 to 15; mountainsides and canyons, 15 to 60 or more

Elevation 1,820 to 2,430 m (6,000 to 8,000 ft)

Mean Annual Precipitation 460 to 760 mm (18 to 30 in)

Winter Precipitation as the Percentage of Annual Precipitation 50 to 60

Mean January Air Temperature 2 to 7 C (35 to 45 F)

Mean July Air Temperature 18 to 27 C (65 to 80 F)

Mean Annual Soil Temperature 6 to 11 C (43 to 52 F)

Frost-Free Days 80 to 170

Area and Percent of State 37,430 ha (92,500 ac), 0.1

Land Uses grazable woodland, wildlife habitat, recreation, water supply and some mining

This association consists of well-drained soils on high mountains. The soils formed in residuum on granite and schist mountains and high basalt mesas.

Mirabal soils make up about 50 percent of the association, Dandrea soils 25 percent, Brolliar soils 10 percent and rock outcrop and minor areas of associated soils 15 percent. These minor soils are mostly small areas of Hogg and Wilcoxson on uplands and narrow areas of Clover Springs and Luth along drainageways.

The dominant native vegetation is ponderosa pine, Gambel oak, Douglas fir and some quaking aspen at high elevations. The grass understory is mostly Arizona fescue, mountain muhly, mountain brome and squirreltail. Most livestock grazing use is during summer months. Timber is harvested in some areas of this association. The principal wildlife species are deer, Wild Turkey, porcupine, Abert's squirrel, Band-tailed Pigeon, black bear and mountain lion. A few dove and rabbits are in the aspen areas.

Factors limiting the potential of these soils for the development of summer homesites and recreation sites are slope, depth to bedrock and potential frost action on all soils, and slow permeability and high shrink-swell on Brolliar and Dandrea soils. Smooth, gently sloping areas of these soils are suitable for campgrounds and picnic sites. Hunting, hiking and winter sports and camping are the main recreational activities.

FH2 Sponseller-Ess-Gordo Association

Moderately deep and deep, medium and moderately fine-textured, moderately sloping to very steep mountain soils.

Soil Classification
Typic Argiborolls
Argic Pachic Cryoborolls

Percent Slope 5 to 60 or more

Elevation 2,000 to 3,790 m (6,600 to 12,500 ft)

Mean Annual Precipitation 460 to 760 mm (18 to 30 in) or more

Winter Precipitation as the Percentage of Annual Precipitation 45 to 65

Mean January Air Temperature -4 to 4 C (25 to 40 F)

Mean July Air Temperature 16 to 27 C (60 to 80 F)

Mean Annual Soil Temperature 3 to 8 C (38 to 47 F)

Frost-Free Days 70 to 125

Area and Percent of State 1,280,670 ha (3,164,500 ac), 4.3

Land Uses grazable woodland, wildlife habitat, recreation and water supply

This association consists of well-drained soils on high mountains. The soils formed in residuum and colluvium weathered from basalt, rhyolite, andesite, cinders, ash-flow tuff and related volcanic rocks.

Sponseller soils make up about 25 percent of the association, Ess soils 25 percent, Gordo soils 25 percent and minor areas of a large number of associated and similar soils 25 percent. The minor associated soils include small areas of Cambern, Baldy, Bushvalley, Brolliar, Mirabal, Siesta, Sizer, Clover Springs, Luth and several uncorrelated soils mapped by the U.S. Forest Service. Small areas of rock outcrop and talus slope on peaks also are included.

The dominant native vegetation is ponderosa pine, Douglas fir, spruce, quaking aspen, juniper and Gambel oak. Grasses occur in small open areas and consist of Arizona fescue, pine dropseed, junegrass, wheatgrass, mountain muhly, squirreltail and brackenfern. Most livestock grazing is during summer months. The potential for timber production is fair to good. The potential for woodland wildlife habitat management also is good. The dominant wildlife species are deer, elk, Wild Turkey, Abert's squirrel, black bear, mountain lion, porcupine, Band-tail Pigeon and a few dove and cottontail rabbits. Several large trout-stocked lakes are in the mapping unit. Recreation uses are hunting, fishing, hiking, camping and skiing.

The major factors limiting the potential of these soils for building sites and recreational facilities are moderately slow permeability, potential frost action, rock fragments on the surface and slope. Smooth, gently sloping areas can be used for campgrounds, picnic sites and other community uses.

FH3 Soldier-Hogg-McVickers Association

Moderately deep and deep, fine-textured, gently sloping to steep mountain soils.

Soil Classification
Typic Glossoboralfs
Mollic Eutroboralfs
Typic Eutroboralfs

Percent Slope mostly 2 to 20; range is 2 to 50

Elevation 2,000 to 2,730 m (6,600 to 9,000 ft)

Mean Annual Precipitation 460 to 760 mm (18 to 30 in)

Winter Precipitation as the Percentage of Annual Precipitation 50 to 60

Mean January Air Temperature -1 to 4 C (30 to 40 F)

Mean July Air Temperature 18 to 21 C (65 to 70 F)

Mean Annual Soil Temperature 5 to 8 C (41 to 47 F)

Frost-Free Days 85 to 125

Area and Percent of State 192,430 ha (475,500 ac), 0.7

Land Uses grazable woodland, forestry, wildlife habitat, recreation and water supply

This association consists of moderately well- and well-drained soils on high mountainous areas on the Coconino Plateau and adjacent areas. The soils formed in residuum weathered from cherty limestone and sandstone, members of the Kaibab and Coconino geologic formations.

Soldier soils make up about 55 percent of the association, Hogg soils 20 percent, McVickers soils 15 percent and minor areas of associated soils and rock outcrop 10 percent. The included minor soils are mostly small areas of Overgaard, Wildcat, Sanchez, Chilson and Palomino on uplands, and narrow areas of Clover Springs and Luth along drainageways.

The dominant native vegetation is ponderosa pine, Douglas and white fir, quaking aspen and some Gambel oak, juniper and ceanothus. Grasses in open areas and as understory include Arizona fescue, pine dropseed, squirreltail, blue grama, junegrass and brackenfern. These areas are some of the best timber-producing soils in the state. Most livestock grazing is during summer months. The potential for woodland wildlife habitat management is good. The dominant wildlife species are deer, elk, black bear, mountain lion, Wild Turkey, Band-tailed Pigeon, dove, porcupine, and small birds and rodents. Lake Mary and Mormon Lake are the largest lakes in the area and are stocked with fish at times.

The dominant factors limiting the potential of these soils for homesites and recreation facilities are slow permeability, high shrink-swell, moderate frost-action potential and depth to bedrock. Some areas have excessive slope.

FH4 Soldier-Lithic Cryoborolls Association

Shallow and deep, medium to fine-textured, gently sloping to steep mountain soils.

Soil Classification
 Typic Glossoboralfs
 Lithic Cryoborolls

Percent Slope mostly 2 to 30; range to 50

Elevation 2,000 to 2,730 m (6,600 to 9,000 ft)

Mean Annual Precipitation 410 to 760 mm (16 to 30 in)

Winter Precipitation as the Percentage of Annual Precipitation 55 to 70

Mean January Air Temperature -4 to 7 C (25 to 45 F)

Mean July Air Temperature 16 to 29 C (60 to 85 F)

Mean Annual Soil Temperature 5 to 8 C (41 to 47 F)

Frost-Free Days 85 to 125

Area and Percent of State 210,440 ha (520,000 ac), 0.7

Land Uses timber production, grazable woodland, wildlife habitat, recreation and water supply

This association consists of moderately well-drained and well-drained soils on the higher part of the Kaibab Plateau. The soils formed in residuum weathered from calcareous sandstone and limestone.

Soldier soils make up about 65 percent of the association, Lithic Cryoborolls 15 percent and rock outcrop and minor areas of associated and similar soils 20 percent. These included soils are mostly small areas of Palomino on uplands and Luth and Clover Springs on narrow floodplains along drainageways.

The dominant native vegetation is ponderosa pine, Douglas and white fir, Gambel oak and quaking aspen, with an understory of Arizona fescue, junegrass, mountain muhly, muttongrass and pine dropseed. Soldier soils are considered to be the best pine-fir producing soils in the state. Most livestock grazing is during summer months. The potential for woodland wildlife management is good. The dominant wildlife species are mule deer, grouse, mountain lion, Wild Turkey, Band-tailed Pigeon, squirrels, porcupine and small birds and rodents.

Factors limiting the potential of these soils for development of homesites and recreation areas are the shallow depth to bedrock and steep slopes of the Lithic Cryoborolls. Also, the Soldier soils have high shrink-swell and very slow permeability, which is poor for use as septic tank absorption fields. Smooth, gently sloping areas can be used for campgrounds and picnic areas.

FH5 Mirabal-Baldy-Rock Outcrop Association

Shallow to deep, gravelly and cobbly, moderately coarse-textured, hilly to very steep mountain soils and rock outcrop.

Soil Classification
Typic Ustorthents
Typic Cryorthents

Percent Slope 5 to 70

Elevation 2,060 to 3,280 m (6,800 to 10,800 ft)

Mean Annual Precipitation 410 to 760 mm (18 to 30 in)

Winter Precipitation as the Percentage of Annual Precipitation 50 to 60

Mean January Air Temperature -1 to 4 C (30 to 40 F)

Mean July Air Temperature 18 to 27 C (65 to 80 F)

Mean Annual Soil Temperature 4 to 8 C (40 to 47 F)

Frost-Free Days 80 to 125

Area and Percent of State 101,380 ha (250,500 ac), 0.3

Land Uses grazable woodland, recreation, wildlife habitat and water supply

This association consists of well-drained soils and rock outcrop on high mountain ranges. The soils formed in residuum weathered from granite, gneiss, schist and other igneous rocks.

Mirabal soils make up about 30 percent of the association, Baldy soils 30 percent, rock outcrop 20 percent and minor areas of associated soils 20 percent. The included minor soils are unnamed and uncorrelated and range from shallow to deep, gravelly to very stony and moderately coarse to moderately fine-textured.

The dominant vegetation is ponderosa and limber pine, Douglas fir, quaking aspen, Gambel and silverleaf oak, mountain brome, Arizona fescue, mountain muhly and junegrass. Some areas are used for timber production. Most livestock grazing is during summer months. Both summer and winter recreation activities are enjoyed in most areas, including hiking, skiing, camping and fishing. The potential for woodland wildlife habitat management is good. The dominant wildlife species are whitetail and mule deer, Wild Turkey, black bear, mountain lion, Abert's and Arizona gray squirrels, porcupine, Band-tailed Pigeon, Gambel's and Scale Quail and small birds and rodents.

Factors limiting the potential of these soils for homesite and campground development are shallow depth to rock and steep slopes. Smooth, gently sloping areas can be used for campgrounds, picnic sites and other community uses.

FH6 Eutroboralfs-Mirabal Association

Shallow to deep, cobbly, moderately coarse and gravelly fine-textured, gently sloping to very steep mountain soils.

Soil Classification
  Typic Eutroboralfs
  Typic Ustorthents

Percent Slope 5 to 75

Elevation 2,060 to 2,760 m (6,800 to 9,100 ft)

Mean Annual Precipitation 410 to 630 mm (16 to 25 in)

Winter Precipitation as the Percentage of Annual Precipitation 45 to 50

Mean January Air Temperature -1 to 4 C (30 to 40 F)

Mean July Air Temperature 18 to 24 C (65 to 75 F)

Mean Annual Soil Temperature 5 to 8 C (42 to 47 F)

Frost-Free Days 70 to 110

Area and Percent of State 62,730 ha (155,000 ac), 0.2

Land Uses grazable woodland, recreation, wildlife habitat and water supply

This association consists of well-drained soils on high mountains. The soils formed in residuum and alluvium weathered from volcanic, granitic and other igneous rocks and sandstone.

The Eutroboralfs make up about 60 percent of the association, Mirabal soils 30 percent and rock outcrop and minor areas of associated soils 10 percent.

The dominant vegetation is ponderosa and pinyon pine, juniper, Douglas and white fir and some spruce and quaking aspen on north slopes at high elevations. The understory is mountain muhly, mountain brome, grama grasses, Arizona fescue, pine dropseed and junegrass. Most livestock grazing is during summer months. Timber is harvested in a few areas. Recreation includes camping, hiking, hunting and fishing. The principal wildlife species are deer, elk, black bear, mountain lion, Wild Turkey, quail, Band-tailed Pigeon, squirrels, porcupines and small birds and rodents.

Factors limiting the potential of these soils for homesite and campground development are high cobble content and shallow depth to rock in the Mirabal soils. Also the Eutroboralfs have slow permeability, which is poor for use as septic tank absorption fields. Both Mirabal and Eutroboralf soils have steep slopes.

FH7 Cryorthents-Eutroboralfs Association

Shallow to deep, moderately coarse to fine-textured, gently sloping to steep, high mountain soils.

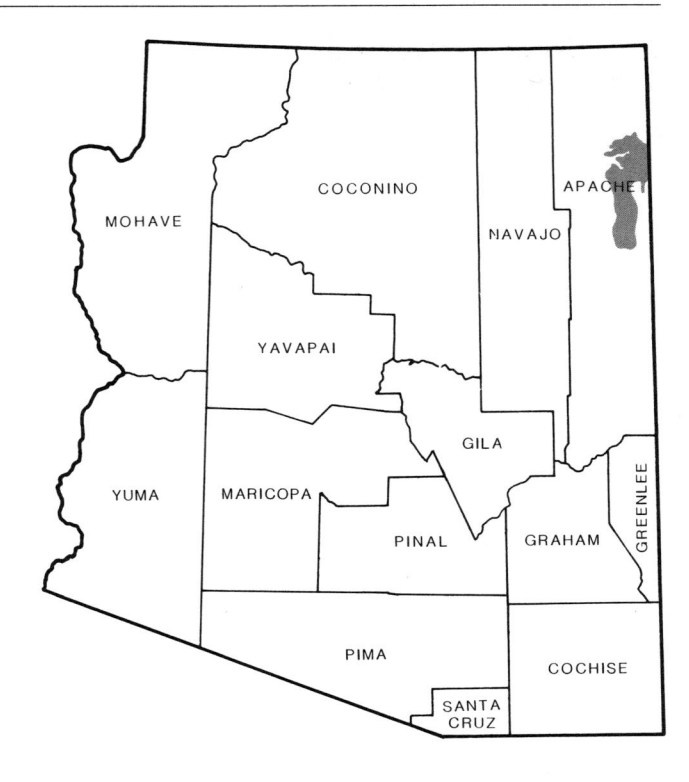

Soil Classification
Cryorthents
Eutroboralfs

Percent Slope 2 to 50

Elevation 2,120 to 2,970 m (7,000 to 9,800 ft)

Mean Annual Precipitation 410 to 560 mm (16 to 22 in)

Winter Precipitation as the Percentage of Annual Precipitation 55 to 60

Mean January Air Temperature -4 to -1 C (25 to 30 F)

Mean July Air Temperature 21 to 24 C (70 to 75 F)

Mean Annual Soil Temperature 5 to 8 C (41 to 47 F)

Frost-Free Days 85 to 110

Area and Percent of State 171,190 ha (423,000 ac), 0.6

Land Uses livestock grazing, wildlife habitat, recreation and some timber production

This association consists of well-drained soils at high elevations on the Chuska and Lukachukai mountains. The soils formed in residuum weathered from sandstone, limestone, shale, conglomerate and volcanic rocks.

Soil series have not been assigned to soils in this mostly unsurveyed area. It is estimated that medium and moderately fine-textured Cryorthents make up about 40 percent of this association, fine and moderately fine-textured Eutroboralfs 40 percent and rock outcrop and shallow or very shallow soils over sandstone or shale bedrock 20 percent.

The dominant vegetation is ponderosa and pinyon pine, juniper and Gambel oak. Understory grasses include mountain muhly, pine dropseed, Arizona fescue, squirreltail and western wheatgrass. This unit is the primary source of commercial timber for the Navajo Indian Reservation in Arizona. Sheep, goats and cattle graze accessible areas during warmer months. The principal wildlife species are deer, mountain lion, Wild Turkey, grouse, quail, squirrel, marmot and small birds and rodents.

Factors limiting the potential of these soils for homesite and campground development are shallow depth to rock and moderately slow to slow permeability, which is poor for use as septic tank absorption fields.

FH8 Gordo-Tatiyee Association

Deep and moderately deep, gravelly, medium to fine-textured, nearly level to rolling soils of the mountain meadows.

Soil Classification
 Argic Pachic Cryoborolls
 Argic Cryoborolls

Percent Slope none to 15

Elevation 2,300 to 3,030 m (7,600 to 10,000 ft)

Mean Annual Precipitation 460 to 760 mm (18 to 30 in)

Winter Precipitation as the Percentage of Annual Precipitation 45 to 50

Mean January Air Temperature -4 to -1 C (25 to 30 F)

Mean July Air Temperature 16 to 21 C (60 to 70 F)

Mean Annual Soil Temperature 3 to 8 C (38 to 47 F)

Frost-Free Days 70 to 110

Area and Percent of State 36,830 ha (91,000 ac), 0.1

Land Uses summer grazing of livestock, water supply, wildlife habitat and recreation

This association consists of well-drained soils in high mountain meadows near Big Lake and Green's Peak. the soils formed in residuum weathered from basalt, cinders and ash.

Gordo soils make up about 45 percent of the association, Tatiyee soils 40 percent and small areas of stony, shallow soils on basalt, and narrow areas of Luth and Clover Springs along drainageways and depressions make up about 15 percent.

Several large trout-stocked lakes are within this unit, and one ski run is nearby. The dominant native vegetation is grass, and includes Arizona fescue, Kentucky bluegrass, mountain muhly, sedges, weeds and native flowers. Some areas have big sagebrush, bitterbrush, bluebunch wheatgrass and mutton bluegrass. The dominant wildlife species using this association are elk, deer, Wild Turkey, badger and small rodents.

Factors limiting the potential of these soils for homesite and campground development are the high shrink-swell and slow permeability of the Tatiyee soils, which is poor for use as septic tank absorption fields. Both Tatiyee and Gordo soils are gravelly and cobbly.

Appendix A

Geologic Time Scale

Geologic time correlates rocks and time. The modern geologic time scale shown in Figure A1 was nearly complete by the end of the 19th century and was based on stratigraphic and fossil studies in northern Europe and the United States. The scale, developed before absolute dating techniques were discovered, is a relative geologic scale that provides a standard of reference for dating rocks throughout the world. It lists the succession of rock depositions that are recognized on and immediately beneath the Earth's surface. The standard stratigraphic column, based on fossil plant and animal assemblages from different European strata, is used to date fossils in strata from other parts of the Earth and is the foundation of the geologic time scale.

The application of radiometric dating techniques began early in the 20th century. The quantitative methods provided by these techniques had the potential for dating divisions of the geologic time scale and for estimating the age of the Earth itself. The age of the Earth now is estimated to be between 5 billion and 4.7 billion years and estimates of the duration of the geologic time scale divisions have been made.

Originally, geologic time scale divisions were based on the natural breaks in the stratigraphic column. The breaks were thought to have resulted from worldwide events of mountain building during which no sedimentation occurred and left gaps in the record of rocks, that is, an unconformity. It is now realized that mountain building events were not necessarily worldwide, but may be limited to a single continent or even part of a continent during one interval of geologic time.

The longest divisions of geologic time are the eras. Most geologic time scales recognize four eras, three of which have been named for the fossils in the associated strata. Thus, the Paleozoic Era refers to "ancient life," the Mesozoic Era to "medieval life" and the Cenozoic Era to "modern life." Rocks older than Paleozoic generally lack diagnostic fossils and are widely known as belonging to the Precambrian Era. This era included about 80 percent of Earth's history, that is, from nearly 5 billion years to 800 million or 700 million years ago.

The eras are divided into periods of time. Rock deposits that relate to or were formed during a certain period of time constitute a system of rocks. Periods are divided further into epochs of time and rock systems into series.

In northern Arizona nearly the entire stratigraphic column is exposed. Near the bottom of the 1.6-km- (1 mi) deep Grand Canyon, a complex of deformed and metamorphosed rocks lies unconformably below the lowest Cambrian sedimentary rocks (Figure A2). The oldest strata in the canyon is the Vishnu Schist that contains intrusive granite and pegmatite dikes. These rocks have been called Archeozoic (early Precambrian). Radiometric dating of the granites indicates an intrusive age of between 1.8 billion and 1.6 billion years placing them in the early Proterozoic (Palmer, 1983). Lying uncomfortably on the Vishnu Schist are strata of the Grand Canyon series, quartzite, limestone, conglomerate and shale. These sequences have been called Proterozoic and probably are late Precambrian (middle or late Proterozoic).

Approximately 910 m (3,000 ft) of Paleozoic sedimentary rock also are exposed in the Grand Canyon. These rocks extend from the Cambrian to the Permian, but rocks of the Ordovician and Silurian periods are missing. They either were never deposited or were deposited and worn away.

ARIZONA SOILS

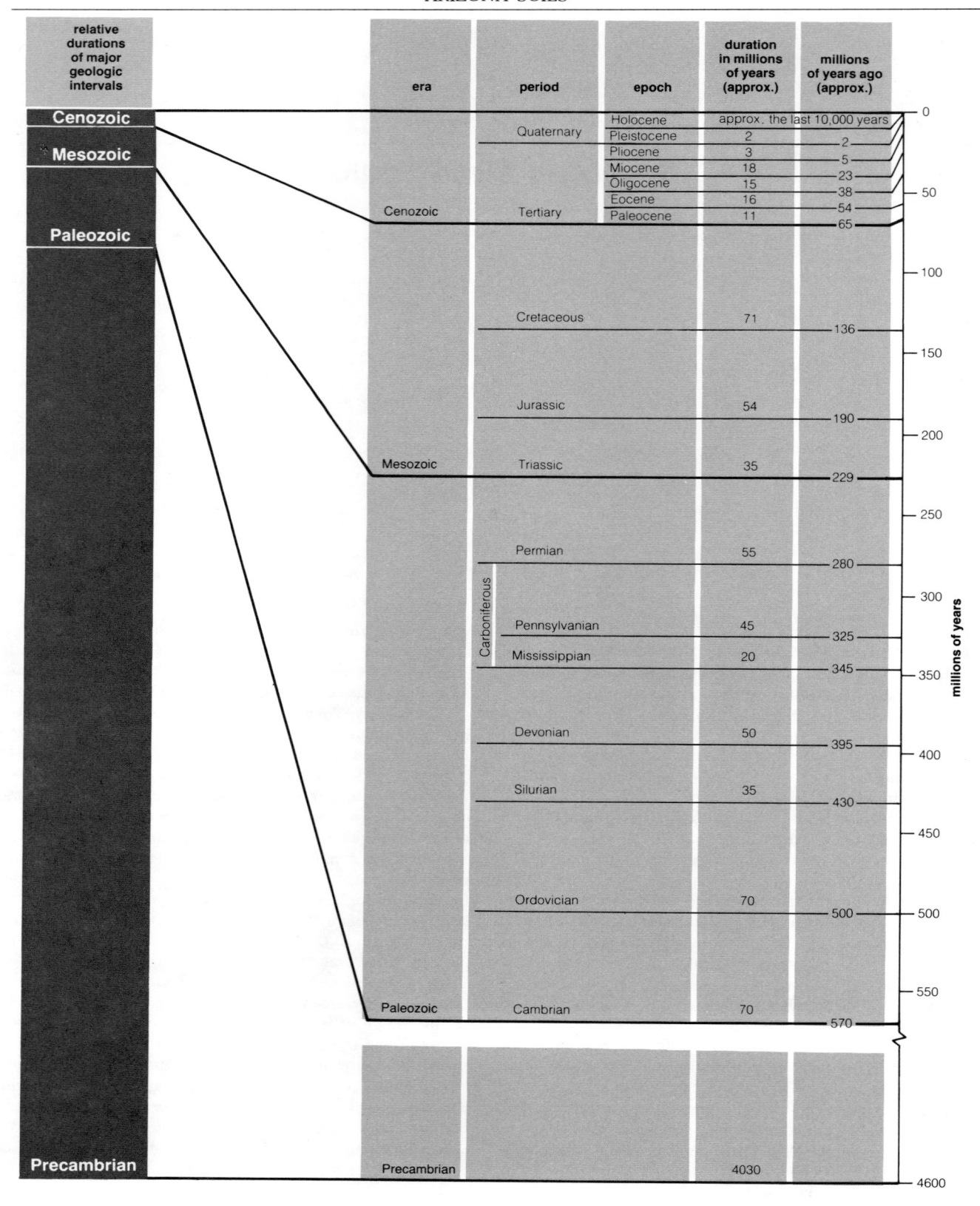

relative durations of major geologic intervals	era	period	epoch	duration in millions of years (approx.)	millions of years ago (approx.)
Cenozoic	Cenozoic	Quaternary	Holocene	approx. the last 10,000 years	0
			Pleistocene	2	2
Mesozoic		Tertiary	Pliocene	3	5
			Miocene	18	23
			Oligocene	15	38
Paleozoic			Eocene	16	54
			Paleocene	11	65
	Mesozoic	Cretaceous		71	136
		Jurassic		54	190
		Triassic		35	229
	Paleozoic	Permian		55	280
		Pennsylvanian (Carboniferous)		45	325
		Mississippian		20	345
		Devonian		50	395
		Silurian		35	430
		Ordovician		70	500
		Cambrian		70	570
Precambrian	Precambrian			4030	4600

FIGURE A1. Modern Geologic Time Scale (after D. L. Eicher, 1976)

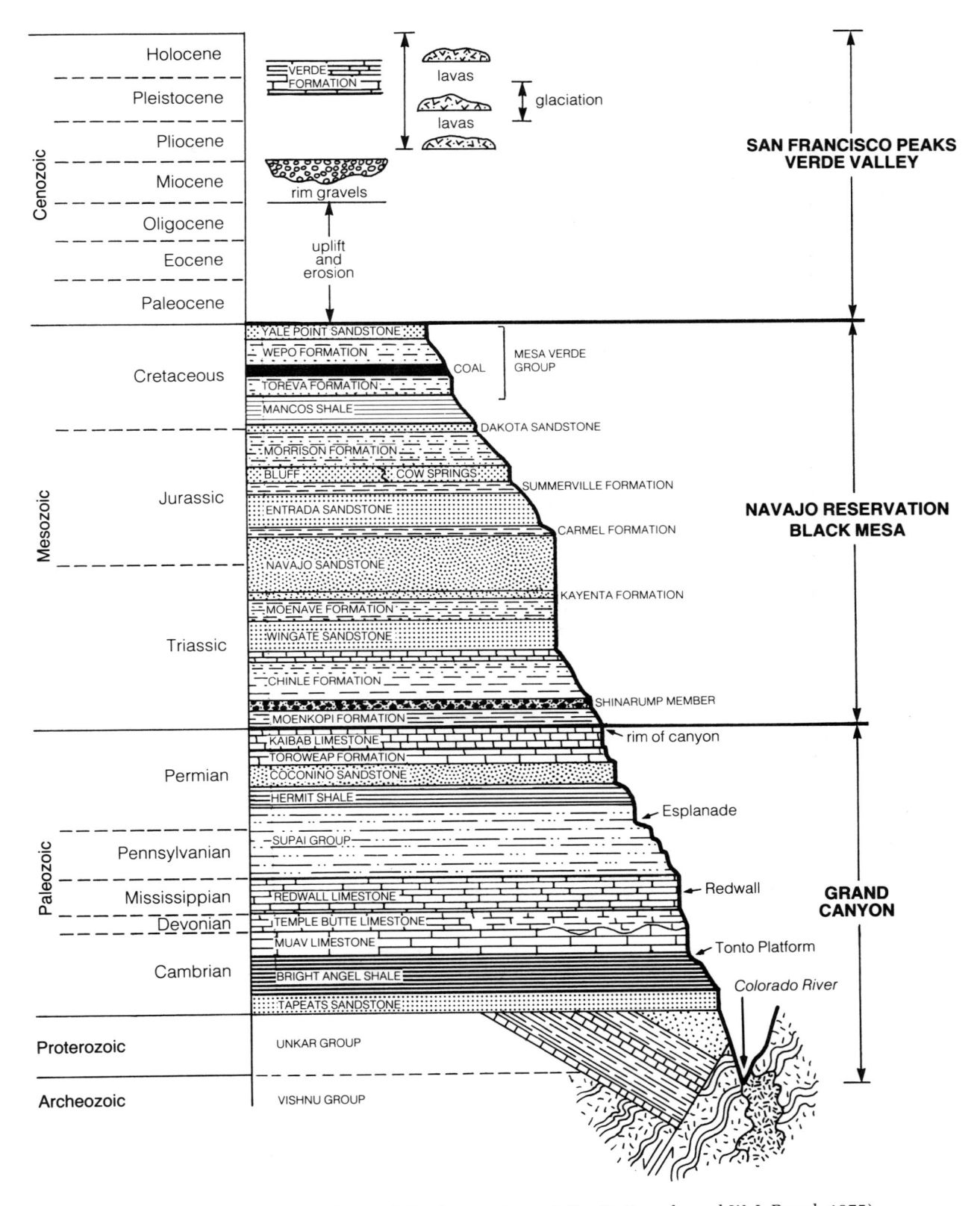

FIGURE A2. Stratigraphic Column of Northern Arizona (after D. Beasely and W. J. Breed, 1975)

The Quaternary Period includes the Pleistocene and Holocene epochs. These epochs are of major importance in evaluating the genesis and distribution of modern soils since the environmental conditions that existed then may have directly influenced the soils. Conversely, the climatic conditions of earlier geologic time periods only indirectly influenced modern soils by affecting the nature of the soil parent material and the evolution of the Earth's physiography.

Glaciation of large portions of North America, northern Europe and other parts of the world is associated with the Pleistocene. The Pleistocene did not experience just one great ice age, but had alternations of glacials, or stadials, and stages of relatively greater warmth, interglacials, or interstadials. Table A outlines the classical glacial divisions of the Pleistocene for North America, four glacial and three interglacial ages. Similar divisions also were developed for European glaciation. But use of modern and refined observational and dating techniques has shown that glacial events were much more complex than the classical divisions indicate. Of particular significance is the investigation of the stratigraphic records of sea floors that has led to new views on the length and frequency of glacial and interglacial episodes. For example, van Dork (1976) suggested 21 glacial cycles during the last 2.3 million years, while Kent, Opdyke and Ewing (1971) favored 16 glacial phases during the last 2.5 million years. Although these researchers do not agree on the exact number of cycles, it is clear that their data provide a very different picture than the classical fourfold sequence of glaciation.

The complexity of the nature of glaciated cycles and the important gaps in the knowledge of geologic time also have given rise to a considerable range of views of the duration and definition of the Pleistocene Epoch (Cooke, 1973). Some researchers have tended to put the boundary between the Pliocene and the Pleistocene at about 2 million years ago based on the appearance of some faunal genera and the disappearance of others (Glass et al, 1967; Zagwijn, 1974). Others place the boundary at between 3 million and 2.5 million years ago based on the marked appearance of mid-latitude glaciers as contrasted with those of the polar regions. Thus, the duration of the Pleistocene now is considered to be between 2 million and 3 million years. The classical view was that it lasted about 1 million years.

Researchers also have disagreed about the boundary between the Pleistocene and Holocene. Different authors have placed it between more than 20,000 and 4,000 years ago. A compromise time of 10,000 years ago for the beginning of the Holocene has been recommended. Using this time, the boundary stratotype (the earliest deposition of rocks different from those associated with the Pleistocene) in the vicinity of Goteborg, Sweden, was further recommended. (Fairbridge, 1983).

TABLE A.

The Classical Subdivision
of the Quaternary for North America
(Modified from Schuchert and Dunbar, 1964).

Epoch	Stages/Ages of Deposition	Estimated Years Before Present
Holocene		10,000
Pleistocene	Wisconsin glacial	125,000
	Sangamon interglacial	250,000
	Illinoian glacial	350,000
	Yarmouth interglacial	650,000
	Kansan glacial	750,000
	Aftonian interglacial	950,000
	Nebraskan glacial	1,000,000

Appendix B

Vegetation: Common and Scientific Names

Common Name	Scientific Name	Common Name	Scientific Name
agave	*Agave* spp.	cholla	*Opuntia* spp.
algerita	*Berberis trifoliata* Moric.	cliffrose	*Cowania mexicana* D. Don
allthorn	*Koeberlinia spinosa* Zucc.	clover	*Trifolium* spp.
almond, desert	*Prunus fasciculata* (Torr.) Gray	coldenia	*Tiquilia* spp.
arrowweed	*Tessaria sericea* (Nutt.) Shiners	cottontop, Arizona	*Trichachne californica* (Benth.) Chase
ash	*Fraxinus pennsylvanica* Marsh. ssp. *velutina* (Torr.) G.N. Miller	cottonwood	*Populus fremontii* Wats.
		creosotebush	*Larrea tridentata* (DC.) Coville
aspen, quaking	*Populus tremuloides* Michx.	curly mesquite	*Hilaria belangeri*
beargrass	*Nolina microcarpa* Wats.	daisy	*Erigeron* spp.
bitterbrush	*Purshia tridentata* (Pursh) DC.	dandelion, mountain	*Agoseris* spp.
blackbrush	*Coleogyne ramosissima* Torr.	desertthorn	*Lycium* spp.
bluegrass	*Poa* spp.	desertwillow	*Chilopsis linearis* (Cav.) Sweet
Kentucky	*P. pratensis* L.	dropseed, pine	*Blepharoneuron tricholepis* (Torr.) Nash
bluestem, cane	*Bothriochloa barbinodis* (Lag.) Herter	dropseed, sand	*Sporobolus cryptandrus* (Torr.) Gray
bluestem, Texas	*Schizachyrium cirratum* (Hack.) Woot. & Standl.	fescue, Arizona	*Festuca arizonica* Vasey
boxelder	*Acer negundo* L.	filaree	*Erodium cicutarium* (L.) L'Her.
brackenfern	*Pteridium aquilinum* (L.) Kuhn	fir, Douglas	*Pseudotsuga menziesii* (Mirb.) Franco
bristlegrass, plains	*Setaria macrostachya* H.B.K.	fir, white	*Abies concolor* (Gordon & Glendinning) Hoopes
brome, mountain	*Bromus marginatus* Nees		
bullgrass	*Muhlenbergia emersleyi* Vasey	fluffgrass	*Erioneuron pulchellum* (H.B.K.) Takeoka
burroweed	*Haplopappus tenuisectus* (Greene) Blake		
bursage	*Ambrosia* spp.	galleta	*Hilaria jamesii* (Torr.) Benth.
white	*A. dumosa* (Gray) Payne	big	*H. rigida* (Thurb.) Benth.
woolly-fruited	*A. eriocentra* (Gray) Payne	grama	*Bouteloua* spp.
		black	*B. eriopoda* (Torr.) Torr.
calliandra	*Calliandra eriophylla* Benth.	blue	*B. gracilis* (H.B.K.) Lag.
catclaw	*Acacia greggii* Gray	Rothrock	*B. rothrockii* Vasey
ceanothus	*Ceanothus* spp.	sideoats	*B. curtipendula* (Michx.) Torr.
chokecherry	*Prunus virginiana* L.	slender	*B. repens* (H.B.K.) Scribn. & Merr.

Common Name	Scientific Name	Common Name	Scientific Name
hairgrass	*Deschampsia* spp.	rabbitbrush	*Chrysothamnus* spp.
horsebrush	*Tetradymia* spp.	ratany, range	*Krameria parvifolia* Benth.
		ricegrass, Indian	*Oryzopsis hymenoides* (Roemer & Schultes) Ricker
ironwood	*Olneya tesota* Gray		
		Russian thistle (tumbleweed)	*Salsola kali* L.
jojoba	*Simmondsia chinensis* (Link) Schneid.		
Joshua tree	*Yucca brevifolia* Engelm.		
junegrass	*Koeleria cristata* (L.) Pers.	sacaton, alkali	*Sporobolus airoides* Torr.
juniper	*Juniperus* spp.	sage, Mojave	*Salvia mohavensis* Greene
alligator	*J. deppeana* Steud.	sagebrush	*Artemisia* spp.
one-seed	*J. monosperma* (Engelm.) Sarg.	big	*A. tridentata* Nutt.
Utah	*J. osteosperma* (Torr.) Little	sand	*A. filifolia* Torr.
		saguaro	*Cereus giganteus* Engelm.
locust	*Robinia neomexicana* Gray	saltbush	*Atriplex* spp.
lovegrass, plains	*Eragrostis intermedia* Hitchc.	fourwing	*Atriplex canescens* (Pursh) Nutt.
		saltcedar	*Tamarix pentandra* Pall.
madrone	*Arbutus arizonica* (Gray) Sarg.	saltgrass	*Distichlis spicata* (L.) Greene
manzanita	*Arctostaphylos pungens* H.B.K.	sandpaper plant	*Petalonyx* spp.
maple	*Acer* spp.	sedge	*Carex* spp.
big-tooth	*A. grandidentatum* Nutt.	senna	*Cassia* spp.
Rocky Mountain	*A. glabrum* Torr.	shadscale	*Atriplex confertifolia* (Torr. & Frem.) Wats.
mariola	*Parthenium incanum* H.B.K.		
mesquite	*Prosopis* spp.	snakeweed	*Gutierrezia* spp.
Mormon-tea	*Ephedra* spp.	sotol	*Dasylirion wheeleri* Wats.
mountainmahogany	*Cercocarpus* spp.	sprangletop, green	*Leptochloa dubia* (H.B.K.) Nees
muhly	*Muhlenbergia* spp.	spruce	*Picea* spp.
bush	*M. porteri* Scribn.	blue	*P. pungens* Engelm.
ring	*M. torreyi* (Kunth) Hitchc.	Engelmann	*P. engelmannii* Parry
spike	*M. wrightii* Vasey	squawbush, pubescent	*Rhus trilobata* Nutt.
muttongrass	*Poa fendleriana* (Steud.) Vasey	squirreltail	*Sitanion hystrix* (Nutt.) J.G. Smith
		sycamore	*Platanus wrightii* Wats.
needlegrass	*Stipa* spp.		
needleandthread	*Stipa comata* Trin. & Rupr.	tamarisk	*Tamarix aphylla* (L.) Karst.
		tanglehead	*Heteropogon contortus* (L.) Beauv.
oak	*Quercus* spp.	tarbush	*Flourensia cernua* DC.
Arizona white	*Q. arizonica* Sarg.	threeawn	*Aristida* spp.
Emory	*Q. emoryi* Torr.	timothy, mountain	*Phleum pratense* L.
Gambel	*Q. gambelii* Nutt.	tobosa	*Hilaria mutica* (Buckl.) Benth.
Mexican blue	*Q. oblongifolia* Torr.	tridens	*Tridens* spp.
silverleaf	*Q. hypoleucoides* Camus	turkshead	*Chlorizanthe rigida* (Torr.) Torr. & Gray
turbinella	*Q. turbinella* Greene	turpentinebush	*Haplopappus laricifolius* Gray
ocotillo	*Fouquieria splendens* Engelm.		
organpipe cactus	*Lemaireocereus thurberi* (Engelm.) Britt. & Rose	vine-mesquite	*Panicum obtusum* H.B.K.
		walnut	*Juglans major* (Torr.) Heller
paloverde	*Cercidium* spp.	wheatgrass	*Agropyron* spp.
blue	*C. floridum* Benth.	bluebunch	*A. spicatum* (Pursh) Scribn. & Smith
foothill	*C. microphyllum* (Torr.) Rose & Johnst.	western	*A. smithii* Rydb.
pine	*Pinus* spp.	whitebrush	*Aloysia wrightii* (Gray) Heller
Apache	*P. engelmannii* Carr	whitethorn	*Acacia constricta* Benth.
bristlecone	*P. aristata* Engelm.	winterfat	*Ceratoides lanata* (Pursh) J.T. Howell
Chihuahua	*P. leiophylla* Schiede & Deppe	wolfberry	*Lycium* spp.
limber	*P. flexilis* James	wolftail	*Lycurus phleoides* H.B.K.
ponderosa	*P. ponderosa* Lawson		
pinyon	*Pinus edulis* Engelm.	yucca, Mohave	*Yucca schidigera* Ruezl.
Mexican	*P. cembroides* Zucc.		
singleleaf	*P. monophylla* Torr. & Frem.	zinnia, desert	*Zinnia acerosa* (DC.) Gray
pricklypear	*Opuntia* spp.		

Appendix C

Animals: Common and Scientific Names

Mammals

Common Name	Scientific Name
antelope, prong-horned	*Antilocapra americana* (Ord)
badger	*Taxidea taxus* (Shreber)
bats*	
free-tailed	*Tadarida* spp (3)
mastif	*Eumops* spp (2)
plain-nosed	*Myotis* spp (9)
bear, black	*Euarctos americanus* (Pallas)
beaver	*Castor canadensis* Kuhle
bobcat	*Lynx rufus* (Shreber)
chipmunk	*Eutamias* spp (4)
coati (or coatimundi)	*Nasua narica* (Linnaeus)
coyote	*Canis latrans* Say
deer	
mule (or black-tailed)	*Odocoileus hemionus* (Rafinesque)
white-tailed (or eastern)	*O. virginianus* (Zimmermann)
elk	*Cervus canadensis* (Erxleben)
fox	
gray	*Urocyon cinereoargenteus* (Shreber)
Kit	*Vulpes macrotis* Merriam
gopher, pocket	*Thomomys* spp (3)
javelina	*Pecari tajacu* (Linnaeus)
mountain lion	*Felis concolor* Linnaeus
mouse (most common)	
cactus	*Peromyscus eremicus* (Baird)
pinyon	*P. truei* (Shufeldt)**
pocket	*Perognathus* spp (10)
muskrat	*Ondatra zibethica* (Linnaeus)

Common Name	Scientific Name
otter, river	*Lutra canadensis* (Shreber)
prairie dog, Gunnison's	*Cynomys gunnisoni*
porcupine	*Erethizon dorsatum* Linnaeus
rabbit	
cottontail	*Sylvilagus* spp (3)
jack	*Lepus* spp (2)
raccoon	*Procyon lotor* (Linnaeus)
rat	
kangaroo	*Dipodomys* spp (5)
pack	*Neotoma albigula* Hartley
wood	*Neotoma* spp (4)
ringtail	*Bassariscus astutus* (Lichenstein)
sheep, bighorn	*Ovis canadensis* Shaw
shrew	
desert	*Notiosorex crawfordi* (Coues)
vagrant (and others)	*Sorex* spp (4)
skunk	
hog-nosed	*Conepatus mesoleucus* Lichenstein
hooded	*Mephitis macroura* (Lichenstein)
striped	*M. mephitis* (Shreber)
spotted	*Spilogale putorius* (Linnaeus)
squirrel	
Abert's	*Sciurus aberti* Woodhouse
Kaibab	Sub *S. aberti*
Apache	*S. apache* J.A. Allen
Arizona gray	*S. arizonensis* (Coues)
red	*Tamiasciurus hudsonicus*
ground	*Citellus* spp (7)
vole	
long-tailed	*Microtus longicaudus* Merriam
Mexican	*M. Mexicanus* (Saussure)
montane	*M. montanus* (Peale)

*14 other species of bats reported in Arizona.
**7 other *Peromyscus* species are known in Arizona.

Birds

Common Name	Scientific Name
Dove	
Mourning	*Zenaida macroura* (Linnaeus)
White-winged	*Z. asiatica* (Linnaeus)
Eagle	
Bald	*Haliaeetus leucocephalus* (Linnaeus)
Golden	*Aquila chrysaetos* (Linnaeus)
Gnatcatcher, Blue-gray	*Polioptila caerulea* (Linnaeus)
Grosbeak, Blue	*Guiraea caerulea* (Linnaeus)
Grouse, Blue (or Dusky)	*Dendragapus obscurus* (Say)
Hawk	
Cooper's	*Accipiter cooperii* (Bonaparte)
Red-tailed (or Common)	*Buteo jamaicensis* (Gmelin)
Jay, Rocky Mountain (also Gray or Canada)	*Perisoreus canadensis* (Linnaeus)
Kestrel, American	*Falco sparverius* Linnaeus
Owl	
Elf	*Micrathene whitneyi* (Cooper)
Great-Horned	*Bubo virginianus* (Gmelin)
Pelican	
Brown	*Pelecanus occidentalis* Linnaeus
White	*P. erythrorhynchos* Gmelin
Pigeon, Band-tailed	*Columba fasciata* Say
Pipit, Water (or American)	*Anthus spinoletta* (Linnaeus)
Pheasant	*Phasianus colchicas*
Quail	
Bobwhite	*Colinus virginianus*
Gambel's (or Desert)	*Lophortyx gambelii* Gambel
Scaled	*Callipepla squamata* (Vigors)
Roadrunner	*Geococcyx californianus* (Lesson)
Sapsucker, Yellow-bellied	*Sphyrapicus varius* (Linnaeus)
Sparrow	
Chipping	*Spizella passerina* (Bechstein)
Vesper	*Pooecetes gramineus* (Gmelin)
Titmouse, Plain	*Parus inornatus* Gambel
Turkey, Wild	*Meleagris gallopavo* Linnaeus
Vulture, Turkey	*Carthartes aura* (Linnaeus)
Warbler, Lucy's	*Vermivora luciae* (Cooper)
Woodpecker	
Acorn	*Melanerpes formicivorus* (Swainson)
Gila	*Centurus uropygialis* Baird
Wren, Cactus	*Campylorhynchus brunneicapillus* (Lafresnaye)

Amphibians and Reptiles

Common Name	Scientific Name
frog, canyon treefrog	*Hyla arenicolor* Cope
lizard	
banded gecko	*Coleonyx variegatus* Baird
desert iguana	*Dipsosaurus dorsalis* Baird and Girard
Gila monster	*Heloderma suspectum* Cope
side-blotched lizard	*Uta stansburiana* Baird and Girard
whiptail lizard	*Cnemidophorus* spp (6)
salamander, tiger	*Ambystoma tigrinum* Green
snake	
blind snake, western	*Leptotyphlops humilis* Baird and Girard
bullsnake (or gopher snake)	*Pituophis melanoleucus* Daudin
burrowing snake, banded	*Chilomeniscus cinctus* Cope
coral snake, Arizona	*Micruroides euryxanthus* Kennicott
garter snake, checkered	*Thamnophis marcianus* Baird and Girard
glossy snake	*Arizona elegans* Kennicott
leaf-nosed snake, saddled	*Phyllorhynchus browni* Stejneger
leaf-nosed snake, spotted	*P. decurtatus* Cope
long-nosed snake	*Rhinocheilus lecontei* Baird and Girard
shovel-nosed snake, Sonora	*Chionactis palarostris* Kaluber
shovel-nosed snake, western	*C. occipitalis* Hallowell
whipsnake (or coachwhip)	*Masticophis flagellum* Shaw
whipsnake, striped	*M. taeniatus* Hallowell
toad	
Colorado River toad	*Bufo alvarius* Girard
Great Plains toad	*B. cognatus* Say
spadefoot toad, Couch's	*Scaphiopus couchi* Baird
spadefoot toad, western	*S. (Spea) hammondi* Baird
tortoise, desert	*Gopherus agassizi* Cooper

Appendix D

Outline of Soil Taxonomy

Order	Suborder	Great group	Order	Suborder	Great group	Order	Suborder	Great group
Alfisols	Aqualfs	Albaqualfs			Plinthustalfs		Fluvents	Cryofluvents
		Duraqualfs			Rhodustalfs			Torrifluvents
		Fragiaqualfs						Tropofluvents
		Glossaqualfs		Xeralfs	Durixeralfs			Udifluvents
		Natraqualfs			Haploxeralfs			Ustifluvents
		Ochraqualfs			Natrixeralfs			Xerofluvents
		Plinthaqualfs			Palexeralfs			
		Tropaqualfs			Plinthoxeralfs		Orthents	Cryorthents
		Umbraqualfs			Rhodoxeralfs			Torriorthents
								Troporthents
	Boralfs	Cryoboralfs	*Aridisols*	Argids	Duragids			Udorthents
		Eutroboralfs			Haplargids			Ustorthents
		Fragiboralfs			Nadurargids			Xerorthents
		Glossoboralfs			Natrargids			
		Natriboralfs			Paleargids		Psamments	Cryopsamments
		Paleboralfs						Quartzipsamments
				Orthids	Calciorthids			Torripsamments
	Udalfs	Agrudalfs			Camborthids			Tropopsamments
		Ferrudalfs			Durorthids			Udipsamments
		Fraglossudalfs			Gypsiorthids			Ustipsamments
		Fragiudalfs			Paleorthids			Xeropsamments
		Glossudalfs			Salorthids			
		Hapludalfs				*Histosols*	Fibrists	Borofibrists
		Natrudalfs	*Entisols*	Aquents	Cryaquents			Cryofibrists
		Paleudalfs			Haplaquents			Luvifibrists
		Rhodudalfs			Hydraquents			Medifibrists
		Tropudalfs			Psammaquents			Sphagnofibrists
					Tropaquents			Tropofibrists
	Ustalfs	Durustalfs						
		Haplustalfs		Arents	Arents		Folists	Borofolists
		Natrustalfs						Cryofolists
		Paleustalfs						Tropofolists

159

ARIZONA SOILS

Order	Suborder	Great group	Order	Suborder	Great group	Order	Suborder	Great group
	Hemists	Borohemists Cryohemists Luvihemists Medihemists Tropohemists		Borolls	Argiborolls Calciborolls Cryoborolls Haploborolls Natriborolls Paleborolls Vermiborolls		Ferrods Humods	Ferrods Cryohumods Fragihumods Haplohumods Placohumods Tropohumods
	Saprists	Borosaprists Cryosaprists Medisaprists Troposaprists		Rendolls Udolls	Rendolls Argiudolls Hapludolls Paleudolls Vermudolls		Orthods	Cryorthods Fragiorthods Haplorthods Placorthods Troporthods
Inceptisols	Andepts	Cryandepts Durandepts Dystrandepts Eutrandepts Hydrandepts Placandepts Vitrandepts		Ustolls	Argiustolls Calciustolls Durustolls Haplustolls Natrustolls Paleustolls Vermustolls	*Ultisols*	Aquults	Albaquults Fragiaquults Ochraquults Paleaquults Plinthaquults Tropaquults Umbraquults
	Aquepts	Andaquepts Cryaquepts Fragiaquepts Halaquepts Haplaquepts Humaquepts Plinthaquepts Tropaquepts	*Mollisols*	Xerolls	Argixerolls Calcixerolls Durixerolls Haploxerolls Natrixerolls Palexerolls		Humults Udults	Haplohumults Palehumults Sombrihumults Tropohumults Fragiudults Hapludults
	Ochrepts	Cryochrepts Durochrepts Drystrochrepts Eutrochrepts Fragiochrepts Ustochrepts Xerochrepts	*Oxisols*	Aquox Humox	Gibbsiaquox Ochraquox Plinthaquox Acrohumox Gibbsihumox Haplohumox Sombrihumox		Ustults	Paleudults Plinthudults Rhodudults Tropudults Haplustults Paleustults Plinthustults Rhodustults Tropustults
	Plaggepts Tropepts	Plaggepts Dystropepts Eutropepts Humitropepts Sombritropepts Ustropepts		Orthox	Acrorthox Eutrorthox Gibbsiorthox Haplorthox Sombriorthox Umbriorthox	*Vertisols*	Xerults Torrerts	Haploxerults Palexerults Torrerts
	Umbrepts	Anthrumbrepts Cryumbrepts Fragiumbrepts Haplumbrepts Xerumbrepts		Torrox Ustox	Torrox Acrustox Eutrustox Gibbsiustox Haplustox Sombriustox		Uderts Usterts Xererts	Chromuderts Pelluderts Chromusterts Pellusterts Chromoxererts Pelloxererts
Mollisols	Albolls Aquolls	Argialbolls Natralbolls Argiaquolls Calciaquolls Cryaquolls Duraquolls Haplaquolls Natraquolls	*Spodosols*	Aquods	Cryaquods Duraquods Fragiaquods Haplaquods Placaquods Sideraquods Tropaquods			

Appendix E

Description of Soil Series

This appendix contains descriptions of the 221 soils mentioned in *Arizona Soils*. The mapping units of the Arizona General Soil Map (Plate 1) that contain a particular soil are listed after the name of that soil. Included in the first phrase describing each soil series is another soil name in parentheses. These are names of Great Soil Groups, a category of the 1938 Soil Classification System (Baldwin et al, 1938) that was used before adoption of *Soil Taxonomy* (Soil Survey Staff, 1975). Not all soils are listed by a soil series, the sixth and lowest category in *Soil Taxonomy* (Soil Survey Staff, 1975). Those soil mapping units are named by the subgroup to which they belong. The subgroup is the fourth-level category in *Soil Taxonomy* (Soil Survey Staff, 1975).

The Cristobol, Deloro and Ligurta series were established formally in 1977, two years after publication of the Arizona General Soil Map (Plate 1), but were included because they had been described and named earlier. The process of formalizing a soil series description and name through the USDA Soil Conservation Service takes from two to three years.

Complete and partial laboratory data were available for 118 of the soil series and are included. Most of the headings of the tables should be understandable to persons who have read the text. A few that may not be are explained here.

O.M.% heads the column that shows the percentage of organic matter in the soil being described.

CEC is the cation exchange capacity of the soil being described. The capacity is measured in meq/100 g, or milliequivalents per 100 grams. Lower values indicate that the soil tends to have lesser amounts of clay and/or organic matter.

In some instances $CaCO_3$%, or percent calcium carbonate, data were not available, but B.S.%, or percent base saturation, data were. Lower percentages of base saturation indicate lower pH values and increased leaching. In one instance, the description of the Purgatory soil, the $CaSO_4 \cdot 2H_2O$%, or percent gypsum, was used instead of either $CaCO_3$% or B.S.%.

Finally, BD is bulk density or volume density and is measured as g/cm^3, or grams per cubic centimeter. Bulk density values indicate levels of aeration or porosity of soil as well as degree of compaction. Generally, the lower the value, the higher the level of aeration or of porosity and the lower the level of compaction. Lower values also may indicate higher permeability.

Blanks in the column indicate that no laboratory data were available. Zeros indicate that laboratory test results showed no values. Very low values are indicated with a "T", or trace.

Soils Descriptions

Abra (MS8, MS9, MS10). Ustollic Calciorthids; fine-loamy, mixed, mesic (Calcisols). Abra soils are deep and well drained. Typically, the surface layer is brownish gray gravelly sandy loam about 8 cm (3 in) thick. The underlying material, to more than 150 cm (60 in) is light brownish gray, light yellowish brown and very pale brown calcareous loam. A zone containing many soft lime masses with more than 15 percent calcium carbonate occurs between depths of 46 and 125 cm (18 and 50 in). Depth to this lime zone ranges from 20 to 71 cm (8 to 28 in). Abra soils are on fan terraces with slopes ranging from 1 to 30 percent. These soils have high available water capacity and moderate permeability. They are moderately alkaline and calcareous throughout. Runoff is slow to medium and the hazard of erosion is slight to moderate. The series was established in Yavapai County in 1957 and the name was taken from a siding on the Santa Fe Railroad.

Agua (TA2). Typic Torrifluvents; coarse-loamy over sandy or sandy-skeletal, mixed (calcareous), thermic (Alluvial). Agua soils are deep and well drained. Typically, they have light brownish gray surface and subsurface layers, about 76 cm (30 in) thick, over stratified sand and gravelly sand to more than 150 cm (60 in). They occur on floodplains with slopes of 0 percent to 2 percent. They have moderate available water capacity. The permeability is moderate in the upper part and rapid in the lower part. They are moderately alkaline and calcareous throughout. Runoff is slow and the hazard of erosion is slight to moderate. They may be subject to piping along entrenched streams. The series was established in Maricopa County in 1969.

Agualt (HA1). Typic Torrifluvents; coarse-loamy over sandy or sandy-skeletal, mixed (calcareous); hyperthermic (Alluvial). Agualt soils are deep and well drained. Typically, they have a brown loam surface layer about 28 cm (11 in) thick. The underlying material is brown loam about 40 cm (16 in) thick. Below this to a depth of 150 cm (60 in) is pale brown sand. Agualt soils occur on alluvial fans and floodplains with slopes of 0 to 3 percent. These soils have high available water capacity in the upper 68 cm (27 in) and very low available water capacity below that. They have moderate permeability to 68 cm (27 in) and rapid permeability below that. They are moderately alkaline and calcareous throughout. Runoff is slow and the hazard of erosion is moderate. The series was established in Maricopa County in 1966 and is the hyperthermic equivalent of the Agua series.

Ajo (HA4). Petrocalcic Paleargids; loamy-skeletal, mixed, hyperthermic (Red Desert). Ajo soils are moderately deep and well drained. Typically, they have a brown very gravelly loam surface layer about 5 cm (2 in) thick. The subsoil is reddish brown and yellowish red clay loam about 55 cm (22 in) thick. This is underlain by a white and pink indurated lime cemented hardpan many centimeters thick. Depth to the hardpan ranges from 50 to 100 cm (20 to 40 in). Ajo soils occur on old fan terraces with slopes ranging from 0 to 10 percent. These soils have low available water capacity and moderately slow permeability. They are moderately alkaline and are noncalcareous in the upper part of the solum and are calcareous throughout the remainder of the profile. Runoff is slow and the hazard of erosion is slight. The series was established on the Organ Pipe Cactus National Monument in 1971 and the name was taken from the town of Ajo in Pima County.

Some properties of the Ajo soils

Horizon	Depth (cm)	O.M.%	Sand%	Silt%	Clay%	CEC meg/100g	$CaCO_3$ %	pH	BD g/cm^3
A	0-5	0.46	54	28	18	17	T	7.5	
Bt	5-33	0.56	41	24	35	30	1	7.7	1.26
Bk	33-61		43	17	40	35	4	7.9	1.30
Bkm	61-		77	18	5	13	21	8.7	1.75

Amos (MH5). Typic Hapludalfs; fine, mixed, mesic (Brown). Amos soils are deep and well drained. Typically, they have a brown clay loam surface layer about 8 cm (3 in) thick. The subsoil is yellowish brown and brown clay about 100 cm (40 in) thick. This is underlain by light yellowish brown and yellowish red weathered stratified shale and siltstone about 18 cm (7 in) thick. Hard shale and siltstone is at a depth of 125 cm (50 in). Amos soils occur on hillslopes with slopes of 5 to 20 percent. These soils have high available water capacity and slow permeability. They are neutral to moderately alkaline and are calcareous in the subsoil and underlying material. Runoff is medium and the hazard of erosion is slight. The series was established on the Fort Apache Indian Reservation in Navajo County in 1966 and the name was taken from a local wash.

Anklam (TS6). Typic Haplargids; clayey-skeletal, mixed, thermic, shallow (Lithosol). Anklam soils are very shallow and shallow and well drained. Typically, they have a brown very gravelly sandy clay loam surface layer about 5 cm (2 in) thick. The upper 13 cm (5 in) of subsoil is reddish brown gravelly light clay. The lower 15 cm (6 in) of subsoil is red very gravelly heavy clay loam. The underlying material to a depth of 150 cm (60 in) is varicolored latite rock that is weathered and fractured with clay in the fractures. Anklam soils occur on hillslopes with slopes of 5 to 50 percent. These soils have low available water capacity and moderately slow permeabilty. They are moderately alkaline and noncalcareous throughout the profile. Runoff is medium to rapid and the hazard of erosion is slight. The series was established in Pima County in 1974.

Antho (HA1, HA5, HA7). Typic Torrifluvents; coarse-loamy, mixed (calcareous), hyperthermic (Alluvial). Antho soils are deep and well drained. Typically, they have light yellowish brown sandy loam surface layers about 20 cm (8 in) thick and similar subsurface layers to more than 150 cm (60 in) that commonly contain slightly coarser or finer textured strata. They occur on floodplains and alluvial fans with slopes of 0 to 3 percent. These soils have moderate available water capacity and moderately rapid permeability. They are moderately alkaline and are usually calcareous throughout. Runoff is slow and the hazard of erosion is slight. The series was established in Maricopa County in 1971 as the hyperthermic equivalent of the Anthony series.

Some properties of the Antho soils

Horizon	Depth (cm)	O.M.%	Sand %	Silt %	Clay %	CEC meg/100g	CaCO$_3$ %	pH	BD g/cm^3
Ap	0-18	0.24	76	15	9	7	T	7.7	1.55
C	18-127	0.14	79	13	8	6	T	8.1	1.70

Anthony (TA2, TA4, TS2, TS3, TS9, TS12, TS19). Typic Torrifluvents; coarse-loamy, mixed (calcareous), thermic (Alluvial). Anthony soils are deep and well drained. They have a brown sandy loam or gravelly sandy loam surface layer about 25 cm (10 in) thick over stratified pale brown and brown fine sandy loam, sandy loam and gravelly sandy loam to more than 150 cm (60 in). Anthony soils are on floodplains and alluvial fans with slopes of 0 to 3 percent. They have moderate available water capacity and moderately rapid permeability. They are moderately alkaline and slightly calcareous throughout. Runoff is slow and the hazard of erosion is moderate. The series was established in the Mesilla Valley, New Mexico, in 1912.

Some properties of the Anthony soils

Horizon	Depth (cm)	O.M.%	Sand %	Silt %	Clay %	CEC meg/100g	CaCO$_3$%	pH	BD g/cm^3
Ap	0-30	0.17	70	20	10	7	2	7.7	1.65
C	30-107	0.15	64	28	8	7	2	7.8	1.52
Btb	107-137		61	18	21	10	T	7.8	1.56

Anway (TS3). Typic Haplargids; fine-loamy, mixed, thermic (Red Desert). Anway soils are deep and well drained. Typically, they have a light yellowish brown loam surface layer about 8 cm (3 in) thick. The subsoil is brown clay loam about 38 cm (15 in) thick. This is underlain by pale brown and light yellowish brown stratified loam, silt loam and fine sandy loam. Anway soils occur on nearly level stream terraces with slopes of 0 to 1 percent. These soils have high available water capacity and moderate to moderately slow permeabilty. They are mildly to moderately alkaline throughout the profile. Runoff is slow and the hazard of erosion is slight. The series was established in the Tucson-Avra Valley area in 1965 and the name was taken from Anway Road.

Apache (MS4, MS7). Lithic Haplustolls; loamy, mixed, mesic (Lithosols). Apache soils are shallow and well drained. Typically, they have a grayish brown cobbly heavy loam surface layer about 20 cm (8 in) thick. The subsoil is brown cobbly light clay loam about 10 cm (4 in) thick. The underlying material is composed of soft to slightly hard white carbonate concretions and hard carbonate coatings on rock fragments and soil material similar to the overlying subsoil about 10 cm (4 in) thick. Hard black basalt coated with white carbonate is at a depth of 40 cm (16 in). Depth to bedrock ranges from 10 to 50 cm (4 to 20 in). Apache soils occur on basalt mesas and lava flows with slopes of 1 to 15 percent. These soils have low available water capacity and moderate permeability. They are moderately alkaline and calcareous throughout the profile. Runoff is medium and the hazard of erosion is slight. The series was established on the Mansher Soil Conservation Demonstration Project in northeastern New Mexico in 1938.

Arada (TA5). Typic Calciorthids; sandy, mixed, thermic (Calcisols). Arada soils are deep and somewhat excessively drained. Typically, they have a pink fine sand surface layer about 15 cm (6 in) thick. The upper 53 cm

(21 in) of underlying material is pink fine sand. The middle 25 cm (10 in) of underlying material is pink gravelly loamy fine sand. The lower portion of the underlying material to a depth of 150 cm (60 in) is stratified pinkish white and very pale brown very gravelly sandy loam and loamy coarse sand. Arada soils occur on nearly level to moderately sloping stream terraces and alluvial fans with slopes of 0 to 8 percent. These soils have low available water capacity and rapid permeability. They are moderate to strongly alkaline and calcareous throughout the profile. Runoff is slow and the hazard of erosion is high. The series was established in Clark County, Nevada, in 1970.

Arizo (TA5, TS2). Typic Torriorthents; sandy-skeletal, mixed, thermic (Alluvial). Arizo soils are deep and excessively drained. Typically, they have light brownish gray gravelly sandy loam surface layers about 20 cm (8 in) thick overlying very gravelly and cobbly sand to 150 cm (60 in) or more. Arizo soils occur on floodplains and alluvial fans that have slopes of 0 to 8 percent. These soils have very low available water capacity and very rapid permeability. They are moderately alkaline and usually are calcareous throughout. Runoff is slow and the hazard of erosion is moderate. The series was established in Clark County, Nevada, in 1971.

Arp (MS6). Ustollic Haplargids; fine, montmorillonitic, mesic (Reddish Brown). Arp soils are moderately deep and well drained. Typically, they have a dark brown gravelly clay loam surface layer about 5 cm (2 in) thick. The subsoil is reddish brown clay about 40 cm (16 in) thick. The underlying material is light yellowish brown weathered schist about 40 cm (16 in) thick. Hard schist bedrock is at a depth of 85 cm (34 in). Depth to the hard schist is 50 to 100 cm (20 to 40 in). Arp soils occur on hillslopes with slopes of 10 to 25 percent. These soils have moderate available water capacity and moderately slow to slow permeability. They are medium acid to neutral and noncalcareous throughout the profile. Runoff is medium to rapid and the hazard of erosion is slight. The series was established in Yavapai County in 1960 and the name was taken from Bill Arp Wash.

Some properties of the Arp soils

Horizon	Depth (cm)	O.M. %	Sand %	Silt %	Clay %	CEC meq/100g	B.S.%	pH	BD g/cm³
A	0-5	1.4	36	37	27	30	92	6.7	1.60
Bt	5-40	1.4	14	33	53	54	93	6.9	1.68
C	40-75	.34	24	50	26	49	97	7.3	1.73
R	75-								

Atascosa (TS6, TS13). Aridic Lithic Argiustolls; loamy-skeletal, mixed, thermic (Lithosols), Atascosa soils are shallow and very shallow and well drained. Typically, they have a dark grayish brown very gravelly sandy loam surface layer about 5 cm (2 in) thick. The subsoil is dark gray very gravelly sandy clay loam about 18 cm (7 in) thick. This is underlain by white rhyolitic conglomerate at a depth of 23 cm (9 in). Depth to bedrock ranges from 10 to 50 cm (4 to 20 in). Atascosa soils occur on hillslopes with slopes ranging from 5 to 60 percent. These soils have low available water holding capacity and moderate permeability. They are neutral to mildly alkaline and noncalcareous throughout the profile. Runoff is rapid and the hazard of erosion is slight. The series was established in Santa Cruz County in 1971.

Avondale (HA1). Typic Torrifluvents; fine-loamy, mixed (calcareous), hyperthermic (Alluvial). Avondale soils are deep and well drained. Typically, they have a brown clay loam surface layer about 30 cm (12 in) thick. The underlying material to a depth of 150 cm (60 in) is pale brown loam and very fine sandy loam. Avondale soils occur on floodplains with slopes of 0 to 3 percent. These soils have high available water capacity and moderate to moderately slow permeability. They are moderately alkaline and calcareous throughout the profile. Runoff is slow and the hazard of erosion is slight to moderate. The series was established in Maricopa County in 1969 and the name was taken from the town of Avondale.

Some properties of the Avondale soils

Horizon	Depth (cm)	O.M. %	Sand %	Silt %	Clay %	CEC meq/100g	CaCO₃ %	pH	BD g/cm³
Ap	0-25	1.80							
C	25-150	0.69	47	35	22				

Badland (MA1, MA3, MA5). Badland consists of moderately steep to very steep, nearly barren areas of highly erodible, multicolored, clayey shales and siltstones interbedded with thin layers of harder sandstone and conglomerate. These shaly areas are dissected by numerous drainageways, leaving small remnants as ridges and low buttes capped by the more resistant sandstone and conglomerate rocks. The dominant slopes are 30 to 50 percent. Vertical relief is from 8 to 60 m (25 to 200 ft) or more.

Baldy (FH2, FH5). Typic Cryorthents; coarse-loamy, mixed, nonacid, mesic (Regosols). Baldy soils are deep and well drained. Typically, there is a 3 cm (1 in) leaf and pine needle litter on the soil surface. The surface mineral layer is light brownish gray cobbly fine sandy loam about 10 cm (4 in) thick. The underlying material to more than 107 cm (42 in) is pinkish gray, very pale brown and yellowish brown cobbly or gravelly fine sandy loams. Gravel, cobble and stone content averages less than 35 percent. Baldy soils are on hillslopes with slopes of 10 to 50 percent. These soils have moderate available water capacity and moderately rapid permeability. They are medium acid throughout and noncalcareous throughout the profile. Runoff is medium and the hazard of erosion is slight to moderate. The series was established on the Fort Apache Indian Reservation in 1965 and the name was taken from Mt. Baldy.

Balon (MS6, MS9). Ustollic Haplargids; fine-loamy, mixed, mesic (Reddish Chestnut). Balon soils are deep and well drained. Typically, they have dark grayish brown gravelly sandy clay loam surface layers about 8 cm (3 in) thick, and dark grayish brown gravelly clay loam subsoils that grade at about 38 cm (15 in) to mottled yellowish brown and pale brown gravelly sandy clay loam and gravelly sandy loam. The underlying material from 91 to more 150 cm (36 to 60 in) is light yellowish brown and very pale brown gravelly sandy loam. Balon soils are on fan terraces with slopes ranging from 0 to 30 percent. These soils have moderate available water capacity and moderately slow permeability. They are medium acid to mildly alkaline and are noncalcareous throughout the profile. Runoff is medium and the hazard of erosion is moderate. The series was established in Yavapai County in 1960.

Some properties of the Balon soils

Horizon	Depth (cm)	O.M.%	Sand%	Silt %	Clay %	CEC meq/100g	B.S.%	pH	BD g/cm^3
A	0-8	1.09	57	22	21	21	93	6.5	1.54
Bt	8-58	0.54	46	28	26	35	96	6.7	1.64
BCt	58-90	0.15	57	28	15	32	96	6.7	1.60
C	90-183	0.12	56	29	15	35	97	6.8	

Bandera (MS4). Torriorthentic Haploborolls; cindery (Lithosols). Bandera soils are shallow and somewhat excessively drained. Typically, they have dark grayish brown and brown gravelly (cindery) loam surface layers about 23 cm (9 in) thick. The underlying material is yellowish brown very gravelly (cindery) loam 8 cm (3 in) thick that grades at about 41 cm (16 in) into very dark gray loose cinders. The gravel in the profile is mainly cinders. Depth to the unweathered cinders ranges from 30 to 66 cm (12 to 26 in). Bandera soils occur on cinder cones that have slopes of 2 to 45 percent. These soils have low available water capacity and moderate permeability. They are slightly acid to mildly alkaline in the soil profile and moderately alkaline and calcareous in the cinders. Runoff is rapid and the hazard of erosion is slight. The series was established in Valencia County, New Mexico, in 1957 and the name was taken from Bandera Crater.

Some properties of the Bandera soils

Horizon	Depth (cm)	O.M.%	Sand %	Silt %	Clay %	CEC meq/100g	CaCO$_3$ %	pH	BD g/cm^3
A	0-23	4.86	37	47	16	41	0	7.4	
C	23-48	3.62	47	42	11	37	6	8.4	
Cr	48-								

Barkerville (MS6, MH2, MH4). Udorthentic Haplustolls; loamy, mixed, mesic, shallow (Lithosols). Barkerville soils are shallow and well drained. Typically, they have a 2.5 cm (1 in) layer of undecomposed and partially decomposed leaves and twigs on the soil surface. The upper surface mineral layer is dark grayish brown cobbly sandy loam about 8 cm (3 in) thick. This is underlain by yellowish brown and light gray highly weathered granite to a depth of 65 cm (26 in). Below this to a depth of 100 cm (40 in) is varicolored slightly weathered granite. Barkerville soils occur on hillslopes with slopes of 10 to 60 percent. These soils have very low available water capacity and moderately rapid permeability. They are neutral to mildly alkaline and noneffervescent throughout the profile. Runoff is rapid and the hazard of erosion is slight. The series was established in Santa Cruz County in 1971.

Some properties of the Barkerville soils

Horizon	Depth (cm)	O.M.%	Sand %	Silt %	Clay %	CEC meq/100g	B.S.%	pH	BD g/cm³
A	0-8	0.77	81	11	8	9	93	7.0	
C	8-63	0.20	85	10	5	13	100	6.7	

Bernardino (TS4, TS5, TS7, TS15). Ustollic Haplargids; fine, mixed, thermic (Reddish Brown). Bernardino soils are deep and well drained. They typically have a dark brown loam surface layer about 5 cm (2 in) thick and a dark reddish brown gravelly clay loam upper subsoil and clay lower subsoil that grades at about 38 cm (15 in) to lime mottled pinkish gray gravelly sandy clay loam to more than 150 cm (60 in). Bernardino soils are on rolling fan terraces with slopes that are dominantly 5 to 15 percent but may range up to 25 percent. These soils have moderate available water capacity and slow permeability. They are moderately alkaline and are usually calcareous throughout with zones of high lime at less than 50 cm (20 in). Runoff is slow to medium and the hazard of erosion is slight to moderate. The series was established in Santa Cruz County in 1971 and the name was taken from the San Bernardino Valley.

Some properties of the Bernardino soils

Horizon	Depth (cm)	O.M.%	Sand %	Silt %	Clay %	CEC meq/100g	CaCO₃%	pH	BD g/cm³
A	0-5	2.77	67	15	18	19	0	7.2	1.03
Bt	5-43	1.75	50	14	36	34	0	7.0	1.47
Bk	43-127	0.41	75	13	12	18	13	7.6	1.41

Bitter Spring (TA5). Typic Haplargids; loamy-skeletal, mixed, thermic (Red Desert). Bitter Spring soils are deep and well drained. Typically, they have a well-developed desert pavement of subangular to rounded varnished gravels on the surface. The surface layer is pink loam about 4 cm (1.5 in) thick. The subsoil is light reddish brown sandy loam and light sandy clay loam about 15 cm (6 in) thick. The underlying material to a depth of 150 cm (60 in) is pink and light reddish brown very gravelly sandy loam. Bitter Spring soils occur on nearly level to moderately sloping fan terraces with slopes of 0 to 8 percent. These soils have low available water capacity and moderately rapid permeability. They are moderate to strongly alkaline with alkalinity decreasing with depth and are calcareous throughout the profile. Runoff is medium and the hazard of erosion is slight. The series was established in Clark County, Nevada, in 1969.

Some properties of the Bitter Spring soils

Horizon	Depth (cm)	O.M.%	Sand %	Silt %	Clay %	CEC meq/100g	CaCO₃%	pH	BD g/cm³
A	0-5	0.43	61	28	11		3	7.9	
Bt	5-55	1.50	43	33	24		7	7.9	
Bk	55-90	0.29	62	29	9		23	8.2	

Bonita (TS4, TS15, TS17, TS18). Typic Chromusterts; fine, montmorillonitic, thermic (Brown soil-Grumusols). Bonita soils are deep and well drained. Typically, they have a dark brown granular cobbly silty clay surface layer about 5 cm (2 in) thick. The subsurface layer is dark brown silty clay about 74 cm (29 in) thick. The underlying material to 150 cm (60 in) is reddish brown very cobbly and gravelly clay loam. Bonita soils occur on nearly level to gently sloping fan terraces with slopes of 0 to 3 percent. These soils have high available water capacity. Permeability is very slow and the upper layers have high shrink-swell that causes them to crack widely and deeply when dry and swell when wet. They are moderately alkaline and are calcareous throughout the profile. Runoff is slow and erosion hazard is slight. Short pipes may develop where water follows cracks. The series was established in Graham County in 1936 and the name was taken from Bonita Canyon.

Some properties of the Bonita soils

Horizon	Depth (cm)	O.M.%	Sand %	Silt %	Clay %	CEC meq/100g	CaCO₃%	pH	BD g/cm³
A	0-28	1.31	13	42	45	39	1	7.6	1.74
C	28-79	1.00	11	42	47	39	2	8.0	1.75
2 Bk	79-160	0.63	30	45	25	28	7	7.7	1.50

Boysag (MS1, MS2, MS5, MS6). Lithic Ustollic Haplargids; clayey, mixed, mesic (Reddish Brown). Boysag soils are shallow and well drained. Typically, these soils have reddish brown and brown fine sandy loam surface layers about 8 cm (3 in) thick. Below this is a layer of yellowish red clay about 20 cm (8 in) thick that grades to pinkish white highly fractured and weathered limy sandstone about 8 cm (3 in) thick. Below this, the bedrock is very pale brown calcareous sandstone having widely spaced fractures. Depth to bedrock ranges from 25 to 50 cm (10 to 20 in). Boysag soils occur on gently undulating hillslopes with slopes ranging from 0 to 8 percent. These soils have low available water capacity and slow permeability. They are slightly acid and noncalcareous in the surface and upper subsoil and moderately alkaline and calcareous in the lower subsoil and bedrock. Runoff is slow to medium and the hazard of erosion is slight to moderate. The series was established in Coconino County in 1962 and the name was taken from an Indian name for an isolated point on the north rim of the Grand Canyon.

Some properties of the Boysag soils

Horizon	Depth (cm)	O.M.%	Sand%	Silt%	Clay%	CEC meq/100g	CaCO$_3$%	pH	BD g/cm^3
A	0-5	1.77	73	21	6	9	1	7.7	1.39
Bt	5-23	1.96	60	16	24	16	6	8.0	1.33
Bk	23-39	3.33	51	20	29	13	36	82	
R	39-								

Brazito (TS2). Typic Torripsamments; mixed, thermic (Alluvial). Brazito soils are deep and well drained. Typically, they have a grayish brown sandy clay loam surface layer about 23 cm (9 in) thick. The underlying material to a depth of 150 cm (60 in) is light brownish gray or light gray sand and fine sand. Brazito soils occur on floodplains and alluvial fans with slopes of 0 to 3 percent. These soils have very low available water capacity and rapid permeability. They are moderately alkaline and calcareous throughout the profile. Runoff is slow and the hazard of erosion is moderate to high. The series was established in Dona Ana County, New Mexico, in 1912.

Brios (HA1, HA7). Typic Torrifluvents; sandy, mixed, hyperthermic (Alluvial). Brios soils are deep and somewhat excessively drained. Typically, they have a brown sandy loam surface layer about 35 cm (14 in) thick. The upper 20 cm (8 in) of underlying material is brown coarse sand. Below this to a depth of 150 cm (60 in) is brown stratified coarse sand and gravelly coarse sand with thin strata of fine sandy loam and sandy loam. Brios soils occur on nearly level to gently sloping floodplains and alluvial fans with slopes of 0 to 3 percent. Available water capacity is low and permeability is rapid. They are moderately alkaline and calcareous throughout the profile. Runoff is slow and the hazard of erosion is high. The series was established in Maricopa County in 1972.

Brolliar (FH1, FH2). Typic Argiborolls; fine, montmorillonitic (Reddish Chestnut). Brolliar soils are moderately deep and well drained. Typically, they have a thin layer of decomposing pine needles on the soil surface. The surface mineral layers are brown and dark brown stony heavy loam about 13 cm (5 in) thick. The subsoil is reddish brown and brown heavy clay loam and clay about 74 cm (29 in) thick. The bedrock from 86 to more than 150 cm (34 to 60 in) is dark gray, extremely hard, fractured basalt. Depth to the bedrock ranges from 50 to 100 cm (20 to 40 in). Brolliar soils occur on mesas with slopes ranging from 0 to 30 percent. These soils have moderate available water capacity, slow permeability and high shrink-swell potential. Reaction is slightly acid to neutral throughout the profile, and these soils are noncalcareous throughout the profile. Runoff is slow to medium and the hazard of erosion is slight. The series was established in the Beaver Creek area of Yavapai County in 1965.

Some properties of the Brolliar soils

Horizon	Depth (cm)	O.M.%	Sand%	Silt%	Clay%	CEC meq/100g	B.S.%	pH	BD g/cm^3
A	0-10	4.40	14	57	29	33	64	5.7	1.54
Bt	10-80	1.72	7	33	60	54	83	5.9	1.75
Cr	80-88	1.29	17	36	57	61	94	6.7	1.79
R	88-								

Bucklebar (TS19). Typic Haplargids; fine-loamy, mixed, thermic (Red Desert). Bucklebar soils are deep and well drained. Typically, they have a brown, sandy loam surface layer about 15 cm (6 in) thick. The subsoil is reddish brown sandy clay loam about 48 cm (19 in) thick. The underlying material to a depth of 150 cm (60 in) is light brown loam and silty clay loam with few carbonate nodules and masses on surfaces of peds. Bucklebar soils occur on gently sloping fan terraces with slopes of 1 to 5 percent. These soils have high available water capacity and moderate permeability. They are mildly to moderately alkaline, noncalcareous in the surface layer and calcareous in the subsoil and underlying material. Runoff is medium and the hazard of erosion is moderate. The series was established in Dona Ana County, New Mexico, in 1971.

Bushvalley (FH2). Argic Lithic Cryoborolls; loamy-skeletal, mixed (Prairie). Bushvalley soils are shallow and very shallow and well drained. Typically, they have a dark brown cobbly sandy loam surface layer about 13 cm (5 in) thick. This is underlain by light brown tuff bedrock at a depth of 25 cm (10 in). Depth to bedrock ranges from 18 to 50 cm (7 to 20 in). Bushvalley soils occur on hillslopes with slopes of 8 to 40 percent. These soils have low available water capacity and moderately slow permeability. They are slightly acid to neutral and noncalcareous throughout the profile. Runoff is rapid and the hazard of erosion is slight. The series was established in Apache County in 1971.

Some properties of the Bushvalley soils

Horizon	Depth (cm)	O.M.%	Sand%	Silt%	Clay%	CEC meq/100g	B.S.%	pH	BD g/cm³
A	0-10	2.14	59	27	14	31	92	6.7	1.45
Bt	10-20								1.80
R	20-								

Cabezon (MS4, MS7, MH2, MH6). Lithic Argiustolls; clayey, montmorillonitic, mesic (Chestnut). Cabezon soils are shallow and well drained. Typically, the surface layer is grayish brown extremely stony loam about 10 cm (4 in) thick. The subsoil is brown cobbly clay about 20 cm (8 in) thick that overlies weathered fractured basalt bedrock at about 30 cm (12 in) and relatively unweathered basalt at about 38 cm (15 in). Depth to bedrock ranges from 25 to 50 cm (10 to 20 in). Cabezon soils are on hillslopes with dominant slopes of 2 to 15 percent and short escarpments up to 45 percent or more. These soils have low available water capacity and slow permeability. They are slightly acid to neutral and are noncalcareous except for small accumulations of lime on the bedrock in places. Runoff is medium and the hazard of erosion is slight to moderate. The series was established in Sandoval County, New Mexico, in 1965.

Cambern (FH2). Argic Pachic Cryoborolls; fine-loamy, mixed (Prairie). Cambern soils are moderately deep and well drained. Typically, they have a dark gray sandy loam and loam surface layer about 25 cm (10 in) thick. The subsoil is dark gray heavy loam and light clay loam about 45 cm (18 in) thick. White tuff and ash is at a depth of 70 cm (28 in). Depth to bedrock ranges from 50 to 100 cm (20 to 40 in). Cambern soils occur on hillslopes with slopes ranging from 1 to 25 percent. These soils have moderate available water capacity and moderately slow permeability. They are neutral to slightly acid and noncalcareous throughout the profile. Runoff is medium and the hazard of erosion is slight. The series was established in Apache County in 1966 and the name was taken from an old townsite.

Some properties of the Cambern soils

Horizon	Depth (cm)	O.M.%	Sand%	Silt%	Clay%	CEC meq/100g	B.S.%	pH	BD g/cm³
A	0-25	3.16	56	30	14	23	88	6.3	1.36
BA	25-35	1.94	47	32	21	26	91	6.7	1.45
Bt	35-55	1.68	39	34	27	32	95	6.8	1.55
R	55-								

Camborthids (TA1, MA6). The Camborthids are shallow to moderately deep and well drained. They are medium to moderately coarse in texture and formed in colluvium and residuum from limestone, sandstone and shale. They occur on hillslopes with slopes ranging from 5 to 30 percent and are similar to the Torriorthents except that they have weak B horizons (Cambic horizons).

Canelo (MH1). Aeric Ochraqualfs; clayey-skeletal, mixed, mesic (Reddish Chestnut). Canelo soils are deep and somewhat poorly drained. Typically, they have a brown gravelly sandy loam surface layer about 13 cm (5 in) thick and a very pale brown, very gravelly sandy loam subsurface layer about 23 cm (9 in) thick. The upper subsoil is mottled pinkish white, reddish yellow, white and light gray very gravelly heavy sandy clay loam and very gravelly clay about 46 cm (18 in) thick. The lower subsoil to a depth of more than 150 cm (60 in) is light red and red very cobbly sandy clay with common pinkish gray mottles. The soils are wet and have a temporary perched water table following rainy seasons. Canelo soils occur on the tops and sides of long narrow fan terraces and have slopes ranging from 1 to 40 percent. They have moderate available water capacity and very slow permeability. They are slightly to extremely acid throughout. Runoff is medium and the hazard of erosion is moderate. The series was established in Santa Cruz County in 1971 and the name was taken from the Canelo Hills.

Some properties of the Canelo soils

Horizon	Depth (cm)	O.M.%	Sand%	Silt%	Clay%	CEC meq/100g	B.S.%	pH	BD g/cm³
A	0-13	1.33	60	32	8	5	80	6.3	1.60
E	13-36	0.58	54	35	11	4	78	6.2	1.35
Btg	36-86	0.46	29	16	55	16	96	4.9	1.47
Bt	86-130		46	12	42	14	75	4.3	

Caralampi (TS4, TS5, TS6, TS7, TS8, TS10). Ustollic Haplargids; loamy-skeletal, mixed, thermic (Reddish Chestnut). Caralampi soils are deep and well drained. They typically have a dark grayish brown very gravelly sandy loam surface layer about 5 cm (2 in) thick. The subsoil is reddish brown very gravelly sandy clay loam about 53 cm (21 in) thick. The substratum to more than 150 cm (60 in) is light reddish brown very gravelly sandy loam. Caralampi soils occupy the higher and steeper fan terraces with slopes ranging from 10 to 30 percent. These soils have low available water capacity and moderately slow permeability. They are slightly acid to moderately alkaline and are usually noncalcareous throughout. Runoff is medium and the hazard of erosion is slight to moderate. The series was established in Santa Cruz County in 1971.

Some properties of the Caralampi soils

Horizon	Depth (cm)	O.M.%	Sand%	Silt%	Clay%	CEC meq/100g	B.S.%	pH	BD g/cm³
A	0-5	1.58	73	11	16	16	81	5.9	1.25
Bt	5-58	1.41	61	14	25	25	98	5.9	1.42
BCt	58-107	0.31	69	18	13	22	100	6.4	1.54
C	107-142		74	17	9	26	100	6.8	1.63

Carrizo (HA1, HA7). Typic Torriorthents; sandy-skeletal, mixed, hyperthermic (Alluvial). Carrizo soils are deep and excessively drained. Typically, they have a brown surface layer about 38 cm (15 in) thick. The underlying material is light grayish brown very gravelly sand to more than 150 cm (60 in). Carrizo soils occur on floodplains and alluvial fans with slopes of 0 to 5 percent. These soils have very low available water capacity and very rapid permeability. They are moderately alkaline and calcareous throughout. Runoff is slow and the hazard of erosion is slight. The series was established in the El Centro area of California in 1918.

Some properties of the Carrizo soils

Horizon	Depth (cm)	O.M.%	Sand%	Silt%	Clay%	CEC meq/100g	CaCO₃%	pH	BD g/cm³
A	0-12	0.34	93	4	3	3	2	8.3	
C	12-113	0.07	95	3	2	2	1	8.5	
2Btb	113-139	0.12	77	8	15	12	T	8.0	
3Cb	139-162	0.20	52	32	16	19	T	8.0	

Casa Grande (HA2). Typic Natrargids; fine-loamy, mixed, hyperthermic (Solonetz). Casa Grande soils are deep and well drained. They typically have thin, reddish yellow sandy loam or loam surface layers and reddish brown sandy clay loam or clay loam subsoils about 5 cm (2 in) thick that are sodic and strongly saline and have a strong lime accumulation in the lower part. The substratum below about 50 to 76 cm (20 to 30 in) is yellowish red and brown loam and sandy loam that is sodic and calcareous to more than 150 cm (60 in). Casa Grande soils occur on generally concave ends of relict basin floors and nearly level low fan terraces with slopes ranging from 0 to 2 percent. They have low available water due to high salt and sodium content. Permeability is slow or very slow. These soils are strongly to very strongly alkaline and calcareous throughout the profile. Runoff is slow and the hazard of erosion is moderate. The series was established in Pinal County in 1936 and the name was taken from the town of Casa Grande.

Some properties of the Casa Grande soils

Horizon	Depth (cm)	O.M.%	Sand%	Silt%	Clay%	CEC meq/100g	CaCO$_3$%	pH	BD g/cm^3
A	0-7	0.29	69	26	5	7	T	8.8	1.70
Btk1	7-36	0.20	49	31	20	18	8	9.3	1.73
Btk2	36-92		48	31	21	12	9	8.8	1.62
Bk	92-165		57	31	12	7	11	9.3	1.73

Casto (MH1). Udic Haplustalfs; loamy-skeletal, mixed, mesic (Reddish Chestnut). Casto soils are deep and well drained. Typically, they have a grayish brown, very gravelly sandy loam surface layer about 3 cm (1 in) thick. The subsoil is reddish brown and reddish gray very gravelly heavy sandy clay loam 69 cm (27 in) thick. The underlying material is pinkish white very gravelly sandy loam to a depth of 150 cm (60 in) or more. Casto soils occur on the sides and tops of narrow fan terraces and steep sides of broad fan terraces with slopes ranging from 8 to 40 percent. These soils have moderate available water capacity and slow permeability. They are medium acid in the upper layers and become moderately alkaline and calcareous below depths of about 61 cm (24 in). Runoff is medium to rapid and the hazard of erosion is slight. The series was established in Santa Cruz County in 1971.

Some properties of the Casto soils

Horizon	Depth (cm)	O.M.%	Sand%	Silt%	Clay%	CEC meq/100g	CaCO$_3$%	pH	BD g/cm^3
A	0-3	2.2	72	15	13	18	0	6.2	1.15
Bt	3-56	0.97	57	16	27	35	0	5.7	1.51
C	56-152		59	26	15	38	4	7.7	1.30

Cave (TA4, TA5, TS12, TS14). Typic Paleorthids; loamy, mixed, thermic, shallow (Calcisols). Cave soils are shallow and well drained. Typically, they have pale brown gravelly sandy loam surface layers and thin, pink, gravelly loam subsurface layers that overlie a white indurated lime cemented hardpan at depths ranging fom 10 to 50 cm (4 to 20 in). The pan is underlain at about 76 cm (30 in) by variably cemented very gravelly sandy loam and loamy sand layers to more than 150 cm (60 in). Cave soils occur on fan terraces with slopes ranging from 0 to 15 percent. These soils have very low available water capacity. Permeability is moderately rapid above the pan and very slow through the pan. They are calcareous throughout and are moderately alkaline. Runoff is moderate and the hazard of erosion is slight. The series was established in the San Simon Valley in 1936.

Some properties of the Cave soils

Horizon	Depth (cm)	O.M.%	Sand%	Silt%	Clay%	CEC meq/100g	CaCO$_3$%	pH	BD g/cm^3
A	0-3	1.24	71	22	7	8	1	7.6	1.50
Bw	3-18	0.78	64	27	9	8	1	7.8	1.55
Bkm	18-104		62	27	11	3	47	8.2	

Cavelt (HA4, HA7). Typic Paleorthids; loamy, mixed, hyperthermic, shallow (Calcisols). Cavelt soils are shallow and well drained. Typically, they have light yellowish brown gravelly loam surface layers and light brown gravelly loam subsurface layers that overlie a white indurated lime cemented hardpan at depth ranging from 10 to 50 cm (4 to 20 in). The pan is underlain at about 115 cm (46 in) by variably cemented gravelly loam layers to more than 150 cm (60 in). Cavelt soils occur on nearly level to undulating fan terraces and have slopes of 0 to 9 percent. Available water capacity is very low and permeability is moderately rapid above the pan and very slow through the pan. They are moderately alkaline and strongly calcareous throughout. Surface runoff is medium and the hazard of erosion is slight. The series was established in Pinal County in 1971 as the hyperthermic equivalent of the Cave series.

Cellar (TA1, TA3, TS6, TS10). Lithic Torriorthents; loamy-skeletal, mixed, nonacid, thermic (Lithosols). Cellar soils are shallow and very shallow and well drained. Typically, they have a pale brown very gravelly sandy loam surface layer about 3 cm (1 in) thick. The underlying material is brown very gravelly loam about 18 cm (7 in) thick. Below this at a depth of 20 cm (8 in) is white granite bedrock. Depth to bedrock ranges from 10 to 50 cm (4 to 20 in).

APPENDICES

Cellar soils occur on hillslopes with slopes of 15 to 50 percent. These soils have low available water capacity and moderate to moderately rapid permeability. They are moderately alkaline and noncalcareous throughout the profile. Runoff is rapid and the hazard of erosion is slight. The series was established in Graham County in 1965.

Some properties of the Cellar soils

Horizon	Depth (cm)	O.M.%	Sand%	Silt%	Clay%	CEC meq/100g	CaCO$_3$%	pH	BD g/cm^3
A	0-3	2.87	69	23	8	12	0	7.2	
C	3-20	0.60	78	13	9	9	0	7.4	
R	20-								

Cherioni (HA6, HA9). Typic Durorthids; loamy-skeletal, mixed, hyperthermic, shallow (Lithosols). Cherioni soils are very shallow and shallow and well drained. Typically, they have light colored gravelly very fine sandy loam surface layers, gravelly and very gravelly fine sandy loam and very fine sandy loam subsurface layers, and an indurated duripan at depths of 13 to 30 cm (5 to 12 in) and hard bedrock at 15 to 51 cm (6 to 20 in). Cherioni soils occur on hillslopes with slopes ranging from 5 to 70 percent. These soils have very low available water capacity and moderate permeability above the pan. They are moderately alkaline and calcareous throughout. Runoff is medium and the hazard of erosion is slight. The series was established in Pima County on the Organ Pipe Cactus National Monument in 1971.

Chevelon (MH3). Udic Haplustalfs; fine-silty, mixed, mesic (Reddish Chestnut). Chevelon soils are moderately deep and well drained. Typically, they have a reddish brown silt loam surface layer about 13 cm (5 in) thick. The subsoil is reddish brown silty clay loam about 63 cm (25 in) thick. Below this at a depth of 76 cm (30 in) is interbedded reddish brown and light gray shale. Depth to bedrock ranges from 50 to 100 cm (20 to 40 in). Chevelon soils occur on gently undulating to moderately steep hillslopes with slopes of 5 to 25 percent. These soils have medium available water capacity and moderately slow permeability. They are neutral to mildly alkaline and noncalcareous throughout the profile. Runoff is medium and the hazard of erosion is moderate. The series was established in Navajo County in 1962.

Some properties of the Chevelon soils

Horizon	Depth (cm)	O.M.%	Sand%	Silt%	Clay%	CEC meq/100g	CaCO$_3$%	pH	BD g/cm^3
A	0-8	1.45	45	40	15	15	10	7.8	
Bt	8-63	0.67	30	41	29		11	7.8	

Chilson (FH3). Lithic Argiborolls; clayey, mixed (Prairie). Chilson soils are shallow and well drained. Typically, they have a dark reddish-gray cobbly clay loam surface layer about 5 cm (2 in) thick. The upper 15 cm (6 in) of subsoil is dark reddish gray gravelly clay. The lower 15 cm (6 in) of subsoil is reddish brown cobbly clay. Below this at a depth of 35 cm (14 in) is dark red fine grained sandstone and shale bedrock. Depth to bedrock ranges from 25 to 50 cm (10 to 20 in). Chilson soils occur on hillslopes with slopes of 20 to 45 percent. These soils have low available water capacity and slow permeability. They are mildly to moderately alkaline and noncalcareous throughout the profile. Runoff is rapid and the hazard of erosion is slight. The series was established in Coconino County in 1965 and the name was taken from stock tank in the area.

Chiminea (TS6). Typic Haplargids; loamy, mixed, thermic, shallow (Lithosols). Chiminea soils are very shallow and shallow and well drained. Typically, they have a brown gravelly fine sandy loam surface layer about 5 cm (2 in) thick. The subsoil is reddish brown gravelly coarse sandy loam about 33 cm (13 in) thick. The underlying material to a depth of 150 cm (60 in) is reddish brown, pink and pinkish gray highly weathered granite (grus). Depth to weathered bedrock ranges from 15 to 50 cm (6 to 20 in). Chiminea soils occur on hillslopes and pediments with slopes of 5 to 25 percent. These soils have low available water capacity and moderate permeability. They are moderately alkaline and noncalcareous throughout the profile. Runoff is medium to rapid and the hazard of erosion is slight. The series was established in Pima County in 1975.

Chiricahua (TA3, TS4, TS6, TS10). Ustollic Haplargids; clayey, mixed, thermic, shallow (Reddish Brown). Chiricahua soils are shallow and well drained. Typically, they have a brown very cobbly loam surface layer about 5 cm (2 in) thick. The upper 5 cm (2 in) of subsoil is reddish brown very gravelly clay loam. The lower 30 cm (12 in) of subsoil is reddish brown gravelly clay. Below this is pink and reddish yellow strongly weathered granite or quartzite with clay soil material in seams and fractures about 13 cm (5 in) thick. Pink and reddish yellow hard quartzite bedrock is at a depth of 53 cm (21 in). Chiricahua soils occur on hillslopes with slopes of 8 to 30 percent. These soils have low available water capacity and slow permeability. They are medium acid to neutral and noncalcareous throughout the profile. Runoff is medium and the hazard of erosion is slight. The series was established in Cochise County in 1936 and the name was taken from the Chiricahua Mountains.

Some properties of the Chiricahua soils

Horizon	Depth (cm)	O.M.%	Sand%	Silt%	Clay%	CEC meq/100g	CaCO$_3$%	pH	BD g/cm^3
A	0-10	1.73	49	23	28	17	0	7.0	1.47
Bt	10-38	1.97	31	20	49	25	T	6.8	1.57
Btk	38-46	1.84	28	18	54	28	6	7.6	1.58
R	46-								

Cibeque (MH3). Calciorthids; fine-loamy, mixed, mesic (Calcisols). Cibeque soils are deep and well drained. Typically, they have a dark yellowish brown gravelly loam surface layer about 23 cm (9 in) thick over a yellowish brown gravelly loam upper substratum 94 cm (37 in) thick. The lower substratum is yellowish brown gravelly sandy loam to more than 150 cm (60 in). Cibeque soils occur on undulating to steep fan terraces with slopes of 5 to 50 percent. These soils have moderate available water capacity and moderate permeability. They are moderately alkaline and contain more than 15 percent calcium carbonate. Runoff is medium to rapid and the hazard of erosion is moderate. The series was established in Gila County in 1966 and the name was taken from an Indian community on the Fort Apache Indian Reservation.

Cipriano (HA4). Typic Durorthids; loamy-skeletal, mixed, hyperthermic, shallow (Calcisols). Cipriano soils are shallow and well drained. Typically, they have a light brown gravelly loam surface layer about 15 cm (6 in) thick. The underlying material is white, fractured, weakly cemented platy fragments and gravel with soil material between plates about 23 cm (9 in) thick. Below this is an indurated silica cemented duripan with laminar cap about 13 cm (5 in) thick. Below the duripan to a depth of 150 cm (60 in) is a layer of white and light brown stratified loam and very gravelly loam. Depth to the duripan ranges from 20 to 50 cm (8 to 20 in). Cipriano soils occur on nearly level to undulating fan terraces with slopes of 0 to 8 percent. These soils have low available water capacity and moderate permeability above the hardpan and very slow permeability through the hardpan. They are moderately alkaline and calcareous throughout the profile. Runoff is slow to medium and the hazard of erosion is moderate. The series was established in Pima County on the Organ Pipe Cactus National Monument in 1971.

Some properties of the Cipriano soils

Horizon	Depth (cm)	O.M.%	Sand%	Silt%	Clay%	CEC meq/100g	CaCO$_3$%	pH	BD g/cm^3
A	0-5	0.43	73	20	7	6	1	8.0	
Bk	5-25	0.32	70	21	9	8	2	8.3	
2Bkqm	25-36		77	16	7		12	8.4	

Claysprings (TA1, MA1, MA2, MA5). Typic Ustorthents; fine, montmorillonitic (calcareous), mesic, shallow (Regosols). Claysprings soils are shallow and well drained. Typically, they have a reddish brown clay surface layer about 10 cm (4 in) thick, underlain by reddish brown and pink clay about 36 cm (14 in) thick. The underlying material to more than 150 cm (60 in) is gray, highly fractured, weathered clay shale that becomes harder below about 86 cm (34 in). Claysprings soils occur on gently rolling hillslopes with slopes of 0 to 10 percent. These soils have low available water capacity and very slow permeability. They are moderately alkaline and are calcareous throughout the profile. Runoff is medium to rapid and the hazard of erosion is moderate to severe. The series was established in Navajo County in 1961 and the name was taken from a village in the county.

Some properties of the Claysprings soils

Horizon	Depth (cm)	O.M.%	Sand%	Silt%	Clay%	CEC meq/100g	CaCO$_3$%	pH	BD g/cm^3
A	0-5	1.70	38	22	40	37	13	7.8	
C	5-23	0.68	18	18	64		14	7.9	
Cr	23-150	0.34	18	20	62		14	7.7	

APPENDICES

Clover Springs (FH1, FH2, FH3, FH4, FH8). Cumulic Haplustolls; fine-loamy, mixed, frigid (Alluvial). Clover Springs soils are deep and well drained. Typically, they have a very dark brown and very dark gray silt loam surface layer about 35 cm (14 in) thick. The underlying material to a depth of 150 cm (60 in) is brown loam. Clover Springs soils occur on floodplains and low stream terraces with slopes of 0 to 2 percent. These soils have high available water capacity and moderate permeability. They are neutral to mildly alkaline and noncalcareous throughout the profile. Runoff is slow and the hazard of erosion is slight to moderate. The series was established in Coconino County in 1965 and the name was taken from a spring.

Clovis (MA1, MA2, MS3). Ustollic Haplargids; fine-loamy, mixed, mesic (Reddish Brown). Clovis soils are deep and well drained. Typically, they have a brown fine sandy loam surface layer about 13 cm (5 in) thick. The upper subsoil is brown and light brown sandy clay loam 38 cm (15 in) thick. The lower subsoil and substratum is light brown and pink, calcareous light sandy clay loam and very fine sandy loam to more than 150 cm (60 in). Clovis soils occur on nearly level to rolling fan terraces with slopes ranging from 0 to 15 percent. These soils have moderate available water capacity and moderate permeability. They are neutral to moderately alkaline and contain more than 15 percent calcium carbonate in the substratum. Runoff is slow to medium and the hazard of erosion is slight to moderate. The series was established in Curry County, New Mexico, in 1956.

Some properties of the Clovis soils

Horizon	Depth (cm)	O.M.%	Sand%	Silt%	Clay%	CEC meq/100g	CaCO$_3$%	pH	BD g/cm^3
A	0-15	0.7	67	21	12	11	0	7.8	1.76
Bt	15-58	0.7	59	13	28	21	0	7.5	1.68
BCk	58-135	0.3	57	19	24	15	16	8.2	1.41
C	135-155	.07	84	12	14	10	5	8.1	1.71

Cogswell (TS11). Pachic Calciustolls; fine, mixed, thermic (Calcisols). Cogswell soils are deep and well drained. Typically, they have a brown clay loam surface layer about 30 cm (12 in) thick and a brown clay subsurface layer about 35 cm (14 in) thick. The underlying material is light brown heavy clay loam about 50 cm (20 in) thick. Below this to a depth of 150 cm (60 in) is a buried subsoil layer which is mottled strong brown, reddish yellow and white clay loam. Cogwswell soils occur on alluvial fans and old playa margins with slopes of 0 to 2 percent. These soils have high available water capacity and slow permeability. They are moderately alkaline and calcareous throughout the profile. Runoff is slow and the hazard of erosion is moderate. The series was established in the Sulphur Springs Valley area in Cochise County in 1948.

Some properties of the Cogswell soils

Horizon	Depth (cm)	O.M.%	Sand%	Silt%	Clay%	CEC meq/100g	CaCO$_3$%	pH	BD g/cm^3
Ap	0-25	1.60	36	33	31	21	4	7.8	
A	25-76	0.74	30	31	39	20	20	8.1	
Bk	76-132								
Btkb	132-152								

Comoro (TS2, TS3, TS8, TS9, TS10, TS11, MH1). Typic Torrifluvents; coarse-loamy, mixed (calcareous), thermic (Alluvial). Comoro soils are deep and well drained. Typically, they have a grayish brown sandy loam surface layer about 35 cm (14 in) thick. The underlying material to a depth of 150 cm (60 in) is grayish brown sandy loam and gravelly sandy loam. Comoro soils occur on floodplains and alluvial fans with slopes of 0 to 3 percent. They have moderate available water capacity and moderately rapid permeability. They are moderately alkaline and calcareous throughout the profile. Runoff is slow to medium and the hazard of erosion is moderate. The series was established in Santa Cruz County in 1930.

Some properties of the Comoro soils

Horizon	Depth (cm)	O.M.%	Sand%	Silt%	Clay%	CEC meq/100g	CaCO₃%	pH	BD g/cm³
Ap	0-25	1.87	61	26	13	15	4	7.7	1.46
A	25-94	0.53	57	30	13	16	4	7.7	1.41
C	95-157		78	18	4	12	T	7.8	1.56

Contine (HA3). Typic Haplargids; fine, mixed, hyperthermic (Red Desert). Contine soils are deep and well drained. Typically, they have a brown clay loam surface layer about 30 cm (12 in) thick. The upper 33 cm (13 in) of the subsoil is reddish brown clay. The lower 55 cm (22 in) of the subsoil is light reddish brown clay loam. The substratum to a depth of 150 cm (60 in) is light reddish brown loam. Contine soils occur on fan terraces with slopes of 0 to 10 percent. They have moderate available water capacity and slow permeability. They are moderately alkaline and calcareous throughout the profile. The calcium carbonate content exceeds 15 percent in some pedons. Runoff is medium and the hazard of erosion is moderate. The series was established in Maricopa County in 1971 as the hyperthermic equivalent of the Continental series.

Some properties of the Contine soils

Horizon	Depth (cm)	O.M.%	Sand%	Silt%	Clay%	CEC meq/100g	CaCO₃%	pH	BD g/cm³
A	0-8	0.73	37	38	25	19	T	7.5	1.40
Bt	8-99	0.37	27	33	40	28	2	7.8	1.38
Bkm	99-104	0.12	13	16	71	6	51	8.5	1.60

Continental (TA3, TA4, TS3, TS12, TS14, TS16). Typic Haplargids; fine, mixed, thermic (Reddish Brown). Continental soils are deep and well drained. They typically have a reddish brown gravelly sandy loam surface layer about 15 cm (6 in) thick. The subsoil is reddish brown gravelly sandy clay loam and clay about 63 cm (25 in) thick. The lower subsoil and substratum is calcareous light reddish brown gravelly and very gravelly sandy clay loam and sandy loam. Continental soils occur on the tops and sideslopes of fan terraces with slopes ranging from 2 to 8 percent. These soils have moderate available water capacity and slow permeability. They are slightly acid to mildly alkaline in the surface layers and moderately alkaline in the subsoil and substratum. They are strongly calcareous below about 76 cm (30 in). Runoff is slow and the hazard of erosion is generally slight. The series was established in Pima County in 1931 and the name was taken from a small town south of Tucson.

Some properties of the Continental soils

Horizon	Depth (cm)	O.M.%	Sand%	Silt%	Clay%	CEC meq/100g	CaCO₃%	pH	BD g/cm³
A	0-15	0.70	72	18	10	6	0	6.7	1.46
Bt	15-78	0.60	53	11	36	25	0	7.4	1.48
2BCk	78-183	0.24	66	13	21	35	8	8.4	1.36

Coolidge (HA5, HA8, HA9, HA10). Typic Calciorthids; coarse-loamy, mixed, thermic (Calcisols). Coolidge soils are deep and well drained. They typically have a light yellowish brown gravelly sandy loam surface layer 33 cm (13 in) thick underlain by yellowish brown sandy loam about 28 cm (11 in) thick. The substratum to more than 150 cm (60 in) is pale brown sandy loam containing many soft and hard lime masses. Coolidge soils occur on nearly level to gently undulating fan terraces with slopes ranging from 0 to 5 percent. These soils have moderate available water capacity and moderately rapid permeability. They are moderately alkaline and calcareous throughout. Runoff is slow or medium and the hazard of erosion is slight to moderate. The series was established in Pinal County in 1936 and the name was taken from the town of Coolidge.

Cordes (MS6, MS8, MS9, MH2, MH6). Cumulic Haplustolls; coarse-loamy, mixed, mesic (Alluvial). Cordes soils are deep and well drained. Typically, they have a grayish brown sandy loam surface layer about 8 cm (3 in) thick. The upper 78 cm (31 in) of underlying material is brown sandy loam. Below this to a depth of 150 cm (60 in) is light brown loamy sand. Cordes soils occur on floodplains and low alluvial fans with slopes of 1 to 5 percent. These soils have moderate available water capacity and moderately rapid to rapid permeability. They are mildly to moderately alkaline and noncalcareous throughout the profile. Runoff is medium and the hazard of erosion is slight. The series was established in Yavapai County in 1971.

Cornville (TS16). Typic Haplargids; fine-loamy, mixed, thermic (Reddish Brown). Cornville soils are deep and well drained. They typically have a brown fine sandy loam surface layer about 8 cm (3 in) thick. The subsoil is yellowish red and reddish brown sandy clay loam about 69 cm (27 in) thick. The substratum to 76 to 150 cm (30 to 60 in) is pinkish white gravelly loam high in calcium carbonate. Cornville soils occur on nearly level to undulating fan terraces with slopes ranging from 0 to 10 percent. These soils have high available water capacity and moderately slow permeability. They are moderately alkaline throughout, are noncalcareous in the upper part, and are strongly calcareous below about 76 cm (30 in). Runoff is medium to slow and the hazard of erosion is slight. The series was established in Yavapai County in 1965 in the Beaver Creek area.

Some properties of the Cornville soils

Horizon	Depth (cm)	O.M.%	Sand%	Silt%	Clay%	CEC meq/100g	CaCO$_3$%	pH	BD g/cm^3
Ap	0-8	2.57	51	33	16	15	8	7.9	
AB	8-42	0.90	54	30	16	14	7	7.9	
Bt	42-104	0.60	38	39	23	18	9	7.9	
Bk	104-152	0.29	30	41	29	22	5	8.2	

Courthouse (TS6). Lithic Torriorthents; loamy, mixed (calcareous), thermic (Reddish Brown). Courthouse soils are shallow and very shallow and well drained. Typically, they have a red stony fine sandy loam surface layer about 8 cm (3 in) thick. The underlying material is light red gravelly fine sandy loam about 15 cm (6 in) thick. Below this is red highly weathered sandstone with fine sandy loam in cracks and seams about 13 cm (5 in) thick. Red extremely hard calcareous sandstone is at a depth of 36 cm (14 in). Courthouse soils occur on hillslopes with slopes of 3 to 35 percent. These soils have low available water capacity and moderate permeability. They are moderately alkaline and calcareous throughout the profile. Runoff is rapid and the hazard of erosion is slight. The series was established in the Beaver Creek area of Yavapai County in 1965.

Cristobol (HA4, HA6, HA7, HA9). Typic Haplargids; loamy-skeletal, mixed, hyperthermic (Solonchak). Cristobol soils are deep and well drained. Typically, the surface is covered with 90 to 95 percent pebbles with desert varnish on top. The surface layer is pale brown extremely gravelly loam about 5 cm (2 in) thick. The upper 20 cm (8 in) of subsoil is red and yellowish red very gravelly clay loam with few fine and medium soft lime masses. The next 38 cm (15 in) of subsoil is yellowish red extremely gravelly clay loam and very gravelly sandy clay loam with common fine and medium soft lime masses and common very fine and fine salt crystals. The lower 25 cm (10 in) of subsoil is reddish yellow very gravelly clay loam with many fine and medium soft lime masses and common very fine salt crystals. The substratum to a depth of 150 cm (60 in) is light brown very gravelly clay loam. Cristobol soils occur on fan terraces with slopes of 0 to 20 percent. These soils have very low available water capacity and moderately slow permeability. They are moderately alkaline, strongly saline and calcareous throughout the profile. Electrical conductivity ranges from 35 to 50 mmhos in the control section. Runoff is rapid and the hazard of erosion is slight. The series was established in Yuma County in 1977.

Cross (MS4, MS7, MH2). Lithic Argiustolls; clayey, montmorillonitic, mesic (Reddish Chestnut). Cross soils are shallow and very shallow and well drained. Typically, they have a dark grayish brown gravelly clay loam surface layer about 8 cm (3 in) thick. The subsoil is dark grayish brown clay about 28 cm (11 in) thick. The underlying material is weakly to strongly cemented basalt gravel and cobble about 13 cm (5 in) thick with a thin 1 cm (.25 in) thick laminar cap in places. Black and very dark gray basalt is at a depth of 48 cm (19 in). Depth to bedrock ranges from 20 to 50 cm (8 to 20 in). Cross soils occur on nearly level to rolling basalt hillslopes with slopes of 2 to 15 percent. These soils have moderate available water capacity and moderately slow to slow permeability. They are moderately alkaline and calcareous in the lower part of the profile above the bedrock. Runoff is medium and the hazard of erosion is slight. The series was established in Yavapai County in 1964 and the name was taken from Cross Mountain.

Crot (TS11). Aquic Natrustalfs; fine-loamy, mixed, thermic (Solonetz). Crot soils are deep and somewhat poorly drained. Typically, they have a light brownish gray sandy loam surface layer about 13 cm (5 in) thick. The subsoil is brown and mottled light gray and yellowish brown sandy clay loam about 38 cm (15 in) thick. The underlying material to a depth of 150 cm (60 in) is light olive gray, olive, and pale yellow, finely stratified sand, fine sandy loam and silt loam. The profile is strongly to very strongly alkaline (pH 9.0 to 10.0). Crot soils occur on alkali flats bordering playas with slopes of 0 to 1 percent. Available water capacity is moderate. Permeability is very slow. In places a seasonal high water table occurs at a depth of 1 to 2 m (3 to 7 ft). The soils are calcareous throughout and may be saline. Runoff is very slow and water stands in numerous small playas until it evaporates. The hazard of erosion is slight. The series was established in Cochise County in the Willcox area in 1971.

Some properties of the Crot soils

Horizon	Depth (cm)	O.M.%	Sand%	Silt%	Clay%	CEC meq/100g	CaCO$_3$%	pH	BD g/cm^3
E	0-13	0.51	73	22	5	6	T	7.8	1.73
Btkn	13-43	0.37	64	14	22	14	8	9.4	1.72
BCgn1	43-51	0.36	80	7	13	10	7	9.7	1.83
BCgn2	51-107	0.05	81	4	15	11	4	9.7	1.82

Cryorthents (FH7). Cryorthents are well drained and shallow to moderately deep. Typically, they have brown gravelly loam or clay loam surface layers and brown or reddish brown loam or clay loam subsurface layers over shale, or interbedded shale and sandstone bedrock at depths of 25 to 75 cm (10 to 30 in). These soils have low available water capacity and moderately slow permeability. The Cryorthents are on the tops and sides of high plateaus and mesas dissected by deep canyons. Slopes dominantly range from 15 to 50 percent. Runoff is medium to rapid and the hazard of erosion is moderate to high.

Dandrea (FH1). Mollic Eutroboralfs; fine, montmorillonitic (Regosols). Dandrea soils are moderately deep and well drained. Typically, they have a thin pine needle and oak leaf litter on the soil surface. The surface mineral layers are a dark grayish brown gravelly heavy loam about 18 cm (7 in) thick. The subsoil is reddish yellow gravelly clay loam and brown gravelly clay about 56 cm (22 in) thick. The bedrock, from 74 to more than 125 cm (29 to 50 in), is brown and reddish yellow highly weathered and fractured schist. Depth to the weathered schist ranges from 50 to 100 cm (20 to 40 in). Dandrea soils occur on hillslopes having 15 to 50 percent slopes. These soils have moderate available water capacity, slow permeability and high shrink-swell potential. Reaction is slightly acid or neutral throughout. Runoff is rapid and the hazard of erosion is high. The series was established in Yavapai County in 1971.

Deloro (TS6). Aridic Haplustalfs; clayey-skeletal, mixed, thermic, shallow (Lithosols). Deloro soils are shallow and well drained. Typically, they have a brown very shaly loam surface layer about 5 cm (2 in) thick. The subsoil is reddish brown extremely shaly heavy clay loam about 23 cm (9 in) thick. The underlying material is variegated red, reddish yellow and weak red highly fractured and weathered phyllite with reddish brown clay loam in fractures. Deloro soils occur on hillslopes with slopes of 8 to 50 percent. The soils have low available water capacity and moderately slow to slow permeability. They are neutral to mildly alkaline and noncalcareous throughout the profile. Runoff is rapid and the hazard of erosion is slight. The series was established in Pinal County in 1977.

Disterheff (MH3). Aridic Paleustalfs; fine, montmorillonitic, mesic (Reddish Chestnut). Disterheff soils are deep and well drained. Typically, they have brown cobbly light sandy clay loam surface layers about 8 cm (3 in) thick and reddish brown calcareous clay loam and clay subsoils about 50 cm (20 in) thick. The underlying material is calcareous pink, cobbly clay loam to 150 cm (60 in). Disterheff soils occur on fan terraces and toeslopes of mesas and plateaus with slopes ranging from 0 to 45 percent. These soils have high available water capacity, slow permeability and high shrink-swell potential. They are mildly to moderately alkaline and are strongly calcareous in the lower subsoil and substratum. Runoff is medium to rapid and the hazard of erosion is slight. The series was established in Navajo County in 1964.

Some properties of the Disterheff soils

Horizon	Depth (cm)	O.M.%	Sand%	Silt%	Clay%	CEC meq/100g	CaCO$_3$%	pH	BD g/cm^3
A	0-8	4.59	64	27	9	20	0	6.8	
BAt	8-18	1.39	54	26	20		0	6.3	
Bt	18-65	0.97	33	17	50		0	6.1	
Bk	65-143	0.36	41	25	34		10	7.4	

Dona Ana (TS9, TS12, TS13, TS14). Typic Haplargids; fine-loamy, mixed, thermic (Red Desert). Dona Ana soils are deep and well drained. Typically, they have a reddish brown fine sandy loam or sandy loam surface layer about 10 cm (4 in) thick. The subsoil is reddish brown calcareous sandy clay loam about 35 cm (14 in) thick. The underlying material to a depth of 150 cm (60 in) is pinkish white and light reddish brown light sandy clay loam with many carbonate nodules and cylindroids. Depth to the calcic horizon ranges from 30 to 75 cm (12 to 30 in). Dona Ana soils occur on nearly level to gently undulating fan terraces with slopes of 0 to 5 percent. These soils have high available water capacity and moderate permeability. They are mildly to moderately alkaline and calcareous throughout the profile. Runoff is medium and the hazard of erosion is moderate. The series was established in Dona Ana County, New Mexico, in 1942 and the name was taken from the county in which it was established.

Some properties of the Dona Ana soils

Horizon	Depth (cm)	O.M.%	Sand%	Silt%	Clay%	CEC meq/100g	CaCO$_3$%	pH	BD g/cm^3
A	0-4	0.61	65	25	10	8	4	7.6	1.49
Bt	4-46	0.61	57	27	16	9	9	7.7	1.42
Bk	46-69		46	33	21	11	18	7.6	1.40
Btkb	69-100		51	31	18	9	16	7.8	1.48
2Bkb	100-152		59	26	15	8	12	8.5	1.54

Dry Lake (TS11). Typic Calciorthids; sandy over loamy, mixed, thermic (Alluvial). Dry Lake soils are deep and well drained. Typically, they have a light brown loamy fine sand surface layer about 18 cm (7 in) thick. The upper 70 cm (28 in) of underlying material is light brown loamy fine sand. Below this to a depth of 150 cm (60 in) is white heavy silt loam. Dry Lake soils occur on nearly level floodplains and alluvial fans adjacent to old playas with slopes of 0 to 2 percent. These soils have medium available water capacity and rapid permeability in the upper part but moderately slow permeability in the lower part of the profile. They are moderate to strongly alkaline and noncalcareous in the surface layer and calcareous in the lower underlying material. Runoff is slow and the hazard of erosion is high. The series was established in Clark County, Nevada, in 1964.

Some properties of the Dry Lake soils

Horizon	Depth (cm)	O.M.%	Sand%	Silt%	Clay%	CEC meq/100g	CaCO$_3$%	pH	BD g/cm^3
Ap	0-23	0.85	89	6	5	6	1	7.6	1.48
A	23-61	0.68	86	7	7	6	2	7.7	1.64
Bk	61-150	0.46	47	26	27	6	36	7.8	1.44

Duncan (TS11). Typic Nadurargids; fine, mixed, thermic (Solonetz). Duncan soils are moderately deep and well drained. Typically, they have a light gray loam surface layer about 13 cm (5 in) thick. The subsoil is brown, light brown and pinkish gray clay about 75 cm (30 in) thick over a very pale brown indurated duripan about 13 cm (5 in) thick. Below this to a depth of 150 cm (60 in) is a buried layer of mottled reddish brown gravelly clay loam. Depth to the duripan ranges between 50 and 100 cm (20 to 40 in). Duncan soils occur on level to nearly level alkali flats bordering playas with slopes of 0 to 2 percent. These soils have high available water capacity and very slow permeability. They are strongly to very strongly alkaline and calcareous throughout the profile with filaments and segregations of calcium carbonate in the subsoil. Runoff is slow and the hazard of erosion is moderate. The series was established in the Sulphur Springs Valley, Cochise County, in 1942 and the name was taken from the town of Duncan.

Some properties of the Duncan soils

Horizon	Depth (cm)	O.M.%	Sand%	Silt%	Clay%	CEC meq/100g	CaCO$_3$%	pH	BD g/cm^3
E	0-13	0.81	49	42	9	10	T	8.2	1.69
Bt	13-28	0.95	23	36	41	33	12	8.4	1.59
Btn	28-79	0.53	22	33	45	36	18	9.6	1.59
Bk	79-89		16	33	51	32	44	9.9	
Bqknm	89-100		71	16	13	11	80	9.9	2.20
Btkb	100-115		27	31	42	31	28	9.0	
Btb	115-150		31	28	41	34	1	8.7	1.52

Dye (MS1, MH4). Lithic Udic Haplustalfs; clayey, mixed, mesic (Lithosols). Dye soils are shallow and well drained. Typically, they have a brown fine sandy loam surface layer about 10 cm (4 in) thick. The upper 15 cm (6 in) of the subsoil is reddish brown sandy clay loam. The lower 23 cm (9 in) of the subsoil is yellowish red clay. Below this at a depth of 48 cm (19 in) is pale brown to light brown sandstone. Depth to bedrock ranges from 25 to 50 cm (10 to 20 in). Dye soils occur on hillslopes with slopes of 5 to 30 percent. These soils have low available water capacity and slow permeability. They are neutral to mildly alkaline and noncalcareous throughout the profile. Runoff is medium to rapid and the hazard of erosion is moderate. The series was established in Coconino County in 1967 and the name was taken from a prominent old ranch.

Eba (TS3, TS12). Typic Haplargids; clayey-skeletal, mixed, thermic (Red Desert). Eba soils are deep and well drained. Typically, they have a reddish brown very gravelly loam and clay loam surface layer about 15 cm (6 in) thick. The subsoil is dark red and red very gravelly clay about 58 cm (23 in) thick. Below this to a depth of 125 cm (50 in) is pink very gravelly light clay loam. Eba soils occur on fan terraces with slopes of 2 to 20 percent. These soils have moderate available water capacity and slow permeability. They are neutral to moderately alkaline and noncalcareous in the upper part of the profile. The zone of calcium carbonate accumulation is in the lower subsoil and underlying material and contains more than 15 percent $CaCO_3$. Runoff is medium and the hazard of erosion is slight. The series was established in Santa Cruz County in 1971.

Ebon (HA3). Typic Haplargids; clayey-skeletal, mixed, hyperthermic (Red Desert). Ebon soils are deep and well drained. Typically, they have a surface layer about 5 cm (2 in) thick. The upper part of the subsoil is reddish brown very cobbly clay loam about 28 cm (11 in) thick. The lower part of the subsoil is yellowish red and reddish brown very cobbly clay about 38 cm (15 in) thick. The substratum to a depth of 130 cm (52 in) is light reddish brown very cobbly sandy clay loam. Ebon soils occur on fan terraces with slopes of 2 to 20 percent. These soils have moderate available water capacity and slow permeability. They are moderately alkaline and calcareous in the lower subsoil and substratum. Runoff is medium and the hazard of erosion is slight. The series was established in Maricopa County in 1971.

Elfrida (TS11, TS13). Pachic Calciustolls; fine-loamy, mixed, thermic (Reddish Brown). Elfrida soils are deep and well drained. Typically, they have a grayish brown silty clay loam surface layer about 55 cm (22 in) thick. The underlying material to a depth of 125 cm (50 in) is pinkish gray and pinkish white clay loam. Elfrida soils occur on nearly level alluvial fans and margins of old lakes with slopes of 0 to 2 percent. These soils have high available water capacity and moderately slow permeability. They are moderately alkaline and calcareous throughout the profile. Depth to the calcic horizon ranges from 40 to 100 cm (16 to 40 in). Runoff is slow and the hazard of erosion is moderate. The series was established in the Sulpher Springs Valley, Cochise County, in 1942 and the name was taken from the town of Elfrida.

Some properties of the Elfrida soils

Horizon	Depth (cm)	O.M.%	Sand%	Silt%	Clay%	CEC meq/100g	CaCO$_3$%	pH	BD g/cm³
Ap	0-56	1.34	16	54	30	22	6	7.5	1.50
Bk	56-125	0.61	33	40	27	13	32	7.8	1.49

Elledge (MH3, MH5). Typic Hapludalfs; fine, mixed mesic (Planosols). Elledge soils are moderately deep and moderately well to well drained. Typically, they have a thin pine-oak litter on the surface. The surface mineral layer is grayish brown cobbly sandy loam about 5 cm (2 in) thick. The subsurface layer is light gray sandy loam 13 cm (5 in) thick. The subsoil is reddish brown and brownish yellow cobbly clay about 58 cm (23 in) thick. Bedrock at about 76 cm (30 in) is light yellowish brown sandstone with widely spaced fractures. Depth to bedrock ranges from 50 to 100 cm (20 to 40 in). Elledge soils occur on hillslopes with slopes ranging from 2 to 20 percent. These soils have moderate available water capacity and slow permeability. They are slightly acid to neutral throughout. Runoff is slow to medium and the hazard of erosion is slight. The series was established in Navajo County in 1962 in the Holbrook-Showlow area.

Some properties of the Elledge soils

Horizon	Depth (cm)	O.M.%	Sand%	Silt%	Clay%	CEC meq/100g	B.S.%	pH	BD g/cm³
A	0-3	5.27	76	16	8	17	74	6.3	
E	3-8	1.92	69	22	9	13	64	5.5	
Bt	8-53	1.26	28	18	54	44	88	5.7	
R	53-								

Ess (FH2). Argic Cryoborolls; loamy-skeletal, mixed (Brown Forest). Ess soils are deep and well drained. Typically, they have a thin pine needle and oak leaf litter on the soil surface. The surface mineral layer is dark brown cobbly loam and silt loam about 28 cm (11 in) thick. The subsoil is brown very cobbly heavy silt loam and light clay loam to a depth of more than 150 cm (60 in). Ess soils are on gently rolling to moderately steep hillslopes with slopes ranging from 2 to 20 percent. These soils have moderate available water capacity and moderate or moderately slow permeability. Reaction ranges from medium acid to mildly alkaline throughout the profile. Runoff is slow or medium and the hazard of erosion is slight. The series was established in the Paunsaugunt area in Utah in 1969.

Estrella (HA1). Typic Torrifluvents; fine-loamy, mixed (calcareous), hyperthermic (Alluvial). Estrella soils are deep and well drained. Typically, they have a brown loam surface layer 28 cm (11 in) thick. The underlying material to a depth of 60 cm (24 in) is light brown loam. Below this to a depth of 150 cm (60 in) is a buried subsoil which is brown and yellowish red clay loam and gravelly clay loam. Depth to the buried argillic horizon ranges from 50 to 90 cm (20 to 36 in). Estrella soils occur on broad alluvial fans and floodplains with slopes of 1 to 5 percent. These soils have high available water capacity and moderate permeability in the upper 60 cm (24 in) and moderately slow permeability in the lower 90 cm (36 in). They are moderately alkaline and calcareous throughout the profile. Runoff is slow and the hazard of erosion is moderate. The series was established in Maricopa County in 1969.

Eutroboralfs (FH6, FH7). Eutroboralfs are deep or moderately deep to very gravelly volcanic conglomerate materials. Typically, they have a 2.5 cm (1 in) pine needle and leaf litter on the soil surface. The surface mineral layers are grayish brown and light brownish gray gravelly loam about 25 cm (10 in) thick. The subsoil is brown and reddish brown gravelly clay about 50 cm (20 in) thick. The substratum is brown and reddish brown very gravelly clay loam or loam. Eutroboralfs occur on hillslopes with slopes ranging from 10 to 60 percent. These soils have moderate available water capacity and slow permeability. They are neutral to medium acid throughout. Runoff is medium and the hazard of erosion is slight to moderate.

Faraway (MS6, MH2, MH6). Lithic Haplustolls; loamy-skeletal, mixed, mesic (Lithosols). Faraway soils are shallow or very shallow and well drained. Typically, the soil profile is a grayish brown very cobbly loam about 20 cm (8 in) thick over light gray rhyolite bedrock having widely spaced fractures. Faraway soils occur on hillslopes with slopes ranging from 10 to 80 percent. They have very low available water capacity and moderate permeability above the bedrock. They are medium acid to mildly alkaline and noncalcareous throughout the profile. Runoff is rapid and hazard of erosion is slight. The series was established in 1971 in Santa Cruz County.

Some properties of the Faraway soils

Horizon	Depth (cm)	O.M.%	Sand%	Silt%	Clay%	CEC meq/100g	B.S.%	pH	BD g/cm^3
A	0-20	1.29	67	25	8	10	70	5.3	1.40
C	20-33	0.68	69	21	10	11	77	5.1	1.84
R	33-								

Forrest (TS3, TS14). Ustollic Haplargids; fine, mixed, thermic (Reddish Brown). Forrest soils are deep and well drained. Typically, they have a brown loam surface layer about 10 cm (4 in) thick. The subsoil is reddish brown and red clay about 63 cm (25 in) thick. The underlying material to a depth of 150 cm (60 in) is reddish yellow and light brown clay loam and loam. Forrest soils occur on fan terraces with slopes of 0 to 5 percent. These soils have high available water capacity and slow permeability. They are slightly acid to moderately alkaline and calcareous in the lower subsoil and underlying material. Depth to the calcic horizon ranges from 50 to 100 cm (20 to 40 in). Runoff is slow to medium and the hazard of erosion is slight. The series was established in Sulphur Springs Valley, Cochise County, in 1948.

Some properties of the Forrest soils

Horizon	Depth (cm)	O.M.%	Sand%	Silt%	Clay%	CEC meq/100g	CaCO$_3$%	pH	BD g/cm^3
A	0-10	1.10	49	34	17	11		6.3	1.71
Bt	10-79	0.80	35	19	46	25	1	7.3	1.81
Ck	79-180	0.46	50	31	19	9	35	7.9	1.88

Fruitland (MA1, MA3, MA6). Typic Torriorthents; coarse-loamy, mixed (calcareous), mesic (Alluvial). Fruitland soils are deep and well drained. Typically, they have pale brown sandy loam surface layers about 18 cm (7 in) thick. The underlying material to more than 150 cm (60 in) is pale brown and light yellowish brown sandy loam. Fruitland soils occur on plateaus and young fan terraces with slopes of 2 to 15 percent. These soils have moderate available water capacity and moderately rapid permeability. They are moderately alkaline and are calcareous throughout the profile. Runoff is slow. The hazard of water erosion is slight and the hazard of wind erosion is moderate. The series was established in San Juan County, New Mexico, in 1965.

Some properties of the Fruitland soils

Horizon	Depth (cm)	O.M.%	Sand%	Silt%	Clay%	CEC meq/100g	CaCO$_3$%	pH	BD g/cm^3
A	0-23	1.09	47	33	18	32	4	7.4	1.35
C	23-75	0.44	51	32	17	43	3	7.7	1.28
Ab	75-98	1.17	62	23	15	23	4	7.7	1.41
Cb	98-163	0.22	60	26	14	23	5	8.1	1.40

Gachado (HA6). Lithic Haplargids; loamy-skeletal, mixed, hyperthermic (Lithosols). Gachado soils are very shallow and shallow and well drained. Typically, they have brown very cobbly loam surface layers, yellowish red gravelly sandy clay loam and heavy clay loam subsoils, and dark gray igneous bedrock at 23 to 50 cm (9 to 20 in). Gachado soils are on low hills and toeslopes of hills and mountains with slopes of 0 to 40 percent. These soils have very low available water capacity and slow permeability. They are moderately alkaline throughout and are calcareous in the lower subsoil. Surface runoff is medium and the hazard of erosion is slight. The series was established in Pima County on the Organ Pipe Cactus National Monument in 1971.

Some properties of the Gachado soils

Horizon	Depth (cm)	O.M.%	Sand%	Silt%	Clay%	CEC meq/100g	CaCO$_3$%	pH	BD g/cm^3
A	0-5	0.36	54	31	15	18	T	7.6	1.30
Bt	5-21	0.34	40	26	34	32	T	7.6	1.42
BCtk	21-33	0.41	44	22	34	28	2	8.3	1.39
R	33-								

Gaddes (MS6, MH2). Ustollic Haplargids; fine-loamy, mixed, mesic (Regosols). Gaddes soils are moderately deep and well drained. Typically, they have a brown gravelly sandy loam surface layer about 5 cm (2 in) thick. The subsoil is reddish brown gravelly sandy clay loam and clay loam about 55 cm (22 in) thick. The underlying material to a depth of 135 cm (54 in) is reddish brown highly weathered granite (grus). Gaddes soils occur on undulating to steep granite hillslopes with slopes of 3 to 35 percent. These soils have moderate available water capacity and moderately slow permeability. They are slightly acid to mildly alkaline and noncalcareous throughout the profile. Runoff is medium and the hazard of erosion is slight. The series was established in Santa Cruz County in 1971.

Some properties of the Gaddes soils

Horizon	Depth (cm)	O.M.%	Sand%	Silt%	Clay%	CEC meq/100g	B.S.%	pH	BD g/cm^3
A	0-5	1.48	67	25	8	10	100	6.8	
Bt	5-60	0.61	46	22	32	45	96	6.8	1.58
Cr	60-	0.19	73	16	11	20	89	7.4	1.70

Gadsden (HA1). Vertic Haplustolls; fine, montmorillonitic, hyperthermic (Alluvial). Gadsden soils are deep and well drained. Typically, they have a brown clay surface layer about 73 cm (29 in) thick. The underlying material to a depth of 150 cm (60 in) or more is brown clay and clay loam. Gadsden soils occur on level to slightly concave low stream terraces and floodplains with slopes of 0 to 1 percent. These soils have high available water capacity and slow permeability. They are moderately alkaline and calcareous throughout the profile. Runoff is slow and the hazard of erosion is moderate. The series was established in the Yuma Valley area in 1957.

Gila (TA2, TA4, TS2, TS12). Typic Torrifluvents; coarse-loamy, mixed (calcareous), thermic (Alluvial). Gila soils are deep and well drained. Typically, they have a grayish brown loam surface layer about 15 cm (6 in) thick. The underlying material to a depth of 165 cm (66 in) is brown loam, silt loam and gravelly sandy loam. Gila soils occur on nearly level to gently sloping floodplains and low alluvial fans with slopes of 0 to 2 percent. These soils have high available water capacity and moderate permeability. They are moderately alkaline and calcareous throughout the profile. Runoff is slow and the hazard of erosion is moderate. The series was established in the Salt River Valley in 1900 and the name was taken from Gila County.

Gilman (HA1, HA2, HA5, HA7). Typic Torrifluvents; coarse-loamy, mixed (calcareous), hyperthermic (Alluvial). Gilman soils are deep and well drained. They typically have pale brown loam and very fine sandy loam surface layers about 33 cm (13 in) thick, and pale brown, stratified loam, very fine sandy loam, loam or silt loam subsurface layers to 150 cm (60 in) or more. Gilman soils occur on nearly level to gently sloping floodplains and low alluvial fans with slopes ranging from 0 to 3 percent. These soils have high available water capacity, moderate permeability and low shrink-swell potential. Reaction is moderately alkaline and they are moderately calcareous throughout. Some areas are saline. Runoff is slow and the hazard of erosion is moderate. The series was established on the Gila River Project in 1936.

Some properties of the Gilman soils

Horizon	Depth (cm)	O.M.%	Sand%	Silt%	Clay%	CEC meq/100g	CaCO$_3$%	pH	BD g/cm^3
Ap	0-38	0.36	68	22	10	17	1	7.8	1.70
C	38-99		50	37	13	17	3	8.1	1.48

Glenbar (HA1). Typic Torrifluvents; fine-silty, mixed (calcareous), hyperthermic (Alluvial). Glenbar soils are deep and well drained. Typically, they have a brown clay loam surface layer about 38 cm (15 in) thick. The underlying material to a depth of 150 cm (60 in) is light brown and pale brown clay loam and silty clay loam. Glenbar soils occur on floodplains and low stream terraces with slopes of 0 to 1 percent. These soils have high available water capacity and moderately slow to slow permeability. They are moderately alkaline and calcareous throughout the profile. Runoff is slow to medium and the hazard of erosion is moderate. The series was established on the Gila River Project in 1937.

Glendale (TA2, TS2, TS9, TS13). Typic Torrifluvents; fine-silty, mixed (calcareous), thermic (Alluvial). Glendale soils are deep and well drained. Typically, they have a light brownish gray loam surface layer about 20 cm (8 in) thick. The underlying material to a depth of 105 cm (42 in) is grayish brown clay loam and silty clay loam. Glendale soils occur on floodplains and low stream terraces with slopes of 0 to 2 percent. These soils have high available water capacity and moderately slow permeability. They are moderately alkaline and calcareous throughout the profile. Runoff is slow to medium and the hazard of erosion is moderate. The series was established in Maricopa County in 1946.

Gordo (FH1, FH8). Argic Pachic Cryoborolls; fine-loamy, mixed (Brown Forest). Gordo soils are moderately deep and deep and well drained. Typically, they have a thin litter of pine needles and aspen leaves on the soil surface. The surface mineral layer is dark brown silt loam about 50 cm (20 in) thick. The subsoil is reddish brown gravelly heavy loam about 50 cm (20 in) thick. This is underlain to more than 150 cm (60 in) by reddish brown highly weathered basalt gravel, cobble and cinders. Depth to weathered parent material ranges from about 66 to 130 cm (26 to 52 in). Gordo soils are on hillslopes with slopes of 5 to 25 percent. These soils have moderate available water capacity and moderate permeability. These soils are medium acid to neutral and noncalcareous. Runoff is slow to medium and the hazard of erosion is moderate. The series was established in Apache County in 1965.

Gothard (TS11, TS13). Typic Natrargids; fine-loamy, mixed, thermic (Solonetz). Gothard soils are deep and moderately well drained. Typically, they have a light brownish gray and light gray fine sandy loam surface layer about 13 cm (5 in) thick. The subsoil is grayish brown, pale brown and pink clay loam and loam about 94 cm (37 in) thick. The underlying material is light gray and olive brown, stratified, sandy loam, loamy sand and sandy clay loam to a depth of 150 cm (60 in) or more. Gothard soils occur on low fan terraces bordering playas with slopes ranging from 0 to 5 percent. These soils have high available water capacity but water is restricted to plants by high sodium content. Permeability is very slow. The soils are strongly to very strongly alkaline (pH 9.0 to 10.0) and may be saline. They are usually calcareous throughout. Runoff is slow and the hazard of erosion is slight to moderate. The series was established in the Sulphur Springs Valley, Cochise County, in 1942.

Some properties of the Gothard soils

Horizon	Depth (cm)	O.M.%	Sand%	Silt%	Clay%	CEC meq/100g	CaCO₃%	pH	BD g/cm³
A	0-13	0.56	61	30	9	9.0	2	9.0	1.63
Bt	13-28	0.65	42	29	29	20.8	6	9.0	1.56
Bk	28-61	0.22	39	37	24	20.3	15	9.9	1.72
2Btkb	61-107	0.12	42	30	28	13.0	24	10.0	1.93
2Ckb	107-130	0.05	70	14	16	14.7	2	9.9	1.74

Grabe (TS2, TS3, TS13, TS14, TS17, MH1). Typic Torrifluvents; coarse-loamy, mixed (calcareous), thermic (Alluvial). Grabe soils are deep and well drained. Typically, they have grayish brown loam surface layers about 40 cm (16 in) thick over grayish brown stratified loam and very fine sandy loam substrata to more than 150 cm (60 in). Grabe soils occur on broad floodplains and alluvial fans with slopes of 0 to 2 percent. These soils have high available water capacity and moderate permeability. They are moderately alkaline and contain small amounts of lime throughout the profile. Runoff is slow and the hazard of erosion is slight. The series was established in Santa Cruz County in 1971.

Some properties of the Grabe soils

Horizon	Depth (cm)	O.M.%	Sand%	Silt%	Clay%	CEC meq/100g	CaCO₃%	pH	BD g/cm³
A	0-100	0.97	43	40	17	19	3	7.6	1.48
C	100-150	0.26	73	19	8	8	3	7.8	1.63

Graham (TS6, TS15, TS17, TS18). Lithic Arguistolls; clayey, montmorillonitic, thermic (Reddish Chestnut). Graham soils are shallow and well drained. Typically, they have a reddish brown cobbly clay loam surface layer about 2.5 cm (1 in) thick. The subsoil is dark reddish brown gravelly clay loam and clay about 33 cm (13 in) thick. This is underlain by basalt bedrock that has a few fractures. Graham soils occur on hillslopes with slopes ranging from 5 to 50 percent. These soils have low available water capacity and slow permeability. They are slightly acid to moderately alkaline and may be slightly calcareous in the lower part. Runoff is medium and the hazard of erosion is slight. The series was established in Graham County in 1936 and the name was taken from the county.

Some properties of the Graham soils

Horizon	Depth (cm)	O.M.%	Sand%	Silt%	Clay%	CEC meq/100g	CaCO₃%	pH	BD g/cm³
A	0-3	4.10	27	39	34		0	6.4	
Bt	3-35	2.33	19	31	50		2	7.5	
R	35-								

Guest (TS2, TS3, TS4, TS11, TS14, TS15, TS17, TS18). Vertic Torrifluvents; fine, mixed (calcareous), thermic (Alluvial). Guest soils are deep and well drained. Typically, they have brown and grayish brown clay surface and subsurface layers from 0 to 150 cm (60 in) or more. Guest soils occur on floodplains and alluvial fans with slopes of 0 to 2 percent. These soils have high available water capacity and slow permeability. They are moderately alkaline and calcareous throughout the profile. Runoff is slow to medium and the hazard of erosion is moderate. The series was established in Yavapai County in 1965.

Some properties of the Guest soils

Horizon	Depth (cm)	O.M.%	Sand%	Silt%	Clay%	CEC meq/100g	CaCO₃%	pH	BD g/cm³
Ap	0-20	1.09	26	34	40	28	1	7.8	1.67
A	20-100	0.61	21	35	44	27	2	7.8	1.83
AC	100-120	0.39	23	33	44	26	5	7.7	1.75
Btkb	120-150		38	20	42	24	19	7.6	1.79

Gunsight (HA4, HA6, HA9). Typic Calciorthids; loamy-skeletal, mixed, hyperthermic (Calcisols). Gunsight soils are deep and well drained. Typically, they have a light brown very gravelly loam or sandy loam surface layer about 5 cm (2 in) thick. The subsoil is pink very gravelly loam about 20 cm (8 in) thick. The underlying material is white and pinkish gray very gravelly loam or sandy loam to more than 150 cm (60 in). Gunsight soils occur on old fan terraces with slopes of 0 to 15 percent. These soils have low available water capacity, moderately rapid permeability and are moderately alkaline and strongly calcareous throughout. The profile generally contains more than 50 percent gravel and cobble and the substratum may be intermittently cemented by calcium carbonate. Runoff is medium and the hazard of erosion is slight. The series was established in Pima County on the Organ Pipe Cactus National Monument in 1971.

Some properties of the Gunsight soils

Horizon	Depth (cm)	O.M.%	Sand%	Silt%	Clay%	CEC meq/100g	CaCO$_3$%	pH	BD g/cm^3
A	0-5	0.26	57	36	7	12	3	8.5	1.02
Bw	5-25	0.27	56	37	7	11	12	8.5	1.03
Ck	25-140		49	37	14	11	27	8.1	1.18

Hantz (TS13). Typic Torrifluvents; fine, mixed (calcareous), thermic (Alluvial). Hantz soils are deep and well drained. Typically, they have a light brownish gray silty clay surface layer about 8 cm (3 in) thick. The underlying material to a depth of 150 cm (60 in) is light brownish gray silty clay. Hantz soils occur on floodplains and alluvial fans with slopes of 0 to 4 percent. These soils have high available water capacity and very slow permeability. They are moderate to strongly alkaline and calcareous throughout the profile. Runoff is medium and the hazard of erosion is moderate. The series was established in Yavapai County in 1965.

Harqua (HA2, HA4, HA9, HA10). Typic Haplargids; fine-loamy, mixed, hyperthermic (Solonchak). Harqua soils are deep and well drained. Typically, they have a desert pavement of varnished gravel on the surface. The surface layer is light brown gravelly loam about 2.5 cm (1 in) thick. The subsoil is yellowish red gravelly clay loam about 34 cm (15 in) thick that is strongly saline. The underlying material to more than 150 cm (60 in) is pink, gravelly sandy loam, loam or clay loam. Harqua soils occur on fan terraces with slopes ranging from 0 to 5 percent. The soils have low available water capacity restricted by salt and slow permeability. The soils are moderately to strongly alkaline and have zones of high lime content at 30 to 60 cm (12 to 24 in). Runoff is slow to medium and the hazard of erosion is slight. The series was established in Pima County on the Organ Pipe Cactus National Monument in 1971.

Some properties of the Harqua soils

Horizon	Depth (cm)	O.M.%	Sand%	Silt%	Clay%	CEC meq/100g	CaCO$_3$%	pH	BD g/cm^3
E	0-8	0.10	39	44	17	22	6	9.7	1.55
Btn	8-30	0.19	41	32	27	21	2	8.4	1.58
2Btk	30-71		59	23	18	9	6	8.3	1.54
3Bk	71-114		46	37	17	10	12	8.7	1.23

Harrisburg (TA2). Typic Paleorthids; coarse-loamy, mixed, thermic (Red Desert). Harrisburg soils are moderately deep and well drained. Typically, they have a yellowish red light fine sandy loam surface layer about 5 cm (2 in) thick. The underlying material to a depth of 90 cm (36 in) is yellowish red fine sandy loam. Below this to a depth of 110 cm (44 in) is an indurated carbonate hardpan. Harrisburg soils occur on mesas with slopes of 1 to 10 percent. Available water capacity is low and permeability is moderately rapid to the hardpan. These soils are mildly alkaline and calcareous throughout the profile. Runoff is slow and the hazard of erosion is moderate. The series was established in Virgin River Valley in southwestern Utah in 1936.

Hathaway (TS4, TS5, TS7, TS8, TS14). Aridic Calciustolls; loamy-skeletal, mixed, thermic (Calcisols). Hathaway soils are deep and well drained. Typically, they have a dark grayish brown very gravelly loam surface layer about 20 cm (8 in) thick underlain by pale brown and pinkish gray, calcareous, very gravelly sandy loam and loamy sand to more than 150 cm (60 in). Hathaway soils occur on long, narrow fan terraces with slopes ranging from 10 to 60 percent. These soils have low available water capacity and moderately rapid permeability. They are moderately alkaline and calcareous throughout with zones of high lime below 18 to 40 cm (7 to 16 in). They have medium runoff and the hazard of erosion is medium. The series was established in Santa Cruz County in 1971.

Some properties of the Hathaway soils

Horizon	Depth (cm)	O.M.%	Sand%	Silt%	Clay%	CEC meq/100g	CaCO$_3$%	pH	BD g/cm^3
A	0-25	2.51	62	19	19	19	10	7.9	1.16
Bk	25-99	1.10	64	22	14	15	19	8.0	1.40
2Bk	99-117		61	27	12	17	13	8.2	1.72

Hogg (FH1, FH3). Mollic Eutroboralfs; fine, mixed, mesic (Western Brown Forest). Hogg soils are moderately deep and deep and well drained. Typically, they have a surface layer of grayish brown fine sandy loam about 8 cm (3 in) thick. The subsoil is brown and reddish brown clay and stony clay about 123 cm (49 in) thick over white limestone bedrock. Depth to the limestone bedrock ranges from about 100 to 150 cm (40 to 60 in). Hogg soils occur on gently undulating plateaus and mesas with slopes ranging from 2 to 10 percent. These soils have high available water capacity and slow permeability. Reaction is mildly to moderately alkaline. Runoff is slow to medium and hazard of erosion is moderate. The series was established in Coconino County in the Beaver Creek area in 1965.

Some properties of the Hogg soils

Horizon	Depth (cm)	O.M.%	Sand%	Silt%	Clay%	CEC meq/100g	CaCO$_3$%	pH	BD g/cm^3
A	0-10	3.91	39	33	28	29	0	6.4	1.70
Bt	10-75	1.28	31	15	54	41	13.0	6.9	1.76

Holtville (HA1). Typic Torrifluvents; clayey over loamy, montmorillonitic (calcareous), hyperthermic (Alluvial). Holtville soils are deep and well drained under natural condition, but perched water tables are common where the soil is irrigated. Typically, they have a light brown silty clay surface layer about 43 cm (17 in) thick. The upper 18 cm (7 in) of underlying material is light brown silty clay. Below this to a depth of 150 cm (60 in) is very pale brown stratified silt loam and loamy very fine sand. Holtville soils occur on floodplains and lakebeds with slopes of 0 to 1 percent. These soils have high available water capacity and slow permeability. They are moderately alkaline and calcareous throughout the profile. Runoff is slow and the hazard of erosion is slight. The series was established in Imperial County, California, in 1918.

House Mountain (TA3, TS6, TS18). Lithic Torriorthents; loamy, mixed, nonacid, thermic (Lithosols). House Mountain soils are shallow and well drained. They typically have a brown cobbly and stony loam surface layer about 5 cm (2 in) thick and brown cobbly and stony light clay loam subsoils about 25 cm (10 in) thick. Dark gray fractured basalt bedrock occurs below about 30 cm (12 in). House Mountain soils occur on hillslopes with slopes ranging from 2 to 50 percent. These soils have very low available water capacity and moderate permeability. They are mildly to moderately alkaline and are usually noncalcareous. Depth to bedrock is 10 to 50 cm (4 to 20 in). Runoff is medium and the hazard of erosion is slight to medium. The series was established in Yavapai County in 1965.

Hubert (MS3). Typic Calciustolls; loamy-skeletal, mixed, mesic (Brown). Hubert soils are deep and well drained. Typically, they have a brown loam and gravelly loam surface layer about 25 cm (10 in) thick. The subsoil is light brownish gray gravelly heavy loam about 13 cm (5 in) thick. The upper 83 cm (33 in) of underlying material is white gravelly loam. The lower portion of the underlying material to a depth of 263 cm (105 in) is pinkish white very gravelly clay loam. Hubert soils occur on nearly level to undulating fan terraces with slopes of 1 to 9 percent. These soils have high available water capacity and moderate permeability. They are moderately alkaline and calcareous throughout the profile. Percent of calcium carbonate in the underlying material (Ck horizons) range from 15 to 40 percent. Runoff is slow to medium and the hazard of erosion is slight. The series was established in Apache County in 1962 and the name was taken from an old homestead.

Indio (HA1). Typic Torrifluvents; coarse-silty, mixed (calcareous), hyperthermic (Alluvial). Indio soils are deep and well to moderately well drained. Typically, they have a light brownish gray very fine sandy loam surface layer about 25 cm (10 in) thick. The underlying material to a depth of 150 cm (60 in) is light brownish gray stratified very fine sandy loam and silt loam. Indio soils occur on lake beds, alluvial fans and floodplains with slopes of 0 to 2 percent. These soils have high available water capacity and moderate permeability. They are moderately alkaline and calcareous throughout the profile. Runoff is slow and the hazard of erosion is moderate. The series was established in Riverside County, California, in 1923 and the name was taken from the town of Indio.

APPENDICES

Ives (MA2, MA4, MA5, MA6, MS3). Typic Torrifluvents; coarse-loamy, mixed (calcareous), mesic (Alluvial). Ives soils are deep and well drained. They typically have brown fine sandy loam surface layers about 20 cm (8 in) thick over brown fine sandy loam thinly stratified with loamy fine sand, loam and silt loam to 150 cm (60 in) or more. Ives soils occur on floodplains and low alluvial fans with slopes of 0 to 3 percent. These soils have moderate available water capacity and moderately rapid permeability. They are moderately alkaline and calcareous throughout. Runoff is slow and the hazard of erosion is slight to moderate. The series was established in the Winslow area, Coconino County, in 1921.

Some properties of the Ives soils

Horizon	Depth (cm)	O.M.%	Sand%	Silt%	Clay%	CEC meq/100g	CaCO$_3$%	pH	BD g/cm^3
A	0-150	0.03	84	10	6	27	2	8.1	

Jacks (MS1, MS5, MS6, MH4, MH5). Udic Haplustalfs; fine, mixed, mesic (Brown). Jacks soils are moderately deep to deep and well drained. Typically, they have a thin layer of pine needles over a reddish brown fine sandy loam surface layer 8 cm (3 in) thick. The subsoil is yellowish red clay and stony clay 98 cm (39 in) thick over massive calcareous sandstone bedrock. Depth to bedrock ranges from 50 to 125 cm (20 to 50 in). Jacks soils occur on hillslopes with slopes ranging from 0 to 45 percent. These soils have moderate available water capacity and slow permeability. They are mildly to moderately alkaline and are usually noncalcareous above the bedrock. Runoff is slow to medium and the hazard of erosion is slight to moderate. The series was established in Coconino County in 1965 in the Beaver Creek area.

Jacques (MS7, MH3, MH4). Cumulic Haplustolls; fine, mixed, mesic (Alluvial). Jacques soils are deep and well drained. Typically, they have a dark grayish brown clay loam surface layer about 25 cm (10 in) thick. The underlying material to a depth of 150 cm (60 in) is dark brown and dark yellowish brown clay. Jacques soils occur on nearly level to gently sloping floodplains and alluvial fans with slopes of 0 to 3 percent. These soils have high available water capacity and slow permeability. They are neutral to mildly alkaline and noncalcareous throughout the profile. Runoff is slow and the hazard of erosion is slight. The series was established in Navajo County in 1962.

Some properties of the Jacques soils

Horizon	Depth (cm)	O.M.%	Sand%	Silt%	Clay%	CEC meq/100g	CaCO$_3$%	pH	BD g/cm^3
A	0-25	1.11	36	30	34	32	1	7.4	
C	25-110	0.95	38	23	39		2	7.5	

Jocity (MA1, MA2, MA4, MA5, MS3, MH5). Typic Torrifluvents; fine-loamy, mixed (calcareous), mesic (Alluvial). Jocity soils are deep and well drained. Typically, they have a reddish gray sandy clay loam surface layer about 23 cm (9 in) thick. The underlying material is reddish gray sandy clay loam about 80 cm (32 in) thick over gray fine sandy loam that extends to 150 cm (60 in). Jocity soils occur on nearly level to undulating alluvial fans and floodplains with slopes of 0 to 5 percent. These soils have high available water capacity and moderately slow permeability. They are moderately alkaline and calcareous throughout the profile. Runoff is medium and the hazard of erosion is moderate. The series was established in Navajo County in 1961 and the name is a coined word taken from Joseph City.

Some properties of the Jocity soils

Horizon	Depth (cm)	O.M.%	Sand%	Silt%	Clay%	CEC meq/100g	CaCO$_3$%	pH	BD g/cm^3
A	0-23	0.26	58	16	26	32	3	8.1	1.46
C	23-148	0.26	49	21	30	35	3	7.9	1.56

Karro (TS11, TS13). Ustollic Calciorthids; fine-loamy, carbonatic, thermic (Calcisols). Karro soils are deep and well drained. Typically, they have light brownish gray and pale brown loam surface layers about 38 cm (15 in) thick. The underlying material is pale brown and light gray calcareous loam and clay loam to 150 cm (60 in) or more and contains more than 40 percent calcium carbonate in the form of soft masses and hard nodules. Karro soils occur on nearly level fan terraces at the margins of lakebeds with slopes of 0 to 5 percent. These soils have high available water capacity and moderately slow permeability. They are moderately to very strongly alkaline and calcareous throughout. Runoff is slow and the hazard of erosion is slight to moderate. The series was established in Cochise County in 1921 in the San Simon area.

Some properties of the Karro soils

Horizon	Depth (cm)	O.M.%	Sand%	Silt%	Clay%	CEC meq/100g	CaCO$_3$%	pH	BD g/cm^3
Ap	0-30	1.26	50	38	12	13	6	7.9	1.34
Bk	30-188	.26	37	46	17	9	52	8.1	1.42

Kimbrough (TS14). Petrocalcic Calciustolls; loamy, mixed, thermic, shallow (Lithosols). Kimbrough soils are shallow and very shallow and well drained. Typically, they have a dark grayish brown gravelly loam surface layer about 20 cm (8 in) thick. Below this to a depth of 90 cm (36 in) is an indurated lime cemented hardpan. Depth to the hardpan ranges from 10 to 50 cm (4 to 20 in). Kimbrough soils occur on fan terraces with slopes of 0 to 8 percent. The soils have low available water capacity and moderate permeability. They are mildly alkaline and calcareous throughout the profile. Runoff is slow to medium and the hazard of erosion is high. The series was established in Lea County, New Mexico, in 1936.

Some properties of the Kimbrough soils

Horizon	Depth (cm)	O.M.%	Sand%	Silt%	Clay%	CEC meq/100g	CaCO$_3$%	pH	BD g/cm^3
A	0-28	2.89	48	35	17	16	16	7.9	1.40
Bkm	28-38	0.65				6	60	8.1	
2C	38-53	1.33	52	32	16	13	26	8.0	
3Bk	53-103	0.10				5	27	8.4	
4C	103-113	0.05	86	10	4	10	1	8.8	

Kofa (HA1). Vertic Torrifluvents; clayey over sandy or sandy-skeletal, montmorillonitic (calcareous), hyperthermic (Alluvial). Kofa soils are deep and well drained. Typically, they have a pale brown clay surface layer about 30 cm (12 in) thick. The upper 40 cm (16 in) of the underlying material is pale brown clay. The lower portion of the underlying material to a depth of 150 cm (60 in) is very pale brown sand. Thickness of the clayey layer over the sand ranges from 55 to 90 cm (22 to 36 in). When dry these soils have cracks 1 cm (0.4 in) or more wide at a depth of 50 cm (20 in). Kofa soils occur on floodplains with slopes of 0 to 1 percent. These soils have moderate available water capacity and slow permeability. They are moderately alkaline and calcareous throughout the profile. Runoff is slow and the hazard of erosion is moderate. The series was established in the Yuma-Wellton area, Yuma County, in 1977 and the name was taken from the Kofa Mountains.

Krentz (TS15). Entic Haplustolls, cindery, thermic (Reddish Brown). Krentz soils are shallow and very shallow and well to somewhat excessively drained. Typically, they have a dark grayish brown gravelly loam surface layer about 10 cm (4 in) thick. The upper 10 cm (4 in) of underlying material is dark grayish brown cindery loam. The lower portion of the underlying material is unconsolidated cinder and ash about 15 cm (6 in) thick. Depth to the layer of cinders ranges from 15 to 50 cm (6 to 20 in). Krentz soils occur on sideslopes of cinder cones with slopes of 1 to 15 percent. These soils have very low available water capacity and moderately rapid permeability. They are slightly to moderately alkaline and calcareous throughout. Runoff is medium and the hazard of erosion is slight. The series was established in the Sulphur Springs Valley area, Cochise County, in 1959 and the name was taken from a crater.

Lagunita (HA1, HA7). Typic Torripsamments; mixed, hyperthermic (Alluvial). Lagunita soils are deep and somewhat excessively drained. Typically they have a pale brown loamy sand surface layer about 20 cm (8 in) thick. The underlying material to a depth of 150 cm (60 in) is pale brown loamy sand. Very fine black sandy biotite flakes are present in thin strata throughout the profile. Lagunita soils occur on floodplains and drainageways with slopes of 0 to 3 percent. These soils have low available water capacity and rapid permeability. They are moderately alkaline and calcareous throughout the profile. Runoff is slow and the hazard of erosion is high. This soil is subject to wind erosion. The series was established in Yuma County in 1978.

APPENDICES

Lampshire (TA3, TS6, TS18). Lithic Haplustolls; loamy-skeletal, mixed, thermic (Lithosols). Lampshire soils are very shallow and shallow and well drained. Typically, they have a grayish brown very cobbly loam profile about 20 cm (8 in) deep over pinkish gray, widely fractured dacite bedrock. Depth to bedrock ranges from 10 to 50 cm (4 to 20 in). Lampshire soils occur on hillslopes with slopes ranging from 50 to 90 percent. These soils have low available water capacity and moderate permeability. They are slightly acid to moderately alkaline and are noncalcareous throughout the profile. The hazard of erosion is slight and runoff is medium to rapid. The series was established in Santa Cruz County in 1971.

Some properties of the Lampshire soils

Horizon	Depth (cm)	O.M.%	Sand%	Silt%	Clay%	CEC meq/100g	$CaCO_3$%	pH	BD g/cm^3
A	0-25	2.23	56	29	15	18	2	7.2	1.11
R	25-								

La Palma (HA2). Typic Durargids; fine-loamy, mixed, hyperthermic (Solonetz). La Palma soils are moderately deep and well drained. They typically have a light brown very fine sandy loam surface layer about 13 cm (5 in) thick over a yellowish red and brown light clay loam subsoil about 33 cm (13 in) thick. Below this is light brown loam containing many durinodes. This grades at about 68 cm (27 in) into an indurated silica and lime cemented duripan many centimeters thick. La Palma soils occur on stream terraces and fan terraces with slopes ranging from 0 to 3 percent. These soils have low available water capacity restricted by alkali and depth and have slow permeability. They are very strongly alkaline (sodic) and calcareous throughout. Runoff is slow and the hazard of erosion is slight. The series was established in Pinal County in 1936.

Some properties of the La Palma soils

Horizon	Depth (cm)	O.M.%	Sand%	Silt%	Clay%	CEC meq/100g	$CaCO_3$%	pH	BD g/cm^3
A	0-12	0.26	65	30	5	7	1	8.8	1.32
Btkn	12-31	0.27	62	20	18	10	10	9.0	1.80
Bctk	31-51	0.14	58	24	18	18	9	8.5	1.69
Bkqm	51-					8	61	9.0	2.23

Latene (TA4, TS9, TS11, TS14, TS16). Typic Calciorthids; coarse-loamy, mixed, thermic (Calcisols). Latene soils are deep and well drained. They typically have a light brown loam surface layer, about 46 cm (18 in) thick underlain by pink gravelly loam or sandy loam to 150 cm (60 in) or more. Latene soils occur on fan terraces with slopes ranging from 0 to 8 percent. Available water capacity is high and permeability is moderate. These soils are moderately alkaline and strongly calcareous throughout. The gravel in the lower layer consists mostly of lime concretions. Surface runoff is slow to medium and the hazard of erosion is slight. The series was established in Valencia County, New Mexico, in 1971.

Laveen (HA2, HA3, HA4, HA5, HA7). Typic Calciorthids; coarse-loamy, mixed, hyperthermic (Calcisols). Laveen soils are deep and well drained. Typically, they have a pale brown loam surface layer about 33 cm (13 in) thick underlain by light brown loam with common soft lime masses about 48 cm (19 in) thick. The underlying material from 80 to more than 150 cm (32 to 60 in) is light brown and pink loam with many soft and hard lime nodules. Laveen soils occur on stream terraces and low fan terraces with slopes ranging from 0 to 3 percent. These soils have high available water capacity and moderate permeability. They are moderately alkaline and calcareous throughout. Surface runoff is slow and the hazard of erosion is slight. The series was established in the Salt River Valley near Phoenix in 1926 and the name was taken from the town of Laveen.

Some properties of the Laveen soils

Horizon	Depth (cm)	O.M.%	Sand%	Silt%	Clay%	CEC meq/100g	$CaCO_3$%	pH	BD g/cm^3
Ap	0-20	0.39	54	34	12	10	5	8.2	1.48
Bw	20-38	0.29	50	36	14	10	7	8.0	1.40
Bk	38-168	0.15	47	36	17	8	20	8.3	1.52

Lehmans (TA3, TS6). Lithic Haplargids; clayey, montmorillonitic, thermic (Lithosols). Lehmans soils are shallow and well drained. Typically, they have a brown gravelly clay loam surface layer about 2.5 cm (1 in) thick. The subsoil is reddish brown and brown gravelly clay and clay about 33 cm (13 in) thick. Pale red, reddish gray and red fractured extremely hard andesite is at a depth of 36 cm (14 in). Depth to bedrock ranges from 25 to 50 cm (10 to 20 in). Lehmans soils occur on hillslopes with slopes of 8 to 60 percent. They have low available water capacity and slow permeability. They are neutral to moderately alkaline and noncalcareous throughout the profile. Runoff is medium to rapid and the hazard of erosion is slight. The series was established in Yavapai County in 1965.

Some properties of the Lehmans soils

Horizon	Depth (cm)	O.M.%	Sand%	Silt%	Clay%	CEC meq/100g	CaCO$_3$%	pH	BD g/cm^3
A	0-4	0.92	67	22	11	9		6.9	1.29
Bt	4-13	1.45	41	14	45	21		6.4	1.18
R	13-								

Ligurta (HA4, HA6, HA7, HA9). Typic Haplargids; fine-loamy mixed, hyperthermic (Solanchak). Ligurta soils are deep and well drained. Typically, the surface is covered with pebbles coated with desert varnish. The surface layer is very pale brown extremely gravelly loam about 5 cm (2 in) thick. The upper 33 cm (13 in) of the subsoil is reddish yellow and yellowish red gravelly clay loam. The lower portion of the subsoil to a depth of 150 cm (60 in) is light reddish brown gravelly clay loam, clay loam and loam. Common very fine and fine salt crystals are present in the subsoil. Electrical conductivity of the saturation extract ranges from 25 to more than 100 mmhos. Ligurta soils occur on fan terraces with slopes of 0 to 6 percent. These soils have very low available water capacity and moderately slow permeability. They are moderately alklaine and calcareous throughout the profile. Runoff is rapid and the hazard of erosion is slight. The series was established in the Yuma-Wellton area, Yuma County, in 1977.

Some properties of the Ligurta soils

Horizon	Depth (cm)	O.M.%	Sand%	Silt%	Clay%	CEC meq/100g	CaCO$_3$%	pH	BD g/cm^3
A	0-5	0.09	32	49	19	15	13	9.0	1.56
Btkz	5-53	0.14	54	28	18	8	11	7.9	1.51
BCtkz	53-94	0.12	72	17	11	7	9	7.7	1.46

Limpia (TS18). Pachic Argiustolls; clayey-skeletal, mixed, thermic (Reddish Chestnut). Limpia soils are deep and well drained. Typically, they have a dark reddish gray very cobbly loam surface layer about 30 cm (12 in) thick. The subsoil to a depth of 150 cm (60 in) is reddish brown very cobbly clay. Limpia soils occur on fan terraces with slopes of 1 to 50 percent. These soils have moderate available water capacity and slow permeability. They are neutral and noncalcareous throughout the profile. Runoff is medium and the hazard of erosion is slight. The series was established in Jeff Davis County, Texas, in 1971.

Lithic Cryoborolls (FH4). Lithic Cryoborolls are shallow soils over limestone or calcareous sandstone bedrock. Typically, they are dark colored and have gravelly medium textured surface layers and cobbly or very cobbly, loamy subsurface layers with bedrock at 15 to 50 cm (6 to 20 in). These soils are intermingled with rock outcrop on the breaks and steep sideslopes of the dissecting canyons and drainageways. Slopes range from 5 to 50 percent. Runoff is medium to rapid and the hazard of erosion is slight to moderate.

Lithic Haplargids (HA6, TA3, MS6). Lithic Haplargids commonly have brown or reddish brown, gravelly or cobbly loam or clay loam surface layers and reddish brown gravelly or cobbly clay loam or clay subsoils over bedrock at depths ranging from about 20 to 50 cm (8 to 20 in). Reaction ranges from slightly acid to neutral on the surface and neutral to moderately alkaline in the subsoil. Representative soils in this subgroup and closely related subgroups are the Arp, Boysag, Chiricahua, Gaddes, Lehmans and Luzena soils. The Lithic Haplargids are usually on the foot slopes and saddles having slopes of 5 to about 45 percent. Runoff is medium to rapid and the hazard of erosion is slight to moderate.

Lithic Torriorthents (TA3, TS6, MS6). Lithic Torriorthents range in color from pale brown to dark brown and light reddish brown to dark reddish brown. Textures of the profiles range from very gravelly to extremely stony sandy loam or loam. Depth to bedrock is usually 10 to 50 cm (4 to 20 in). Reaction ranges from slightly acid to moderately alkaline and from noncalcareous on igneous rocks to strongly calcareous on limestone and other sedimentary rocks. Representative soils in this subgroup are the Cellar, House Mountain, Moano and Moenkopie soils. Also included are the darker Faraway and Tortugas soils (Lithic Haplustolls) and the Barkerville and Mokiak soils on decomposing granite (Typic Ustorthents and Aridic Arguistolls). The Lithic Torriorthents and associated soils are on the steeper hillslopes and mountainslopes with slopes dominantly 15 to more than 60 percent. Runoff is rapid and the hazard of erosion is moderate.

Lomitas (HA6, HA9). Lithic Camborthids; loamy-skeletal, mixed, hyperthermic (Lithosols). Lomitas soils are shallow and well drained. They typically have a surface cover of 50 to 75 percent gravel, cobbles and stones, and have very stony or very cobbly loam surface layers, and very gravelly loam subsoils with bedrock at 30 to 50 cm (12 to 20 in). They occur on hillslopes with slopes of 5 to 40 percent. These soils have very low available water capacity and moderate permeability. They are moderately alkaline and calcareous throughout. Surface runoff is medium to rapid and the hazard of erosion is slight. The series was established on the Organ Pipe Cactus National Monument, Pima County, in 1971.

Some properties of the Lomitas soils

Horizon	Depth (cm)	O.M.%	Sand%	Silt%	Clay%	CEC meq/100g	CaCO$_3$%	pH	BD g/cm^3
A	0-5	0.65	53	38	9	11	1	8.2	1.33
Bw	5-25	0.53	48	40	12	12	5	8.2	1.45
Bk	25-33	0.68	49	37	14	8	31	8.3	1.65
R	33-								

Lonti (MS6, MS9, MS10). Ustollic Haplargids; fine, mixed, mesic (Reddish Brown). Lonti soils are deep and well drained. They typically have a grayish brown gravelly sandy loam surface layer about 5 cm (2 in) thick. The subsoil is dark brown gravelly sandy clay loam 8 cm (3 in) thick and reddish brown gravelly light clay and clay loam 100 cm (40 in) thick. The underlying material from a depth of about 114 to more than 150 cm (45 to 60 in) is pink and light reddish brown, calcareous, gravelly sandy clay loam. Lonti soils occur on nearly level to rolling tops of fan terraces and the steep sideslopes of fan terraces with slopes ranging from 1 to 45 percent. These soils have moderate available water capacity and slow permeability. They are slightly acid to neutral in the upper layers and moderately alkaline and calcareous below depths ranging from 45 to 120 cm (18 to 48 in). Runoff is slow to medium and the hazard of erosion is slight to medium. The series was established in Yavapai County in 1971.

Some properties of the Lonti soils

Horizon	Depth (cm)	O.M.%	Sand%	Silt%	Clay%	CEC meq/100g	B.S.%	pH	BD g/cm^3
A	0-5	2.80	59	23	18	16	91	6.9	1.56
Bt	5-103	0.48	52	13	35	28	93	7.0	1.48
C	103-158	0.05	64	17	19	24	100	8.2	1.55
Ck	158-185	0.05	65	15	20	21	100	8.1	1.61

Luth (FH1, FH2, FH3, FH4, FH8). Typic Haplustolls; fine, mixed, frigid (Alluvial). Luth soils are deep and well drained. Typically, they have a dark brown silt loam surface layer about 10 cm (4 in) thick. The underlying material to depths of 178 cm (71 in) or more is brown and dark brown clay loam or clay. Luth soils occur on floodplains and low stream terraces with slopes of 0 to 2 percent. These soils have high available water capacity and moderately slow to slow permeability. They are neutral to mildly alkaline and noncalcareous throughout the profile. Runoff is slow and the hazard of erosion is slight. The series was established in Coconino County in 1965 and the name was taken from a stock pond.

Luzena (MS6, MH2). Lithic Argiustolls; clayey, montmorillonitic, mesic (Reddish Chestnut). Luzena soils are shallow and well drained. They typically have a brown or grayish brown cobbly loam or clay loam surface layer about 13 cm (5 in) thick and reddish brown clay or gravelly clay subsoils over hard bedrock at 18 to 50 cm (7 to 20 in). Luzena soils occur on hillslopes with slopes ranging from 5 to 60 percent. These soils have low or very low available water capacity and slow permeability. They are slightly acid to neutral in reaction and are noncalcareous throughout the profile. Runoff is medium to rapid and the hazard of erosion is moderate. The series was established in Cochise County in 1936.

Some properties of the Luzena soils

Horizon	Depth (cm)	O.M.%	Sand%	Silt%	Clay%	CEC meq/100g	CaCO$_3$%	pH	BD g/cm^3
A	0-8	2.62	60	23	17	14	0	5.9	1.25
Bt	8-43	1.79	23	15	62	37	1	7.2	1.63
R	43-								

Lynx (MS1, MS2, MS5, MS6, MS7, MS8, MS9, MS10, MH2, MH3, MH4, MH5, MH6). Cumulic Haplustolls; fine-loamy, mixed, mesic (Alluvial). Lynx soils are deep and well drained. Typically, they have a grayish brown loam surface layer about 5 cm (2 in) thick. Below this, and extending to a depth of more than 150 cm (60 in), is brown and dark grayish brown light clay loam that may be thinly stratified with gravelly sandy loam in the lower part. Lynx soils occur on floodplains and low alluvial fans with slopes of 0 to 5 percent. These soils have high available water capacity and moderately slow permeability. They are neutral to moderately alkaline throughout and may be slightly calcareous in the lower part of the profile. Runoff is slow. Erosion hazard is generally slight, but some areas adjacent to entrenched drainageways are gullied. Some areas are subject to rare or frequent flooding in wet seasons. The series was established in the Beaver Creek area, Yavapai County, in 1965.

Some properties of the Lynx soils

Horizon	Depth (cm)	O.M.%	Sand%	Silt%	Clay%	CEC meq/100g	CaCO$_3$%	pH	BD g/cm^3
A	0-150	2.55	16	68	16		7	8.2	

Mabray (TS6). Lithic Haplustolls; loamy-skeletal, carbonatic, thermic (Lithosols). Mabray soils are shallow and very shallow and well drained. Typically, they have dark grayish brown very gravelly or very cobbly loam profiles overlying extremely hard, widely fractured, limestone bedrock at about 30 cm (12 in). Depth to bedrock ranges from 10 to 50 cm (4 to 20 in). Mabray soils occur on steep hillslopes of limestone or marble with slopes ranging from 15 to 70 percent. These soils have very low available water capacity and moderate permeability. They are moderately alkaline and contain more than 40 percent calcium carbonate. Runoff is medium to rapid and the hazard of erosion is slight. The series was established in Santa Cruz County in 1971.

Some properties of the Mabray soils

Horizon	Depth (cm)	O.M.%	Sand%	Silt%	Clay%	CEC meq/100g	CaCO$_3$%	pH	BD g/cm^3
A	0-23	2.82	48	34	18	21	7	7.6	1.25
Bk	23-33	1.87	47	31	22	32	67	7.6	1.35
R	33-								

Maripo (HA1). Typic Torrifluvents; coarse-loamy over sandy or sandy-skeletal, mixed (calcareous), hyperthermic (Alluvial). Maripo soils are deep and well drained. Typically, they have a brown sandy loam surface layer about 33 cm (13 in) thick. The upper 53 cm (21 in) of the underlying material is pale brown sandy loam. The lower portion to a depth of 150 cm (60 in) is brown gravelly sand. Depth to the gravelly sand or sand horizons ranges from 50 to 100 cm (20 to 40 in). Maripo soils occur on floodplains and alluvial fans with slopes of 0 to 3 percent. These soils have moderate available water capacity and moderately rapid permeability. They are neutral to strongly alkaline and calcareous throughout the profile. Runoff is slow to medium and the hazard of erosion is moderate. The series was established in Maricopa County in 1972.

Martinez (MH1). Udic Haplustalfs; very fine, kaolinitic, mesic (Reddish Brown). Martinez soils are deep and well drained. Typically, they have brown gravelly loam and clay loam surface layers about 15 cm (6 in) thick. The subsoil is dark brown and yellowish brown heavy clay 73 cm (29 in) thick. The substratum from 88 to more than 150 cm (35 to 60 in) is mottled yellowish brown and red gravelly sandy clay. Martinez soils occur on high old fan terraces with slopes of 0 to 3 percent. These soils have high available water capacity and very slow permeability.

They are strongly acid to mildly alkaline in the upper layers and become moderately alkaline and may be slightly calcareous in the lower layers. Runoff is slow and the hazard of erosion is slight. The series was established in Santa Cruz County in 1971.

Some properties of the Martinez soils

Horizon	Depth (cm)	O.M.%	Sand%	Silt%	Clay%	CEC meq/100g	B.S.%	pH	BD g/cm³
A	0-15	1.84	44	30	26	12	62	5.4	1.5
Bt	15-89	0.94	19	8	71	27	98	7.7	1.58
BCt	89-117		49	12	39	21	100	8.1	1.55

McAllister (TS3, TS13). Ustollic Haplargids; fine-loamy mixed, thermic (Reddish Brown). McAllister soils are deep and well drained. Typically, they have a brown loam surface layer about 30 cm (12 in) thick. The subsoil is brown and light brown clay loam about 88 cm (35 in) thick. The underlying material to a depth of 180 cm (72 in) is light brown fine sandy loam. McAllister soils occur on fan terraces with slopes of 0 to 2 percent. These soils have high available water capacity and moderately slow permeability. They are moderate to strongly alkaline and calcareous throughout the profile. Depth to calcic horizon (with more than 15 percent $CaCO_3$) ranges from 50 to 90 cm (20 to 36 in). Runoff is slow to medium and the hazard of erosion is moderate. The series was established in the Sulphur Springs Valley area, Cochise County, in 1948.

McVickers (FH3). Typic Cryoboralfs; fine, montmorillonitic (Gray Wooded). McVickers soils are moderately deep to deep and well drained. They typically have 5 cm (2 in) of decomposing pine litter on the soil surface. The surface mineral layers are brown and very pale brown very fine sandy loam about 38 cm (15 in) thick. The upper subsoil is pink loam 18 cm (7 in) thick underlain by reddish brown and reddish yellow clay 63 cm (25 in) thick over sandstone bedrock at 130 cm (52 in). Depth to bedrock ranges from 75 to 175 cm (30 to 70 in). McVickers soils occur on undulating to rolling plateaus with slopes of 2 to 15 percent. These soils have high available water capacity, slow permeability and high shrink-swell potential. Runoff is medium and the hazard of erosion is slight to moderate. The series was established in Coconino County in the Long Valley area in 1960 and the name was taken from a spring in the area.

Some properties of the McVickers soils

Horizon	Depth (cm)	O.M.%	Sand%	Silt%	Clay%	CEC meq/100g	B.S.%	pH	BD g/cm³
E	0-30	1.53	47	46	7	10	68	6.2	1.55
BA	30-35	0.43	53	33	13	8	77	6.1	1.7
Bt	35-78	0.41	38	24	38	27	87	5.8	1.85
BCt	78-108	0.12	52	26	21	18	93	6.2	1.76
R	108-								

Millard (MH3). Typic Haplustalfs; fine-loamy, mixed, mesic (Brown). Millard soils are deep and well drained. Typically, they have a reddish brown gravelly loam surface layer about 25 cm (10 in) thick. The subsoil is light reddish brown and yellowish red light clay loam about 55 cm (22 in) thick. Below this to a depth of 150 cm (60 in) the substratum is light reddish brown and reddish yellow gravelly loam. Millard soils occur on fan terraces with slopes of 2 to 10 percent. These soils have high available water capacity and moderately slow permeability. They are mildly alkaline and calcareous in the subsoil and substratum. Runoff is medium and the hazard of erosion is slight. The series was established in the Richfield area of Utah in 1947.

Some properties of the Millard soils

Horizon	Depth (cm)	O.M.%	Sand%	Silt%	Clay%	CEC meq/100g	$CaCO_3$%	pH	BD g/cm³
A	0-23	1.50	65	26	9	13	0	6.9	
Bt	23-105	0.77	76	8	16		0	7.2	
BC	105-150	0.10	84	1	15		1	6.9	

Millett (MS3). Ustollic Haplargids; fine-loamy, mixed, mesic (Reddish Brown). Millett soils are deep and well drained. Typically, they have a reddish brown gravelly sandy loam surface layer about 5 cm (2 in) thick. The subsoil is reddish brown gravelly sandy clay loam and light clay loam about 15 cm (6 in) thick. The underlying material to a depth of 100 cm (40 in) is pinkish gray gravelly loam and reddish brown very gravelly sandy clay loam. Thickness of the solum is less than 50 cm (20 in). Millett soils occur on fan terraces with slopes of 5 to 25 percent. These soils have high available water capacity and moderate permeability. They are moderately alkaline and calcareous throughout the profile. Runoff is medium and the hazard of erosion is moderate. The series was established in Navajo County in 1961.

Some properties of the Millett soils

Horizon	Depth (cm)	O.M.%	Sand%	Silt%	Clay%	CEC meq/100g	CaCO₃%	pH	BD g/cm³
A	0-10	1.04	71	14	15	9	1	7.8	
Bt	10-38	0.88	59	17	24	15	T	7.8	1.59
Bk1	38-69	0.56	83	9	8	2	18	7.9	
2Bk	69-144	0.05	93	4	3	3	6	8.3	

Mirabal (FH1, FH2, FH5, FH6). Typic Ustorthents; loamy-skeletal, mixed, nonacid, frigid (Gray Wooded). Mirabal soils are moderately deep and well drained. Typically, they have a thin litter of pine needles and leaf litter on the soil surface. The surface mineral layer is grayish brown, strongly weathered very cobbly and stony sandy loam about 58 cm (23 in) thick over granite bedrock. Depth to bedrock ranges from 50 to 88 cm (20 to 35 in). Mirabal soils occur on mountainslopes with slopes ranging from 5 to 70 percent. Intermixed with Mirabal soils are small areas of shallow and very shallow soils and rock outcrop. These soils have low available water capacity and moderately rapid permeability. They are medium acid to neutral throughout the profile. Runoff is medium to rapid and the hazard of erosion is slight to moderate. The series was established in Valencia County, New Mexico, in the Zuni Mountain area in 1964.

Some properties of the Mirabal soils

Horizon	Depth (cm)	O.M.%	Sand%	Silt%	Clay%	CEC meq/100g	B.S.%	pH	BD g/cm³
A	0-8	2.8	75	18	7	12	47	5.0	1.16
C	8-65	0.75	60	24	16	10	65	4.8	1.63
R	65-								

Moano (TA3, MS6, MH2). Lithic Torriorthents; loamy, mixed, nonacid, mesic (Lithosols). Moano soils are shallow to very shallow and well drained. Typically, they have a brown gravelly loam surface layer about 23 cm (9 in) thick. The underlying material to a depth of 35 cm (14 in) is olive to olive brown hard fractured phyllite and schist with thin tongues of soil material in the fractures. Pale yellow extremely hard schist bedrock is at a depth of 35 cm (14 in). Depth to bedrock ranges from 15 to 40 cm (6 to 16 in). Moano soils occur on gently rolling to steep hillslopes with slopes of 8 to 60 percent. These soils have low available water capacity and moderate permeability. They are neutral to moderately alkaline and noneffervescent throughout the profile. Runoff is medium to rapid and the hazard of erosion is slight. The series was established in Yavapai County in 1960 and the name was taken from Rancho Moano.

Some properties of the Moano soils

Horizon	Depth (cm)	O.M.%	Sand%	Silt%	Clay%	CEC meq/100g	B.S.%	pH	BD g/cm³
A	0-8	4.49	40	42	18	24	89	7.1	0.91
AC	8-23	2.43	39	37	24	32	89	6.9	1.39
Cr	23-35								

Moapa (TA5). Typic Torripsamments; mixed, thermic (Regosols). Moapa soils are moderately deep and excessively drained. Typically, they have a light yellowish brown fine sand surface layer about 10 cm (4 in) thick. The underlying material is light yellowish brown fine sand about 65 cm (26 in) thick. White sandstone is at a depth of 75 cm (30 in). Moapa soils occur on hillslopes with slopes of 8 to 30 percent. These soils have very low available water capacity and very rapid permeability to the bedrock. Runoff is slow and the hazard of erosion is high. These soils are subject to wind erosion. The series was established in Clark County, Nevada, in 1970.

Moenkopie (TA1, MA1, MA2, MA3, MS1, MS2, MS3, MS6). Torriorthents; loamy, mixed (calcareous), mesic (Lithosols). Moenkopie soils are shallow and very shallow and well drained. Typically, they have a reddish brown loamy sand surface layer about 5 cm (2 in) thick. The next layer is reddish brown sandy loam about 18 cm (7 in) thick. The bedrock at about 23 cm (9 in) is hard, widely fractured sandstone. Moenkopie soils occur on hillslopes and plateaus with slopes ranging from 1 to 15 percent. These soils have very low available water capacity and moderately rapid permeability. They are moderately alkaline and calcareous. Depth to bedrock ranges from 13 to 50 cm (5 to 20 in). They have slow to moderate runoff and the hazard of erosion is slight to moderate. The series was established in Navajo County in the Winslow area in 1921.

Some properties of the Moenkopie soils

Horizon	Depth (cm)	O.M.%	Sand%	Silt%	Clay%	CEC meq/100g	CaCO$_3$%	pH	BD g/cm^3
A	0-26	0.65	76	12	12	5	14	8.6	1.59
C	26-48	0.51	60	26	14	5	20	8.5	1.47
R	48-								

Mohall (HA2, HA3, HA5, HA8). Typic Haplargids; fine-loamy, mixed, hyperthermic (Red Desert). Mohall soils are deep and well drained. Typically, they have reddish yellow or light brown sandy loam, loam or clay loam surface layers about 25 cm (10 in) thick and reddish brown or brown clay loam subsoils about 75 cm (30 in) thick over brown loam or sandy loam to more than 150 cm (60 in). Mohall soils occur on nearly level to gently undulating fan terraces with slopes of 0 to 5 percent. These soils have high available water capacity and moderately slow permeability. They are moderately alkaline throughout and have zones of high lime accumulation below 50 or 60 cm (20 to 24 in). Runoff is slow and the hazard of erosion is slight. The series was established in Maricopa County in 1971 as the hyperthermic equivalent of the Mohave series.

Some properties of the Mohall soils

Horizon	Depth (cm)	O.M.%	Sand%	Silt%	Clay%	CEC meq/100g	CaCO$_3$%	pH	BD g/cm^3
A	0-13	.73	60	24	16	12	T	7.6	1.61
Bt	13-46	.24	50	23	27	18	T	8.2	1.64
Btk	46-109	.09	36	34	30	20	13	8.9	1.57
BCtk	109-180		32	44	24	17	16	9.4	1.67

Mohave (TA4, TS12). Typic Haplargids; fine-loamy, mixed, thermic (Red Desert). Mohave soils are deep and well drained. Typically, they have brown and light yellowish brown sandy loam surface layer about 28 cm (11 in) thick. The subsoil is brown heavy loam and clay loam about 110 cm (44 in) thick. The underlying material to a depth of 150 cm (60 in) or more is reddish brown loamy coarse sand. Mohave soils occur on fan terraces with slopes of 1 to 5 percent. These soils have high available water capacity and moderately slow permeability. They are neutral to moderately alkaline and noncalcareous in the surface horizons. Depth to carbonates ranges from 30 to 90 cm (12 to 36 in) and depth to the calcic horizon with more than 15 percent CaCO$_3$ ranges from 50 to 100 cm (20 to 40 in). Runoff is slow and the hazard of erosion is slight. The series was established in the Middle Gila River Valley in 1917.

Some properties of the Mohave soils

Horizon	Depth (cm)	O.M.%	Sand%	Silt%	Clay%	CEC meq/100g	CaCO$_3$%	pH	BD g/cm^3
A	0-8	1.43	48	27	25	20	1	7.7	1.45
Bt	8-46	0.36	50	19	31	17	4	7.6	1.50
Btk	46-81		44	26	30	16	9	7.9	1.50
BCk	81-102		44	23	33	19	15	8.3	1.45
2Ck	102-140		51	19	30	11	46	8.6	1.45

Mokiak (MS6, MH2). Aridic Argiustolls; loamy-skeletal, mixed, mesic (Regosols). Mokiak soils are moderately deep and well darined. Typically, they have a brown very cobbly and cobbly sandy loam surface layer about 28 cm (11 in) thick. The subsoil is yellowish brown and light yellowish brown sandy clay loam about 68 cm (27 in) thick overlying fractured gneiss and schist bedrock at a depth of 95 cm (38 in). Mokiak soils occur on hillslopes with slopes of 50 to 70 percent. These soils have low available water capacity and moderate permeability. They are mildly alkaline and noncalcareous throughout the profile. Runoff is rapid and the hazard of erosion is slight. The series was established in Washington County, Utah, in 1972.

Mormon Mesa (TA5). Typic Paleorthids; loamy, carbonatic, thermic, shallow (Calcisols). Mormon Mesa soils are shallow and well drained. They typically have light reddish brown fine sandy loam surface and subsurface layers underlain at about 40 cm (16 in) by a pink and white indurated lime cemented hardpan meters thick. Morman Mesa soils occur on tops of old fan terraces with slopes of 0 to 8 percent. They have very low available water capacity. Permeability is moderately rapid above the pan and very slow through the pan. These soils are very strongly calcareous, more than 40 percent calcium carbonate, and are moderate to strongly alkaline. Runoff is slow to medium and the hazard of erosion is slight to moderate. The series was established in Clark County, Nevada, in 1939.

Navajo (MA1, MA4, MA5). Vertic Torrifluvents; fine, mixed (calcareous), mesic (Alluvial). Navajo soils are deep and well drained. Typically, they have a reddish brown silty clay surface layer about 13 cm (5 in) thick. This is underlain by reddish brown silty clay and clay substrata to more than 150 cm (60 in). Navajo soils occur on floodplains, alluvial fans and playas with slopes of 0 to 5 percent. These soils have high available water capacity and very slow permeability. They are moderately alkaline and calcareous throughout and are commonly saline. Runoff is slow and the hazard of erosion is slight except along entrenched streams where they are subject to piping and gullying. They are subject to very brief seasonal flooding. The series was established in Navajo County in 1921 and the name was taken from that county.

Some properties of the Navajo soils

Horizon	Depth (cm)	O.M.%	Sand%	Silt%	Clay%	CEC meq/100g	CaCO$_3$%	pH	BD g/cm^3
A	0-13	0.87	3	44	53	18	1	7.1	1.68
C	13-150	0.39	2	38	60	27	2	7.6	1.78

Nickel (TA2, TA4, TA5, TS3, TS4, TS6, TS9, TS12, TS14, TS16). Typic Calciorthids; loamy-skeletal, mixed, thermic (Calcisols). Nickel soils are deep and well drained. They typically have a pale brown very gravelly sandy loam surface layer about 18 cm (7 in) thick underlain to 150 cm (60 in) or more by light brownish gray very gravelly sandy loam that is weakly cemented by calcium carbonate. Nickel soils occur on fan terraces with slopes ranging from 8 to 30 percent. These soils have low available water capacity and moderately rapid permeability. The soils are moderately alkaline and strongly calcareous throughout. Runoff is medium and erosion hazard is slight. The series was established in Clark County, Nevada, in 1939.

Some properties of the Nickel soils

Horizon	Depth (cm)	O.M.%	Sand%	Silt%	Clay%	CEC meq/100g	CaCO$_3$%	pH	BD g/cm^3
A	0-10	1.05	74	19	7	7	5	8.6	1.43
Bw	10-25	0.97	65	23	12	8	5	8.5	1.47
Bk	25-100	0.70	50	31	19	7	36	8.5	1.30
C	100-122		72	15	13	6	11	8.7	

Nolam (TS5, TS7). Ustollic Haplargids; loamy-skeletal, mixed, thermic (Red Desert). Nolam soils are deep and well drained. Typically, they have light brown very gravelly fine sandy loam surface layer about 5 cm (2 in) thick. The subsoil is red and reddish brown very gravelly sandy clay loam about 38 cm (15 in) thick. The upper part of the underlying material is calcareous pink and light brown very gravelly sandy loam about 58 cm (23 in) thick. The lower part of the underlying material to a depth of 198 cm (79 in) is brown and very pale brown very gravelly loamy sand and very gravelly sand. Nolam soils occur on rolling fan terraces with slopes of 2 to 15 percent. These soils have low available water capacity and moderate permeability. They are mildly to moderately alkaline and calcareous in the subsoil and underlying material. The upper boundary of the calcic horizon occurs within 100 cm (40 in) of the surface. Runoff is medium and the hazard of erosion is slight. The series was established in Dona Ana County, New Mexico, in 1972.

Oracle (TS6, TS8, TS10). Ustollic Haplargids; loamy, mixed, thermic, shallow (Reddish Brown). Oracle soils are shallow and well drained. Typically, they have a dark brown very gravelly loam surface layer about 13 cm (5 in) thick. The subsoil is reddish brown fine gravelly clay loam that becomes very gravelly in the lower part. The substratum below about 48 cm (18 in) to more than 150 cm (60 in) is strongly weathered, highly fractured coarse grained granite. Oracle soils occur on hillslopes with slopes ranging from 15 to 25 percent. They have low available water capacity and moderately slow permeability. They are slightly acid to mildly alkaline and are noncalcareous throughout the profile. Runoff is medium and the hazard of erosion is slight. The series was established in Pinal County in 1952 and the name was taken from the town of Oracle.

Some properties of the Oracle soils

Horizon	Depth (cm)	O.M.%	Sand%	Silt%	Clay%	CEC meq/100g	B.S.%	pH	BD g/cm^3
A	0-23	1.21	73	15	12	8	85	6.3	
Bt	23-81	0.66	68	7	25	16	98	7.1	
Cr	81-168								

Overgaard (MH5, FH3, FH6). Typic Cryoboralfs; fine, mixed (Gray Wooded). Overgaard soils are deep and well drained. Typically, they have a thin layer of pine needle and oak leaf litter on the surface. The surface mineral layers are grayish brown over light brownish gray gravelly loam about 25 cm (10 in) thick. The subsoil is brown and reddish brown gravelly light clay 80 cm (32 in) thick. The underlying material is mottled brown and reddish brown very gravelly clay loam to more than 150 cm (60 in). Overgaard soils occur on undulating terraces and hillslopes with slopes ranging from 2 to 50 percent. These soils have moderate available water capacity and slow permeability. They are medium acid to neutral throughout and are noncalcareous. Runoff is slow to medium and the hazard of erosion is slight. The series was established in Navajo County in 1962.

Some properties of the Overgaard soils

Horizon	Depth (cm)	O.M.%	Sand%	Silt%	Clay%	CEC meq/100g	CaCO$_3$%	pH	BD g/cm^3
A	0-23	1.22	75	16	9	19	0	6.4	
Bt	23-68	1.00	45	18	37		0	6.4	
C	68-170	0.49	52	22	26		0	6.9	

Pachic Argiustolls (MH6). Pachic Argiustolls typically have brown loam surface layers about 15 cm (6 in) thick and brown clay loam and clay subsoils to 150 cm (60 in) deep or more. These soils have high available water capacity and slow permeability. They are slightly acid to neutral in the upper layers and become moderately alkaline and calcareous below about 75 cm (30 in). These soils are on undulating to rolling fan terraces with dominant slopes of 2 to 8 percent. Runoff is slow and the hazard of erosion is slight.

Palma (MA2, MA3, MS3). Ustollic Haplargids; coarse-loamy, mixed, mesic (Reddish Brown). Palma soils are deep and well drained. They typically have a brown loamy sand surface layer about 8 cm (3 in) thick. The subsoil is reddish brown sandy loam and brown sandy loam about 45 cm (18 in) thick. The underlying material to a depth of more than 100 cm (40 in) is light brown calcareous fine sandy loam. Palma soils occur on nearly level to undulating fan terraces with slopes of 0 to 8 percent. These soils have high available water capacity and moderately rapid permeability. They are moderately alkaline throughout and have zones of high lime in the substratum. Runoff is slow and the hazard of erosion is slight. The hazard of wind erosion is moderate. The series was established in Colorado in 1970.

Some properties of the Palma soils

Horizon	Depth (cm)	O.M.%	Sand%	Silt%	Clay%	CEC meq/100g	CaCO$_3$%	pH	BD g/cm^3
A	0-8	0.39	93	3	4	5	0	8.0	1.73
Bt	8-53	0.44	86	3	11	10	0	7.5	1.67
Bk	53-103	0.20	88	3	9	8	3	8.3	1.60
2Bkb	103-183	0.28	81	6	13	10	4	8.5	1.77

Palomino (FH3, FH4). Lithic Cryoboralfs; loamy-mixed (Lithosols). Palomino soils are shallow and well drained. Typically, they have a 5 cm (2 in) layer of pine needles on the surface. The surface mineral layer is brown extremely stony fine sandy loam about 8 cm (3 in) thick. The subsoil is reddish brown extremely stony sandy clay loam about 30 cm (12 in) thick. Strong brown sandstone is at a depth of 38 cm (15 in). Depth to bedrock ranges from 30 to 50 cm (12 to 20 in). Palomino soils occur on hillslopes with slopes of 2 to 30 percent. These soils have low available water capacity and moderate to moderately slow permeability. They are medium acid to neutral and noncalcareous throughout the profile. Runoff is medium and the hazard of erosion is slight. The series was established in Coconino County in 1965 and the name was taken from Palomino Lake.

Palos Verdes (TS7, TS9, TS19). Haplic Durargids; loamy, mixed, thermic, shallow (Red Desert). Palos Verdes soils are shallow and well drained. Typically, they have a brown gravelly sandy loam surface layer about 8 cm (3 in) thick. The subsoil is reddish brown gravelly sandy clay loam about 30 cm (12 in) thick. The substratum is yellowish red and reddish brown gravelly sandy loam about 10 cm (4 in) thick. The next layer is a pinkish white and pink strongly cemented duripan about 48 cm (19 in) thick. Below this to a depth of 150 cm (60 in) or more is pinkish white and pink gravelly loamy coarse sand. Depth to the duripan ranges from 25 to 50 cm (10 to 20 in). Palos Verdes soils occur on fan terraces with slopes of 1 to 15 percent. These soils have low available water capacity and moderately slow permeability to the pan and very slow through the pan. They are neutral to moderately alkaline and calcareous in the substratum and in the duripan and underlying material. Runoff is medium and the hazard of erosion is slight. The series was established in Pima County in 1931.

Some properties of the Palo Verdes soils

Horizon	Depth (cm)	O.M.%	Sand%	Silt%	Clay%	CEC meq/100g	CaCO$_3$%	pH	BD g/cm^3
A	0-3	0.66	65	28	7	6.7	0	7.9	1.69
BA	3-8	0.53	61	26	13	8.3	0	8.0	
Bt	8-38	0.41	54	14	32	10	1	7.5	1.54
Bkn	38-48		50	27	23	10	28	8.3	1.53
Bkqm	48-97		65	21	14	6	29	8.5	1.44
Bkq	97-163		78	12	10	6	10	8.5	1.67

Pantano (TS6). Typic Calciorthids; loamy-skeletal, mixed, thermic, shallow (Calcisols). Pantano soils are shallow and very shallow and well drained. Typically, they have a pale brown extremely gravelly loam surface layer about 2.5 cm (1 in) thick. The subsurface layer is brown very gravelly loam about 23 cm (9 in) thick. The upper 15 cm (6 in) of the underlying material is white and pale brown calcareous extremely gravelly loam. Below this to a depth of 150 cm (60 in) is gray highly fractured schist bedrock. Depth to highly fractured rock ranges from 25 to 50 cm (10 to 20 in). Pantano soils occur on pediments and hillslopes with slopes of 2 to 60 percent. These soils have very low available water capacity and moderate permeability. They are moderately alkaline and calcareous throughout the profile. Runoff is medium to rapid and the hazard of erosion is slight. The series was established in Pima County in 1974 and the name was taken from a large wash.

Partri (MS8). Aridic Argiustolls; fine, mixed, mesic (Reddish Chestnut). Partri soils are deep and well drained. Typically, they have a brown loam surface layer 5 cm (2 in) thick. The subsoil is reddish brown clay loam and clay about 65 cm (26 in) thick. The underlying material to more than 150 cm (60 in) is white, weakly lime cemented gravelly clay loam. Depth to the zone of high lime is 50 to 90 cm (20 to 36 in). Partri soils occur on nearly level to gently undulating fan terraces with slopes of 1 to 5 percent. They have high available water capacity and slow permeability. They are moderately alkaline throughout and contain more than 15 percent lime in the substratum. Runoff is slow and the hazard of erosion is slight. The series was established in Yavapai County in 1971.

Pastura (MS1, MS4, MS6, MS8, MS10). Ustollic Paleorthids; loamy, mixed, mesic, shallow (Lithosols). Pastura soils are shallow and well drained. Typically, the upper 28 cm (11 in) is pale brown and brown gravelly loam underlain by a pinkish white indurated lime hardpan several inches thick. The upper laminated layer is underlain by variably lime cemented gravelly and cobbly materials to more than 150 cm (60 in). Depth to the pan ranges from 20 to 50 cm (8 to 20 in). Pastura soils occur on fan terraces and toeslopes of limestone hills with slopes of 1 to 15 percent. These soils have very low available water capacity and moderate permeability to the pan. They are moderately alkaline and calcareous throughout. Runoff is slow and the hazard of erosion is slight to moderate. The series was established in Torrance County, New Mexico, in 1970.

Some properties of the Pastura soils

Horizon	Depth (cm)	O.M.%	Sand%	Silt%	Clay%	CEC meq/100g	CaCO₃%	pH	BD g/cm³
A	0-10	1.90	42	36	22	24	11	7.8	
Bw	10-33	1.82	35	36	29	24	21	7.6	
Bk	33-50	1.38	37	35	28	25	23	7.8	
Bkm	50-	0.75	37	36	27	9	64	8.0	

Penthouse (TS16). Ustollic Haplargids; fine, mixed, thermic (Reddish Brown). Penthouse soils are deep and well drained. They typically have a brown cobbly clay loam surface layer about 8 cm (3 in) thick. The subsoil is reddish brown clay about 60 cm (24 in) thick. The substratum, 68 to 150 cm (27 to 60 in), is pink, very cobbly clay loam that is weakly cemented and high in calcium carbonate. Penthouse soils occur on the tops of old fan terraces with slopes of 2 to 8 percent. These soils have moderate available water capacity and very slow permeability. They are mildly alkaline in the surface and moderately alkaline in the lower layers and contain more than 15 percent calcium carbonate in the substratum. Runoff is medium and the hazard of erosion is moderate. The series was established in Yavapai County in 1965 and the name was taken from Penthouse Tank.

Perryville (HA4, HA5, HA9). Typic Calciorthids; coarse-loamy, carbonatic, hyperthermic (Calcisols). Perryville soils are deep and well drained. Typically, they have a surface layer of very pale brown gravelly loam about 20 cm (8 in) thick. This is underlain by light brown gravelly loam or sandy loam to 150 cm (60 in) or more containing 20 to 35 percent gravel and lime nodules and more than 40 percent calcium carbonate. Perryville soils occur on fan terraces with slopes of 0 to 3 percent. The soils have low available water capacity and moderate permeability. They are moderately alkaline and calcareous throughout. Runoff is slow and the hazard of erosion is slight. The series was established on the Organ Pipe National Monument area, Pima County, in 1971.

Some properties of the Perryville soils

Horizon	Depth (cm)	O.M.%	Sand%	Silt%	Clay%	CEC meq/100g	CaCO₃%	pH	BD g/cm³
Ap	0-23	0.38	43	40	17	9	19	7.9	1.32
Bk	23-97	0.12	38	45	17	7	40	7.9	0.90

Pima (TS2, TS3, TS4, TS13, TS14, TS15, TS17, TS18). Typic Torrifluvents; fine-silty, mixed, thermic (Alluvial). Pima soils are deep and well drained. They typically have a dark grayish brown clay loam surface layer about 66 cm (26 in) thick over stratified grayish brown loam and fine sandy loam to more than 150 cm (60 in). Pima soils occur on floodplains with slopes of 0 to 2 percent. The soils have high available water capacity and moderately slow permeability. They are moderately alkaline and slightly calcareous throughout. Runoff is slow and the hazard of erosion is generally slight. The series was established in the Middle Gila River Valley area in 1917.

Some properties of the Pima soils

Horizon	Depth (cm)	O.M.%	Sand%	Silt%	Clay%	CEC meq/100g	CaCO₃%	pH	BD g/cm³
Ap	0-30	2.43	5	57	38	35	3	7.5	1.25
A	30-119	1.24	15	53	32	30	4	7.5	1.30
C	119-152		34	43	23	22	2	7.5	1.38

Pimer (HA1). Typic Torrifluvents; fine-silty, mixed (calcareous), hyperthermic (Alluvial). Pimer soils are deep and well drained. Typically, they have a brown light clay loam surface layer about 38 cm (15 in) thick. The underlying material to a depth of 150 cm (60 in) is brown clay loam and silty clay loam. Pimer soils occur on floodplains with slopes of 0 to 1 percent. These soils have high available water capacity and moderately slow permeability. They are moderately alkaline and calcareous throughout. Runoff is slow and the hazard of erosion is slight. This series was established in Maricopa County in 1971 as the hyperthermic equivalent of the Pima series.

Some properties of the Pimer soils

Horizon	Depth (cm)	O.M.%	Sand%	Silt%	Clay%	CEC meq/100g	CaCO₃%	pH	BD g/cm³
Ap	0-38	1.15	13	58	29	31	2	7.7	1.65
C	38-178	0.24	11	65	24	26	4	7.8	1.44

Pinal (HA4). Typic Durorthids; coarse-loamy, mixed, hyperthermic (Calcisols). Pinal soils are shallow and well drained. They typically have a light brown gravelly very fine sandy loam surface layer about 5 cm (2 in) thick. The subsurface layer is light brown gravelly very fine sandy loam about 43 cm (17 in) thick. This is underlain by a lime and silica cemented duripan from about 48 to more than 75 cm (19 to 30 in). Pinal soils occur on nearly level to undulating fan terraces with slopes of 0 to 9 percent. These soils have low available water capacity and moderate permeability in the upper part of the profile and very slow permeability in the duripan. They are moderately alkaline and calcareous throughout. Runoff is medium and the hazard of erosion is slight. The series was established in the Middle Gila River Valley area in 1917 and the name was taken from Pinal County.

Pinaleno (TA3, TA4, TS3, TS4, TS6, TS9, TS12, TS14, TS19). Typic Haplargids; loamy-skeletal, mixed; thermic (Red Desert). Pinaleno soils are deep and well drained. Typically, they have a brown very gravelly sandy loam surface layer about 5 cm (2 in) thick. The subsoil is reddish brown very gravelly sandy clay loam about 70 cm (28 in). The underlying material from about 76 to more than 150 cm (30 to 60 in) is reddish brown and pinkish gray very gravelly sandy loam and loamy sand. Pinaleno soils occur on fan terraces with slopes of 0 to 25 percent. These soils have low available water capacity and moderate or moderately slow permeability. They are mildly to moderately alkaline throughout and usually have zones of high lime in the lower subsoil and underlying material. Runoff is slow to medium and the hazard of erosion is slight to moderate. The series was established in Maricopa County in 1969.

Some properties of the Pinaleno soils

Horizon	Depth (cm)	O.M.%	Sand%	Silt%	Clay%	CEC meq/100g	CaCO$_3$%	pH	BD g/cm^3
A	0-3	0.80	71	18	11	11	T	7.6	1.02
Bt	3-32	0.73	63	10	27	20	2	8.1	1.45
Btk	32-61	0.65	61	19	20	12	19	8.1	
Bk	61-160	0.26	75	14	11	11	15	8.2	1.10

Pinamt (HA3, HA6, HA8). Typic Haplargids; loamy-skeletal, mixed, hyperthermic (Red Desert). Pinamt soils are deep and well drained. They typically have light brown very gravelly or very cobbly sandy loam or loam surface layers about 15 cm (6 in) thick, and yellowish red very gravelly sandy clay loam subsoils about 75 cm (30 in) thick over very pale brown calcareous very gravelly sandy loam substrata to 150 cm (60 in) or more. Pinamt soils occur on fan terraces with slopes of 0 to 20 percent. Available water capacity is low and permeability is moderately slow. The soils are moderately alkaline throughout and are slightly calcareous in the surface layers. They have a zone of high lime in the lower horizons. Surface runoff is slow to medium and the hazard of erosion is slight. The series was established in Maricopa County in 1971 as the hyperthermic equivalent of the Pinaleno series.

Some properties of the Pinamt soils

Horizon	Depth (cm)	O.M.%	Sand%	Silt%	Clay%	CEC meq/100g	CaCO$_3$%	pH	BD g/cm^3
A	0-7	0.36	63	19	8	8	T	7.6	1.60
BA	7-22	0.20	74	17	9	8	T	7.8	1.64
Bt	22-35	0.26	64	23	13	11	T	7.8	1.60
2Bt	35-81	0.43	63	16	21	17	1	8.2	1.60
3BCt	81-117	0.12	73	10	17	15	T	8.1	1.62
3Ck	117-170	0.09	70	12	8	9	10	8.4	1.50

Poley (MS2, MS6, MS8, MS10). Ustollic Haplargids; fine, mixed, mesic (Reddish Brown). Poley soils are deep and well drained. Typically, they have a reddish brown gravelly sandy loam surface layer about 5 cm (2 in) thick. The subsoil is yellowish red clay and sandy clay loam about 56 cm (22 in) thick. The substratum is white, weakly cemented, very cobbly coarse sandy loam. Depth to the zone of high lime is 50 to 90 cm (20 to 36 in). Poley soils occur on fan terraces with slopes of 1 to 8 percent. These soils have moderate available water capacity and slow permeability. They are moderately alkaline throughout and contain more than 15 percent calcium carbonate in the substratum. Runoff is medium and the hazard of erosion is slight. The series was established in Yavapai County in 1964.

Purgatory (MA2). Typic Gypsiorthids; fine-loamy, gypsic, mesic (Regosols). Purgatory soils are moderately deep and well drained. Typically, they have a yellowish red gravelly fine sandy loam surface layer about 5 cm (2 in) thick. The upper 30 cm (12 in) of underlying material is variegated pink and yellowish red sandy loam with common gypsum crystals. The lower 50 cm (20 in) of underlying material is variegated light gray clay loam with common gypsum crystals. Below this to a depth of 125 cm (50 in) is variegated thin bedded shale. Depth to bedrock layers from 50 to 100 cm (20 to 40 in). Purgatory soils occur on mesas with slopes of 0 to 8 percent. These soils have moderate available water capacity and moderate permeability. They are moderately alkaline and calcareous throughout. Runoff is slow to medium and the hazard of erosion is high. The series was established in the Virgin River area on the Utah-Arizona border in 1936.

Some properties of the Purgatory soils

Horizon	Depth (cm)	O.M.%	Sand%	Silt%	Clay%	CEC meq/100g	2H$_2$O%	pH	BD g/cm^3
A	0-3	1.39	64	30	6	9	25	7.5	1.10
By	3-58	0.44	22	63	15	6	53	7.8	1.30
Cy	58-118	0.07	5	82	13	5	30	8.3	1.40
C	118-143	0.10	16	78	6	4	9	8.6	1.40

Purner (MS1). Lithic Haplustolls; loamy, mixed, mesic (Lithosols). Purner soils are shallow and very shallow and well drained. Typically, they have reddish brown gravelly loam surface layers about 23 cm (9 in) thick over white, strongly lime cemented gravelly loam about 15 cm (6 in) thick. The bedrock is dark gray and pinkish gray limestone that is capped with a lime hardpan. Depth to bedrock ranges from 18 to 46 cm (7 to 18 in). Purner soils occur on gently rolling limestone hillslopes with slopes of 2 to 10 percent. These soils have very low available water capacity and moderate permeability. They are moderately alkaline and calcareous throughout. Runoff is medium and the hazard of erosion is slight. The series was established in Yavapai County in 1964 and the name was taken from a well-known ranch.

Retriever (TS6). Lithic Torriorthents; loamy, carbonatic, thermic (Lithosols). Retriever soils are shallow and very shallow and well drained. Typically, they have a pale brown gravelly loam surface layer about 5 cm (2 in) thick. The underlying material is light brown gravelly loam and gravelly clay loam about 30 cm (12 in) thick. Below this at a depth of 35 cm (14 in) is white extremely hard limestone. Depth to limestone ranges from 18 to 50 cm (7 to 20 in). Retriever soils have low available water capacity and moderate permeability. They are moderately alkaline and calcareous throughout. Runoff is medium to rapid and the hazard of erosion is slight. The series was established in the Beaver Creek area in Yavapai County in 1965.

Rillino (TA2, TS9, TS14, TS16). Typic Calciorthids; coarse-loamy, mixed, thermic (Calcisols). Rillino soils are deep and well drained. They typically have a light brown gravelly sandy loam surface layer about 28 cm (11 in) thick. The underlying layer is light brown and pinkish gray weakly lime cemented gravelly loam about 95 cm (38 in) thick. Below this to 150 cm (60 in) or more is light brown stratified gravelly sandy loam and very gravelly loamy sand. Rillino soils occur on undulating fan terraces with slopes of 1 to 15 percent. These soils have moderate available water capacity and moderate permeability. They are moderately alkaline and calcareous throughout. Runoff is medium and the hazard of erosion is moderate. The series was established in Santa Cruz County in 1973 as the thermic equivalent of the Rillito series.

Some properties of the Rillino soils

Horizon	Depth (cm)	O.M.%	Sand%	Silt%	Clay%	CEC meq/100g	CaCO$_3$%	pH	BD g/cm^3
A	0-13	1.05	52	30	18	17	9	7.7	
Bw	13-25	1.02	49	26	25	21	16	7.6	
2Bk1	25-46	1.17	35	36	29	21	27	7.6	
2Bk2	46-152	0.37	60	23	17	17	30	7.7	

Rillito (HA3, HA4, HA5, HA6, HA7, HA8, HA9, HA10). Typic Calciorthids; coarse-loamy, mixed, hyperthermic (Calcisols). Rillito soils are deep and well drained. They typically have a thin, light brown gravelly sandy loam surface layer about 8 cm (3 in) thick and a pink gravelly subsurface layer over underlying materials of white and pinkish gray gravelly loam, sandy loam or loamy sand from about 30 to more than 150 cm (12 to 60 in). Rillito soils occur on nearly level to gently undulating fan terraces with slopes of 0 to 3 percent. These soils have low to moderate available water capacity and moderate or moderately rapid permeability. They are moderately alkaline and strongly calcareous throughout. The substratum averages 15 to 35 percent gravel and hard lime nodules that may be weakly cemented by calcium carbonate. Runoff is slow to medium and the hazard of erosion is slight. The series was established in Pima County in 1945 and the name was taken from the Rillito River.

Some properties of the Rillito soils

Horizon	Depth (cm)	O.M.%	Sand%	Silt%	Clay%	CEC meq/100g	B.S.%	pH	BD g/cm^3
A	0-6	0.19	66	28	6	9	6	8.5	1.16
Bw	6-35	0.29	60	29	11	10	12	8.4	1.19
Bk	35-122	0.29	51	32	17	7	37	8.3	1.4

Rimrock (TS15, TS17, TS18). Typic Chromusterts; fine, montmorillonitic, thermic (Brown Grumusols). Rimrock soils are moderately deep to deep and well drained. Typically, they have reddish brown granular clay surface layers about 5 cm (2 in) thick. This is underlain by reddish brown clay about 80 cm (32 in) thick. Basalt bedrock with widely spaced fractures occurs at about 85 cm (34 in). Rimrock soils occur on nearly level to rolling basalt flows with slopes of 0 to 10 percent. These soils have moderate available water capacity and very slow permeability. They crack widely and deeply when dry and swell when wet. They are moderately alkaline and are calcareous throughout. Runoff is slow and the hazard of erosion is slight although short pipes may develop where water follows cracks. The series was established in Yavapai County in 1960 and the name was taken from the Rimrock Post Office near where the soil was originally mapped.

Rock Outcrop (HA4, HA6, TA1, TA3, TA5, TS4, TS6, TS10, TS15, TS18, MA2, MA3, MA6, MS1, MS2, MS4, MS5, MS6, MS7, MH2, MH4, MH5, MH6, FH1, FH2, FH3, FH4, FH5, FH6, FH7). Rock outcrop occurs mainly as steep or very steep peaks, ledges and escarpments. In some areas it occurs as low ridges or boulder piles or on pediment surfaces with less than 15 percent slope. rock outcrop is barren of vegetation and has rapid runoff. The rock type is highly variable and may be granite, gneiss, andesite, rhyolite, tuff, basalt, limestone, sandstone, shale or conglomerate.

Romero (TS6, TS10). Typic Ustorthents; loamy-skeletal, mixed, thermic, shallow (Regosols). Romero soils are shallow and very shallow and well drained. Typically, they have a dark grayish brown very gravelly sandy loam and fine sandy loam surface layer about 25 cm (10 in) thick. The upper 18 cm (7 in) of underlying material is strongly weathered granite (grus) with soil material from the surface layer in the fractures. Below this to a depth of 180 cm (72 in) is pale brown, very pale brown and light yellowish brown weathered granite (grus). Depth to highly fractured granite ranges from 13 to 50 cm (5 to 20 in). Romero soils occur on gently rolling to steep hillslopes and pediments with slopes of 5 to 60 percent. These soils have low available water capacity and moderately rapid permeability. They are neutral to slightly acid and noncalcareous throughout. Runoff is medium and the hazard of erosion is slight. The series was established in Pima County in 1975.

Rond (MS5, MS6, MH4). Typic Argiustolls; fine, mixed, mesic (Reddish Chestnut). Rond soils are deep and well drained. Typically, they have a brown gravelly loam surface layer about 8 cm (3 in) thick. The upper 28 cm (11 in) of the subsoil is dark reddish gray gravelly heavy clay loam. The lower 85 cm (34 in) of the subsoil is reddish brown and yellowish red gravelly clay. The underlying material is mottled weak red gravelly clay loam about 15 cm (6 in) thick. Below that is grayish brown limestone bedrock. Depth to bedrock is greater than 100 cm (40 in). Rond soils occur on hillslopes and fan terraces with slopes of 2 to 30 percent. These soils have high available water capacity and slow permeability. They are neutral to moderately alkaline and noncalcareous in the upper part of the profile and calcareous in the lower subsoil and underlying material. Runoff is medium and the hazard of erosion is moderate. The series was established on the Fort Apache Indian Reservation in Gila County in 1967.

Rositas (HA10). Typic Torripsamments; mixed, hyperthermic (Alluvial). Rositas soils are deep and somewhat excessively drained. Typically, they have very pale brown fine sand surface layers, about 23 cm (9 in) thick over similar colored fine sand underlying layers to more than 150 cm (60 in) that are thinly laminated with strata of loamy fine sand 0.1 to 0.5 cm (0.04 to 0.2 in) thick. Rositas soils occur on sand dunes with slopes of 0 to 2 percent. These soils have low available water capacity and rapid permeability. They are moderately alkaline and calcareous throughout. Surface runoff is very slow and the hazard of water erosion is slight. The hazard of wind erosion is severe if the natural surface and cover are disturbed. The series was established in Imperial County, California, in 1918.

Some properties of the Rositas soils

Horizon	Depth (cm)	O.M.%	Sand%	Silt%	Clay%	CEC meq/100g	CaCO$_3$%	pH	BD g/cm^3
A	0-53	0.63	86	11	3		T	8.0	
C	53-165	0.17	84	12	4		1.0	7.8	

Roundtop (MS5, MH4, MH5). Typic Argiustolls; fine, mixed, mesic (Reddish Chestnut). Roundtop soils are moderately deep and well drained. Typically, they have a dark reddish gray gravelly clay loam surface layer about 8 cm (3 in) thick. The subsoil is reddish brown gravelly heavy clay loam and gravelly clay about 84 cm (33 in) thick. The bedrock is grayish brown fractured cherty limestone. Depth to bedrock ranges from 50 to 100 cm (20 to 40 in). Roundtop soils occur on rolling plains and hillslopes with slopes ranging from 2 to 30 percent. These soils have moderate available water capacity and slow permeability. They are neutral to moderately alkaline in the upper layers and moderately alkaline and calcareous below about 38 cm (15 in). Runoff is medium and the hazard of erosion is moderate. The series was established in Gila County in 1964 and the name was taken from a local mountain.

Rudd (MS4, MS7). Lithic Calciustolls; loamy-skeletal, mixed, mesic (Calcisols). Rudd soils are shallow and very shallow and well drained. Typically, they have a grayish brown very gravelly loam surface layer about 5 cm (2 in) thick. The next layer is dark grayish brown gravelly heavy loam 20 cm (8 in) thick underlain by grayish brown, very gravelly, calcareous loam about 8 cm (3 in) thick. The bedrock at about 33 cm (13 in) is dark gray basalt with lime coatings on rock faces and in the fractures. Depth to bedrock ranges from about 15 to 50 cm (6 to 20 in). Rudd soils are on nearly level to gently rolling basalt mesas with slopes of 0 to 8 percent, but a few escarpments are up to 45 percent. These soils have low available water capacity and moderate permeability. They are moderately alkaline and calcareous throughout. Runoff is medium and the hazard of erosion is slight. The series was established in Apache County in 1971 and the name was taken from a local creek.

Some properties of the Rudd soils

Horizon	Depth (cm)	O.M.%	Sand%	Silt%	Clay%	CEC meq/100g	CaCO$_3$%	pH	BD g/cm^3
A	0-25	3.33	46	28	26	22	19	8.1	
Bk	25-33	4.13	45	27	28	19	40	8.0	1.19
R	33-								

Rune (MS6, MS8). Cumulic Haplustolls; fine, mixed, mesic (Alluvial). Rune soils are deep and well drained. Typically, the upper 5 cm (2 in) of the surface layer is reddish brown loam. The lower 53 cm (21 in) of the surface layer is reddish brown clay loam. The underlying material to a depth of 150 cm (60 in) is reddish brown clay. Rune soils occur on floodplains and alluvial fans with slopes of 0 to 8 percent. These soils have high water holding capacity and slow permeability. They are moderately alkaline and calcareous throughout. Runoff is slow and the hazard of erosion is moderate. The series was established in Yavapai County in 1962.

St. Thomas (TA5, TS6). Lithic Torriorthents; loamy-skeletal, carbonatic, thermic (Lithosols). St. Thomas soils are shallow and very shallow and well drained. Typically, they have a very pale brown cobbly loam surface layer about 5 cm (2 in) thick. The underlying material to a depth of 30 cm (12 in) is very pale brown cobbly loam. Below this is light gray extremely hard limestone with a thin capping of secondary calcium carbonate. Depth to bedrock ranges from 10 to 48 cm (4 to 19 in). St. Thomas soils occur on hillslopes with slopes of 15 to 50 percent. These soils have low available water capacity and moderate permeability. They are moderately alkaline and calcareous throughout. Runoff is rapid and the hazard of erosion is slight. The series was established in Clark County, Nevada, in 1970.

Sanchez (FH3). Lithic Eutroboralfs; loamy-skeletal, mixed (Lithosols). Sanchez soils are shallow and well drained. Typically, a 5 cm (2 in) layer of pine needles and twigs covers the soil surface. The surface layer is pinkish gray gravelly sandy loam about 5 cm (2 in) thick. The upper 13 cm (5 in) of subsoil is pinkish gray sandy clay loam. The lower 25 cm (10 in) of subsoil is pinkish gray channery clay loam and channery sandy clay loam. Below this at a depth of 43 cm (17 in) is pinkish gray fine grained sandstone. Depth to bedrock ranges from 28 to 50 cm (11 to 20 in). Sanchez soils occur on hillslopes with slopes of 5 to 25 percent. These soils have low available water capacity and moderately slow permeability. They are medium acid to neutral and noncalcareous throughout. Runoff is medium and the hazard of erosion is slight. The series was established in the Zuni Mountain area in McKinley County, New Mexico, in 1964.

Schrap (TS6). Ustic Torriorthents; loamy-skeletal, mixed (nonacid), thermic (Lithosols). Schrap soils are shallow and very shallow and well drained. Typically, they have a brown very channery clay loam surface layer about 8 cm (3 in) thick. The upper 35 cm (14 in) of underlying material is gray weathered shale. The lower portion of underlying material to a depth of 68 cm (27 in) is yellowish brown, brown, reddish yellow and olive shale bedrock. Depth to the weathered shale ranges from 8 to 30 cm (3 to 12 in). Schrap soils occur on hillslopes with slopes of 5 to 50 percent. These soils are neutral to mildly alkaline and noncalcareous throughout. Runoff is medium to rapid and the hazard of erosion is slight. The series was established in Santa Cruz County in 1971.

Shalet (TA1, MA1, MA2, MA3, MA5). Typic Torriorthents; loamy, mixed (calcareous), mesic, shallow (Lithosols). Shalet soils are shallow and very shallow and well drained. Typically, they have a reddish brown light clay loam surface layer about 10 cm (4 in) thick. The next layer is yellowish red clay loam about 20 cm (8 in) thick that is underlain by weathered, highly fractured shale to more than 150 cm (60 in). Depth to the shale ranges from 10 to 38 cm (4 to 15 in). Shalet soils occur on hillslopes with slopes of 2 to 20 percent. These soils have very low available water capacity and slow permeability. They are strongly alkaline and calcareous throughout and commonly are high in gypsum. Runoff is rapid and erosion hazard is moderate. The series was established in Washington County, Utah, in 1972.

Sheppard (MA1, MA3, MS3). Typic Torripsamments; mixed, mesic (Alluvial). Sheppard soils are deep and somewhat excessively drained. Typically, they have a surface layer of reddish yellow fine sand about 30 cm (12 in) thick. This is underlain by reddish yellow loamy fine sand to 150 cm (60 in) or more. Sheppard soils occur on gently undulating to rolling sand dunes with slopes of 2 to 15 percent. These soils have low available water capacity and rapid permeability. They are moderately alkaline and are slightly calcareous throughout. Runoff is slow. The hazard of water erosion is slight, but the hazard of wind erosion is high. The series was established in Uinta River Valley area of Utah in 1925.

Some properties of the Sheppard soils

Horizon	Depth (cm)	O.M.%	Sand%	Silt%	Clay%	CEC meq/100g	CaCO$_3$%	pH	BD g/cm^3
A	0-25	0.39	96	2	2	2		6.9	
C	25-175	0.25	95	2	3	3		7.2	

Showlow (MS8, MH3, MH5, MH6). Aridic Argiustolls; fine, montmorillonitic, mesic (Reddish Chestnut). Showlow soils are deep and well drained. Typically, they have brown and dark grayish brown gravelly loam surface layers about 8 cm (3 in) thick over a reddish brown clay loam and clay subsoil 70 cm (28 in) thick. The underlying material is reddish brown gravelly and very gravelly, calcareous, sandy clay loam to 150 cm (60 in). Showlow soils occur on nearly level to rolling fan terraces with slopes of 0 to 20 percent. These soils have high available water capacity, slow permeability and high shrink-swell potential. Reaction is neutral in the upper part and mildly to moderately alkaline and calcareous below about 60 cm (24 in). Runoff is medium and the hazard of erosion is slight. The series was established in Navajo County in 1962 in the Holbrook-Showlow area.

Some properties of the Showlow soils

Horizon	Depth (cm)	O.M.%	Sand%	Silt%	Clay%	CEC meq/100g	CaCO$_3$%	pH	BD g/cm^3
A	0-8	2.75	41	43	16	20	0	6.9	1.56
Bt1	8-60	1.20	28	34	38	44	0	7.1	1.64
Bt2	60-78	0.73	22	20	58	59	T	7.5	1.68
Bk	78-130	0.24	71	5	24	27	1	7.6	1.68

Siesta (FH2). Mollic Cryoboralfs; fine, montmorillonitic (Reddish Prairie). Siesta soils are moderately deep and deep and well drained. Typically, they have a 2.5 cm (1 in) layer of partially decomposed pine needles on the soil surface. The surface mineral layer is dark reddish gray cobbly silt loam about 5 cm (2 in) thick. The subsurface layer is reddish brown light clay loam about 8 cm (3 in) thick. The upper 40 cm (16 in) of the subsoil is reddish brown clay. The lower 63 cm (25 in) of the subsoil is red gravelly clay and gravelly clay loam. Vesicular basalt and cinders are at a depth of 115 cm (46 in). Depth to bedrock ranges from 60 to 150 cm (24 to 60 in). Siesta soils occur on cinder cones with slopes of 2 to 20 percent. These soils have high available water capacity and slow permeability. They are neutral to moderately alkaline and noncalcareous throughout the profile. Runoff is slow to medium and the hazard of erosion is slight. The series was established in the Beaver Creek area of Coconino County in 1965.

Some properties of the Siesta soils

Horizon	Depth (cm)	O.M.%	Sand%	Silt%	Clay%	CEC meq/100g	$CaCO_3$%	pH	BD g/cm³
A	0-15	3.52	19	55	26	21	1	6.0	
BA	15-30	2.29	17	53	30	22	1	5.9	
Bt	30-200	0.84	12	26	62	46	2	7.3	

Signal (TS5, TS17). Aridic Paleustolls; clayey-skeletal, montmorillonitic, thermic (Red Desert). Signal soils are deep and well drained. Typically, they have a brown very gravelly loam surface layer about 5 cm (2 in) thick. The subsoil is dark brown very gravelly clay loam over reddish brown very gravelly clay about 75 cm (30 in) thick. The underlying material, from 80 to 150 cm (32 to 60 in) is brown and pink, calcareous, very gravelly sandy loam and loamy sand. Signal soils occur on old fan terraces with slopes of 4 to 15 percent They have moderate available water capacity and slow permeability. Runoff is medium and the hazard of erosion is slight. These soils are medium acid to mildly alkaline in the upper part and moderately alkaline and calcareous below about 55 cm (22 in). The series was established in Graham County in 1937.

Some properties of the Signal soils

Horizon	Depth (cm)	O.M.%	Sand%	Silt%	Clay%	CEC meq/100g	B.S.%	pH	BD g/cm³
A	0-15	1.50	55	20	25	22		6.1	
Bt	15-51	1.27	38	12	50	40		7.1	
BCk	51-119	0.07	73	7	20	31	2	8.2	

Sizer (FH2). Argic Cryoborolls; fine-loamy over fragmental, mixed (Reddish Chestnut). Sizer soils are shallow and moderately deep and well drained. Typically, they have a 2.5 cm (1 in) layer of pine litter on the soil surface. The surface mineral layer is dark grayish brown and brown gravelly silt loam about 20 cm (8 in) thick. The subsoil is reddish brown gravelly light clay loam about 25 cm (10 in) thick. Below this to a depth of 150 cm (60 in) are gray and dark brown cinders with very gravelly sandy loam material filling the fractures and spaces between the cinders. Depth to the cinders ranges from 35 to 100 cm (14 to 40 in). Sizer soils occur on cinder cones with slopes of 6 to 30 percent. They have moderate available water capacity and moderately slow permeability. They are neutral to mildly alkaline and noneffervescent throughout. Runoff is medium and the hazard of erosion is slight. The series was established on the Fort Apache Indian Reservation in 1965.

Soldier (FH3, FH4). Glossic Cryoboralfs; fine, montmorillonitic (Planosols). Soldier soils are deep and moderately well to well drained. Typically, they have a 5 cm (2 in) litter of pine needles on the soil surface. The upper surface mineral layer is brown cobbly loam, about 5 cm (2 in) thick underlain by light brownish gray and very pale brown cobbly loam and very cobbly sandy loam 43 cm (17 in) thick. The upper subsoil is reddish brown clay and cobbly clay about 48 cm (19 in) thick. The lower subsoil from 75 to more than 150 cm (30 to 60 in) is mottled reddish and yellowish brown and reddish yellow cobbly clay. Soldier soils occur on undulating to steep fan terraces with slopes of 2 to 25 percent. These soils have moderate available water capacity. They have moderately rapid infiltration rates in the upper parts and very slow permeability and high shrink-swell in the lower parts. They are neutral to very strongly acid throughout the profile. Runoff is medium and the hazard of erosion is slight to moderate. The series was established in Coconino County in 1960 and the name was taken from Soldier Canyon.

Some properties of the Soldier soils

Horizon	Depth (cm)	O.M.%	Sand%	Silt%	Clay%	CEC meq/100g	B.S.%	pH	BD g/cm³
A	0-10	8.25	26	62	12	31	44	5.5	1.16
E	10-38	3.2	25	58	17	23	36	5.4	1.30
Bt	38-70	0.85	29	18	53	31	49	4.4	1.57
BC	70-110+	0.53	43	23	34	20	35	4.5	1.63

Sonoita (TS3, TS19). Typic Haplargids; coarse-loamy, mixed, thermic (Red Desert). Sonoita soils are deep and well drained. They typically have brown or reddish brown sandy loam or gravelly sandy loam surface layers about 10 cm (4 in) thick over reddish brown gravelly sandy loam and light sandy clay loam subsoils to more than 150 cm (60 in). Sonoita soils occur on fan terraces with slopes of 0 to 8 percent. These soils have moderate available water capacity and moderate permeability. They are medium acid to mildly alkaline and noncalcareous in the upper layers and moderately alkaline and may or may not be calcareous in the lower layers. Runoff is slow to medium and the hazard of erosion is slight. The series was established in Santa Cruz County in 1930 and the name was taken from the town of Sonoita.

Some properties of the Sonoita soils

Horizon	Depth (cm)	O.M.%	Sand%	Silt%	Clay%	CEC meq/100g	CaCO₃%	pH	BD g/cm³
A	0-23	0.53	74	19	7	7.8	0	6.5	1.63
Bt	23-91	0.39	69	17	14	9.0	0	7.2	1.55
BCtk	91-135		66	20	14	9.3	2	8.3	1.71

Sontag (TS15). Typic Argiustolls; fine, montmorillonitic, thermic (Reddish Brown). Sontag soils are deep and well drained. Typically, they have a reddish brown gravelly clay loam surface layer about 5 cm (2 in) thick. The subsoil is dark reddish brown clay about 70 cm (28 in) thick. The underlying material to a depth of 110 cm (44 in) is yellowish red and reddish yellow cobbly clay loam. Sontag soils occur on fan terraces with slopes of 2 to 15 percent. These soils have high available water capacity and slow permeability. They are slightly acid to mildly alkaline and noncalcareous in the upper part and calcareous in the lower subsoil and underlying material. Runoff is medium and the hazard of erosion is moderate. The series was established in Cochise County in 1965.

Sponseller (FH2). Argic Cryoborolls; fine-loamy, mixed (Brown Forest). Sponseller soils are deep and well drained. Typically, they have a thin layer of decomposing pine litter and cinder gravel on the soil surface. The surface mineral layer is reddish brown gravelly loam about 10 cm (4 in) thick. The subsoil is reddish brown clay loam and gravelly and cobbly clay loam 96 cm (38 in) thick. This is underlain from 105 to more than 150 cm (42 to 60 in) by decomposing cinders, basalt cobble and fractured bedrock. Depth to the parent rock ranges from 100 to 150 cm (40 to 60 in). Sponseller soils occur on basalt flows and cinder cones with slopes of 3 to 40 percent. These soils have moderate to high available water capacity and moderately slow permeability. Reaction ranges from medium acid to neutral. Runoff is slow to medium and the hazard of erosion is moderate. The series was established in Navajo County in 1962.

Some properties of the Sponseller soils

Horizon	Depth (cm)	O.M.%	Sand%	Silt%	Clay%	CEC meq/100g	CaCO₃%	pH	BD g/cm³
A	0-10	2.14	28	53	19	11	0	6.8	
Bt1	10-28	1.31	24	45	31		1	6.2	
Bt2	28-105	1.19	29	35	36		T	6.5	
Cr	105-								

Springerville (MS1, MS4, MS7, MS9). Udic Chromusterts; fine, montmorillonitic, mesic (Grumusols). Springerville soils are deep and well drained. Typically, they have brown stony or cobbly silty clay surface layers 10 cm (4 in) thick over brown silty clay subsurface layers 90 cm (36 in) thick underlain by basalt bedrock. Depth to bedrock ranges from 100 to 175 cm (40 to 70 in). Springerville soils occur on toeslopes and terraces having gently undulating microrelief due to shrinking and swelling (gilgai). Slopes are dominantly 1 to 5 percent. These soils have high available water capacity, slow to very slow permeability and high shrink-swell potential. They are mildly to moderately alkaline and are usually slightly calcareous throughout. Runoff is slow and the hazard of erosion is slight. The series was established in Navajo County in 1961 and the name was taken from the town of Springerville.

Some properties of the Springerville soils

Horizon	Depth (cm)	O.M.%	Sand%	Silt%	Clay%	CEC meq/100g	CaCO$_3$%	pH	BD g/cm^3
A	0-5	1.40	6	41	53	52	0	6.6	1.78
C	5-115	0.87	7	39	54	54	5	7.0	1.88
R	115-								

Stellar (TS14). Ustollic Haplargids; fine, mixed, thermic (Red Desert). Stellar soils are deep and well drained. Typically, they have a pinkish gray clay loam surface layer about 8 cm (3 in) thick. The subsoil is reddish brown clay about 50 cm (20 in) thick. The substratum is pinkish white and light reddish brown clay loam about 13 cm (5 in) thick. The upper 23 cm (9 in) of the underlying material is calcareous pink and light brown clay loam. The lower part of the underlying material to a depth of 150 cm (60 in) is calcareous light brown and pinkish white sandy clay loam. A calcic horizon is present between 50 and 100 cm (20 to 40 in). Stellar soils occur on fan terraces and toeslopes of alluvial fans with slopes of 0 to 5 percent. These soils have high available water capacity and slow permeability. They are mildly to moderately alkaline and calcareous throughout the lower subsoil, substratum and underlying material. Runoff is slow and the hazard of erosion is slight. The series was established on the Desert Soil-Geomorphology Project in Dona Ana County, New Mexico, in 1971.

Some properties of the Stellar soils

Horizon	Depth (cm)	O.M.%	Sand%	Silt%	Clay%	CEC meq/100g	CaCO$_3$%	pH	BD g/cm^3
A	0-23	1.96	24	42	34	29	1	7.9	1.44
Bt	23-69	1.21	29	26	45	28	T	8.0	1.50
BCtk	69-152	0.44	62	16	22	13	3	8.2	1.55

Stewart (TS11). Typic Durorthids; loamy, mixed, thermic, shallow (Solonchak). Stewart soils are shallow and very shallow and somewhat poorly drained. Typically, they have a surface layer of light gray loam about 15 cm (6 in) thick. The subsoil is brown fine sandy loam about 8 cm (3 in) thick. Below this is a light colored, indurated lime-silica cemented hardpan about 33 cm (13 in) thick. This is underlain by finely stratified light gray and pale yellow fine sand and loam to 150 cm (60 in) or more. Stewart soils occur on low terraces bordering playas with slopes of 0 to 5 percent. The profile is very strongly alkaline (pH 9.4 to 10.0). Available water capacity is very low and permeability is very slow. Runoff is very slow and small playas and hummocky areas are common. Water stands in the playas until it evaporates. Water erosion is slight but there is a hazard of wind erosion in disturbed areas. The series was established in the Sulphur Springs Valley area in Cochise County in 1941.

Some properties of the Stewart soils

Horizon	Depth (cm)	O.M.%	Sand%	Silt%	Clay%	CEC meq/100g	CaCO$_3$%	pH	BD g/cm^3
E	0-15	0.60	55	35	10	14	4	9.5	1.64
Bn	15-23	0.39	56	25	18	35	4	9.7	1.56
Bnqm	23-56		62	31	7	24	1	9.8	1.63
Bnq	56-107		62	27	11	15	2	9.6	1.71

Superstition (HA10). Typic Calciorthids; sandy, mixed, hyperthermic (Calcisols). Superstition soils are deep and somewhat excessively drained. Typically, they have light brown loamy sand or sand surface and subsurface layers about 58 cm (23 in) thick. Below this to more than 150 cm (60 in) is light brown loamy sand or sand containing calcium carbonate in the form of soft lime masses and hard concretions. Superstition soils occur on terrace and mesa tops with slopes of 0 to 2 percent. These soils have low available water capacity and rapid permeability. They are moderately alkaline and calcareous throughout. Surface runoff is very slow. The water erosion hazard is slight but the hazard of wind erosion is severe. The series was established in the El Centro area of California in 1918.

Some properties of the Superstition soils

Horizon	Depth (cm)	O.M.%	Sand%	Silt%	Clay%	CEC meq/100g	CaCO$_3$%	pH	BD g/cm^3
A	0-16	0.03	95	4	1	2	2	8.2	
Bw	16-89	0.02	91	6	3	3	3	8.1	
2Bk	89-124		83	10	7	6	7	8.4	
2C	124-180		94	3	3	3	5	8.3	

Tatiyee (FH8). Argic Cryoborolls; clayey-skeletal, montmorillonitic (Prairie). Tatiyee soils are deep and well drained. Typically, they have very dark grayish brown gravelly loam and gravelly clay loam surface layers about 30 cm (12 in) thick. The subsoil is brown very gravelly clay grading to very gravelly clay loam to depths of 150 cm (60 in) or more. Tatiyee soils occur on nearly level to undulating fan terraces with slopes of 0 to 8 percent. These soils have moderate to high available water capacity and slow permeability. They are medium acid in the upper part and neutral in the lower part. Runoff is slow and the hazard of erosion is slight. The series was established in the Paunsaugunt area in Utah in 1969.

Telephone (MH4, MH5). Lithic Ustorthents; loamy-skeletal, mixed, nonacid, mesic (Lithosols). Telephone soils are shallow or very shallow and well drained. Typically, there is a thin pine needle and oak leaf litter on the soil surface. The surface mineral layer is light brownish gray very cobbly sandy loam about 10 cm (4 in) thick. This is underlain by light gray very cobbly sandy loam about 33 cm (13 in) thick and light gray sandstone bedrock with widely spaced fractures. Depth to the bedrock ranges from 15 to 50 cm (6 to 20 in). Telephone soils occur on hillslopes with slopes of 15 to 75 percent. These soils have very low available water capacity and moderate permeability above the bedrock. They are medium to slightly acid, contain 35 to 80 percent rock fragments and are noncalcareous. Runoff is rapid and the hazard of erosion is moderate. The Telephone series was established in Santa Cruz County in 1971.

Thunderbird (MS1, MS4, MS7, MH6). Aridic Argiustolls; fine, montmorillonitic, mesic (Reddish Prairie). Thunderbird soils are moderately deep and well drained. Typically, they have a grayish brown cobbly clay loam surface layer 5 cm (2 in) thick over a grayish brown clay subsoil 65 cm (26 in) thick. This is underlain by brown gravelly clay loam 8 cm (3 in) thick and basalt bedrock at about 78 cm (31 in). Depth to bedrock ranges from 50 to 100 cm (20 to 40 in). Thunderbird soils occur on mesa tops and hillslopes with slopes of 2 to 15 percent. These soils have moderate available water capacity, slow or very slow permeability and high shrink-swell potential. They are neutral to moderately alkaline in the upper layers, moderately alkaline in the lower subsoil and calcareous just above the bedrock. Runoff is slow to medium and the hazard of erosion is slight to moderate. The series was established in Yavapai County in 1967 and the name was taken from a mythical bird by that name.

Some properties of the Thunderbird soils

Horizon	Depth (cm)	O.M.%	Sand%	Silt%	Clay%	CEC meq/100g	CaCO$_3$%	pH	BD g/cm^3
A	0-5	4.85	22	57	21	23	0	6.8	
Bt	5-70	1.34	10	43	47	40	0	7.4	1.62
Bk	70-78	0.63	36	38	26	54	9	8.0	
R	78-								

Tobler (TA2). Typic Torrifluvents; coarse-loamy, mixed (calcareous), thermic (Alluvial). Tobler soils are deep and well drained. Typically, they have a red fine sandy loam to silty clay loam surface layer about 33 cm (13 in) thick. Below this to a depth of 150 cm (60 in) is red fine sandy loam. Tobler soils occur on alluvial fans and floodplains with slopes of 1 to 5 percent. These soils have high available water capacity and moderately rapid permeability. They are moderately alkaline and calcareous throughout the profile. Runoff is slow and the hazard of erosion is moderate. The series was established in Utah in 1973.

Toltec (HA2). Entic Durorthids; coarse-loamy, mixed, hyperthermic (Calcisols). Toltec soils are moderately deep and well drained. Typically, they have a pale brown loam surface layer about 30 cm (12 in) thick. The underlying material is light brown loam and very fine sandy loam to a depth of 90 cm (36 in) over a pinkish gray strongly cemented silica-lime duripan about 35 cm (14 in) thick. Below the duripan to a depth of 150 cm (60 in) is pinkish gray fine sandy loam. Depth to the duripan ranges from 50 to 100 cm (20 to 40 in). Toltec soils occur on fan terraces with slopes of 0 to 3 percent. These soils have moderate available water capacity and moderate permeability to the duripan and slow permeability in the duripan. They are mildly to strongly alkaline and calcareous throughout. Runoff is slow and the hazard of erosion is slight. The series was established in Pinal County in 1936.

Tonopah (TA5). Typic Calciorthids; sandy-skeletal, mixed, thermic (Calcisols). Tonopah soils are deep excessively drained. Typically, they have light brown very cobbly sandy loam surface layers about 15 cm (6 in) thick. Below this to a depth of 150 cm (60 in) is light brown extremely gravelly sand. A calcic horizon is present between 45 and 75 cm (18 and 30 in). Tonopah soils occur on fan terrace remnants with slopes of 0 to 15 percent. These soils have very low available water capacity and very rapid permeability. They are moderate to strongly alkaline and calcareous throughout. Runoff is slow to medium and the hazard of erosion is moderate. The series was established in Clark County, Nevada, in 1923.

Torrifluvents (HA3, HA4, HA6, TA5, TS1, TS2, TS5, TS6, TS7, TS9, TO10, TS12, TS16, TS19, MA1, MA1, MA5, MA6, MS5). Torrifluvents typically have pale brown sandy loam or loamy sand surface layers overlying stratified sandy loam, loamy sand, and thin silty or gravelly layers to 150 cm (60 in) or more. These soils have moderate to low available water capacity and moderately rapid to rapid permeability. They are on floodplains and lower alluvial fans with slopes of 0 to 3 percent. Runoff is slow and the hazard of water erosion is generally slight except for piping and bank cutting along entrenched streambeds. They are moderately susceptible to wind erosion. These soils are subject to very brief seasonal flooding.

Torripsamments (TS1). Torripsamments have pale brown loamy sand surface layers overlying stratified layers of loamy sand, fine sand and loamy fine sand to 150 cm (60 in) or more. These soils have low available water capacity and rapid permeability. They occur on alluvial fans and sand dunes with slopes of up to 5 percent. Many areas are hummocky and subject to wind erosion in disturbed or overgrazed areas. Runoff is slow and the hazard of water erosion is slight.

Tortugas (MS1, MS2, MS6, MH2, MH4, MH5). Lithic Haplustolls; loamy-skeletal, carbonatic, mesic (Lithosols). Tortugas soils are shallow and very shallow and well drained. Typically, they have grayish brown very cobbly loam profiles about 30 cm (12 in) thick over gray limestone with widely spaced fractures. Depth to bedrock ranges from 15 to 50 cm (6 to 20 in). Tortugas soils occur on gently rolling to very steep hillslopes with slopes of 5 to 70 percent. Available water capacity is very low and permeability is moderate. These soils are moderately alkaline and calcareous. Runoff is medium to high and erosion hazard is slight to moderate. The series was established on the Gila River Project in 1936.

Tours (MA1, MA2, MA4, MA5, MA6, MS2, MS3, MS6, MH3, MH4, MH5). Typic Torrifluvents; fine-silty, mixed (calcareous), mesic (Alluvial). Tours soils are deep and well drained. They typically have a reddish brown clay loam surface layer about 25 cm (10 in) thick. This is underlain to more than 150 cm (60 in) by reddish brown, stratified, clay loam, light clay loam and silty clay loam. Tours soils occur on alluvial fans and floodplains with slopes of 0 to 2 percent. These soils have high available water capacity and moderately slow permeability. They are moderately alkaline and calcareous throughout. Some areas are saline. Runoff is slow to medium and the hazard of erosion is slight to moderate. They are subject to gullying along entrenched drainageways and to very brief seasonal flooding. The series was established in the Beryl-Enterprise area in Utah in 1958.

Some properties of the Tours soils

Horizon	Depth (cm)	O.M.%	Sand%	Silt%	Clay%	CEC meq/100g	CaCO$_3$%	pH	BD g/cm^3
A	0-3	1.84	16	36	47	15	11	8.0	
C	3-151	0.80	12	55	33	12	13	7.9	1.51

Trail (MA2, MA4, MA5, MS3). Typic Torrifluvents; sandy, mixed, mesic (Alluvial). Trail soils are deep and somewhat excessively drained. They typically have a reddish brown loamy fine sand surface layer about 20 cm (8 in) thick over yellowish red loamy fine sand substrata that is thinly stratified with sandy loam and fine sand to 150 cm (60 in) or more. Trail soils occur on floodplains and alluvial fans with slopes of 0 to 5 percent. These soils have moderate available water capacity and rapid permeability. They are mildly to moderately alkaline and are calcareous throughout. Runoff is slow. The hazard of water erosion is slight and the hazard of wind erosion is moderate to severe. They are subject to very brief seasonal flooding. The series was established in Navajo County in 1957.

Some properties of the Trail soils

Horizon	Depth (cm)	O.M.%	Sand%	Silt%	Clay%	CEC meq/100g	CaCO₃%	pH	BD g/cm³
A	0-8	0.102	93	2	5	4	2	8.6	
C	8-147	0.30	85	10	5	7	4	8.7	1.45

Tremant (HA2, HA3, HA5, HA8). Typic Haplargids; fine-loamy, mixed, hyperthermic (Red Desert). Tremant soils are deep and well drained. They typically have thin, brown, gravelly loam or sandy loam surface layers and reddish brown gravelly clay loam or gravelly sandy clay loam subsoils. The substratum from about 58 to 150 cm (23 to 60 in) is pink gravelly loam high in lime. Tremant soils occur on nearly level to undulating fan terraces with slopes of 0 to 5 percent. These soils have a moderate available water capacity and moderately slow permeability. They are moderately alkaline and calcareous throughout and have zones of high lime in the lower subsoil and substratum. Runoff is medium and the hazard of erosion is generally slight. The series was established in Maricopa County in 1971 as the hyperthermic equivalent of the Tres Hermanos series.

Tres Hermanos (TA3, TA4, TS3, TS9, TS12, TS14, TS19). Typic Haplargids; fine-loamy, mixed, thermic (Red Desert). Tres Hermanos soils are deep and well drained. They typically have a light brown gravelly sandy loam or sandy clay loam surface layer about 8 cm (3 in) thick. The subsoil is reddish brown calcareous gravelly light clay loam about 60 cm (24 in) thick. The substratum to more than 150 cm (60 in) is pink very gravelly loam high in lime. Tres Hermanos soils occur on fan terraces with slopes of 2 to 15 percent. These soils have moderate available water capacity, moderate permeability and moderate shrink-swell. They are moderately alkaline and calcareous throughout. Runoff is medium to rapid and the hazard of erosion is moderate. The series was established in Sierra County, New Mexico, in 1942.

Some properties of the Tres Hermanos soils

Horizon	Depth (cm)	O.M.%	Sand%	Silt%	Clay%	CEC meq/100g	CaCO₃%	pH	BD g/cm³
A	0-3	0.54	65	23	12	9	5	8.5	1.56
Bt	3-43	0.60	58	25	17	11	11	8.1	1.46
Btk	43-114		46	31	23	8	29	8.8	1.48
BCtk	114-147		62	21	17	7	17	8.4	1.50
2C	147-160		40	32	28	16	17	8.5	1.45

Trix (HA1). Typic Torrifluvents; fine-loamy, mixed (calcareous), hyperthermic (Alluvial). Trix soils are deep and well drained. Typically, they have a light brown clay loam surface layer about 35 cm (14 in) thick. The underlying material is light brown clay loam about 25 cm (10 in) thick. Below this is a buried argillic horizon. The upper 33 cm (13 cm) of the buried horizon is light reddish brown clay loam. The lower part to a depth of 150 cm (60 cm) is light brown loam. Depth to the buried argillic horizon ranges from 50 to 100 cm (20 to 40 in). Trix soils occur on floodplains and low stream terraces with slopes of 0 to 1 percent. These soils have high available water capacity and moderately slow permeability. They are moderately alkaline and calcareous throughout. Runoff is slow and the hazard of erosion is slight. The series was established in Maricopa County in 1969.

Tubac (TS3, TS19). Typic Paleargids; fine, mixed, thermic (Reddish Brown). Tubac soils are deep and well drained. Typically, they have reddish brown gravelly sandy loam and loam surface layers about 35 cm (14 in) thick underlain abruptly by dark red clay subsoils about 43 cm (17 in) thick. This is underlain to 150 cm (60 in) or more by reddish brown calcareous gravelly sandy clay loam. Tubac soils are on fan terraces with slopes of 0 to 5 percent. These soils have high available water capacity and slow permeability. They are medium acid to neutral in the surface layers and moderately alkaline in the subsoil and substratum. They are usually noncalcareous in the surface and upper subsoil, but become calcareous below about 50 cm (20 in). Runoff is slow and the hazard of erosion is slight. The series was established in Santa Cruz County in 1930 and the name was taken from the town of Tubac.

Some properties of the Tubac soils

Horizon	Depth (cm)	O.M.%	Sand%	Silt%	Clay%	CEC meq/100g	CaCO$_3$%	pH	BD g/cm^3
A	0-28	0.61	61	28	11	8	0	6.4	1.52
Bt	28-53	0.44	36	12	52	31	1	7.7	1.45
Btk	53-112	0.07	61	15	24	19	T	7.7	1.50
C	112-178	0.03	83	6	11	14	T	7.6	1.4

Tusayan (MS2). Typic Calciorthids; loamy-skeletal, carbonatic, mesic (Calcisols). Tusayan soils are moderately deep and well drained. Typically, they have a brown gravelly sandy loam and gravelly loam surface layer about 25 cm (10 in) thick. The subsoil is light brownish gray and brown gravelly clay loam and very gravelly loam. The underlying material to a depth of 73 cm (29 in) is light brown very gravelly loam. Below this is light brown calcareous sandstone. Depth to bedrock ranges from 50 to 100 cm (20 to 40 in). Tusayan soils occur on hillslopes and fan terraces with slopes of 0 to 8 percent. These soils have low available water capacity and moderate permeability. They are moderate to strongly alkaline and calcareous throughout. Runoff is medium and the hazard of erosion is slight. The series was established in Coconino County in 1975.

Some properties of the Tusayan soils

Horizon	Depth (cm)	O.M.%	Sand%	Silt%	Clay%	CEC meq/100g	CaCO$_3$%	pH	BD g/cm^3
A	0-25	0.99					11	8.4	
Bw	25-41	1.08	58	23	19		25	8.3	
Bk1	41-69	0.92	61	24	15		34	8.4	
Bk2	69-74	0.65	59	27	14		54	8.5	
R	74-								

Valencia (HA1, HA8, TS18). Typic Torrifluvents; coarse-loamy, mixed (calcareous), hyperthermic (Alluvial). Valencia soils are deep and well drained. Typically, they have a brown sandy loam surface layer about 25 cm (10 in) thick and a light brown subsurface layer about 40 cm (16 in) thick over a buried substratum that is brown sandy clay loam and clay loam to a depth of 150 cm (60 in). Valencia soils occur on alluvial fans with slopes of 0 to 2 percent. These soils have moderate available water capacity and moderately rapid permeability in the upper part and moderately slow permeability in the buried part. The soils are moderately alkaline and are mildly calcareous throughout. Runoff is slow and the hazard of erosion is slight to moderate. The series was established in 1969 in Maricopa County.

Vecont (HA2, HA3). Typic Haplargids; fine, mixed, hyperthermic (Red Desert). Vecont soils are deep and well drained. They typically have brown clay loam or clay surface layers about 35 cm (14 in) thick, reddish brown clay subsoils about 68 cm (27 in) thick and a reddish brown clay loam substratum to more than 150 cm (60 in). Vecont soils occur on low alluvial fans and stream terraces with slopes of 0 to 2 percent. The soils have high available water capacity and slow permeability. They are moderately alkaline and weakly to strongly calcareous throughout. Surface runoff is slow and the hazard of erosion is generally slight. The series was established in Maricopa County in 1971.

Some properties of the Vecont soils

Horizon	Depth (cm)	O.M.%	Sand%	Silt%	Clay%	CEC meq/100g	CaCO$_3$%	pH	BD g/cm^3
A	0-28	0.92	36	23	41	23	1	8.7	1.76
Bt	28-74	0.30	34	24	42	23	2	9.1	1.90
BCtk	74-178	0.17	30	26	43	23	6	9.2	1.80

Venezia (MS7). Lithic Haplustolls; loamy, mixed, mesic (Lithosols). Venezia soils are shallow and very shallow and well drained. Typically, they have a brown very stony loam surface layer about 5 cm (2 in) thick. The underlying material to a depth of 25 cm (10 in) is brown heavy loam. Dark gray extremely hard basalt is at a depth of 25 cm (10 in). Depth to bedrock ranges from 13 to 40 cm (5 to 16 in). Venezia soils occur on hillslopes with slopes of 5 to 40 percent. These soils have low available water capacity and moderate permeability. They are neutral to moderately alkaline and noncalcareous throughout. Runoff is medium to rapid and the hazard of erosion is slight.

The series was established in Yavapai County in 1965 and the name was taken from a small settlement between Prescott and Crown King.

Vint (HA1, HA7). Typic Torrifluvents; sandy, mixed, hyperthermic (Alluvial). Vint soils are deep and well drained. They typically have pale brown loamy fine sand surface layers about 40 cm (16 in) thick over pale brown loamy fine sand to more than 150 cm (60 in) that is very thinly stratified with silt loam, loam or very fine sandy loam. Vint soils occur on floodplains with slopes of 0 to 3 percent. These soils have low available water capacity, rapid permeability and are moderately alkaline and calcareous. Runoff is slow and the hazard of erosion is slight, but they are subject to wind erosion. The series was established in Maricopa County in 1971 as the hyperthermic equivalent of the Vinton series.

Some properties of the Vint soils

Horizon	Depth (cm)	O.M.%	Sand%	Silt%	Clay%	CEC meq/100g	CaCO$_3$%	pH	BD g/cm^3
Ap	0-36	0.41	87	9	4	8	T	8.1	1.27
C	36-114		92	5	3	8	T		
2C	114-157		97	2	1	6	T		

Vinton (TA2, TS2). Typic Torrifluvents; sandy, mixed, thermic (Alluvial). Vinton soils are deep and well drained. They typically have brown loamy sand surface layers 15 to 40 cm (6 to 16 in) thick and yellowish brown loamy sand or loamy fine sand subsurface layers to 150 cm (60 in) or more. Vinton soils occur on floodplains with slopes of 0 to 2 percent. These soils have low available water capacity and rapid permeability. They are moderately alkaline and calcareous throughout. They have slow runoff and a low water erosion hazard, but they are subject to wind erosion. The series was established in New Mexico in 1940.

Waldroup (MS7). Udic Rhodustalfs; fine, montmorillonitic, mesic (Brown). Waldroup soils are deep and moderately deep and well drained. Typically, they have a reddish brown silty clay loam and gravelly silty clay loam surface layer about 18 cm (7 in) thick. The subsoil is dark reddish brown and weak red clay and gravelly clay about 70 cm (28 in) thick. The upper 8 cm (3 in) of underlying material is reddish brown and gray very gravelly clay loam. The lower portion of underlying material to a depth of 125 cm (50 in) is mottled red, gray and yellowish red weathered cinders. Waldroup soils occur on cinder cones and basalt flows with slopes of 2 to 50 percent. These soils have moderate available water capacity and slow permeability. They are neutral to moderately alkaline and calcareous in the underlying material. Runoff is medium to rapid and the hazard of erosion is slight. The series was established in Yavapai County in 1965.

Welring (MS2). Lithic Ustic Torriorthents; loamy-skeletal, carbonatic, mesic (Lithosols). Welring soils are shallow and well drained. Typically, they have a brown very gravelly loam surface layer about 8 cm (3 in) thick. The underlying material is pale brown gravelly loam and very gravelly loam about 40 cm (16 in) thick. Fractured limestone bedrock is at a depth of 48 cm (19 in). Depth to bedrock ranges from 25 to 50 cm (10 to 20 in). Welring soils occur on hillslopes with slopes of 30 to 70 percent. These soils have low available water capacity and moderate permeability. They are mildly to moderately alkaline and calcareous below the surface layer. Runoff is rapid and the hazard of erosion is moderate. The series was established in Washington County, Utah, in 1972.

White House (TS4, TS5, TS7, TS8, TS10, TS15, MH1). Ustollic Haplargids; fine, mixed, thermic (Reddish Chestnut). White House soils are deep and well drained. They typically have a surface layer of brown gravelly loam, about 8 cm (3 in) thick over a reddish brown and dark red clay upper subsoil that grades at about 65 cm (26 in) to yellowish red, lime mottled, gravelly sandy clay loam to more than 150 cm (60 in). White House soils occur on fan terraces with slopes of 0 to 35 percent. These soils have high available water capacity and slow permeability. They are medium acid to neutral in the upper layers and moderately alkaline and calcareous below about 50 cm (20 in). Runoff is slow to medium and the hazard of erosion is moderate. The series was established in Pima County in 1931.

Some properties of the White House soils

Horizon	Depth (cm)	O.M.%	Sand%	Silt%	Clay%	CEC meq/100g	CaCO$_3$%	pH	BD g/cm^3
A	0-8	1.88	46	38	16	12	0	5.3	1.42
Bt	8-56	1.46	22	22	56	30	0	6.8	1.56
Btk	56-124	0.29	37	24	39	45	8	8.1	1.52
BCk	124-157		54	18	28	43	1	7.9	1.45
Ck	157-198		52	24	24	43	1	7.8	1.36

Whitlock (TA4, TS9, TS12). Typic Calciorthids; coarse-loamy, mixed, thermic (Calcisols). Whitlock soils are deep and well drained. Typically, they have a light brown sandy loam surface layer about 30 cm (12 in) thick. The upper 43 cm (17 in) of underlying material is pink sandy loam. The lower portion of underlying material to a depth of 150 cm (60 in) or more is light brown sand. Depth to the calcic horizon, which has more than 15 percent calcium carbonate, ranges from 25 to 70 cm (10 to 28 in). Whitlock soils occur on fan terraces with slopes of 0 to 5 percent. These soils have moderate available water capacity and moderately rapid permeability. They are moderately alkaline and calcareous throughout. Runoff is medium to slow and the hazard of erosion is moderate. The series was established in Graham County in 1936 and the name was taken from the Whitlock Mountains.

Wildcat (FH3). Aquic Eutroboralfs; fine, montmorillonitic (Planosols). Wildcat soils are moderately deep to deep and somewhat poorly drained. Typically they have a 1 cm (0.4 in) layer of partially decomposed pine needles and leaves on the surface. The surface mineral layer is light brownish gray and light gray gravelly fine sandy loam and loam about 18 cm (7 in) thick. The subsoil is mottled brown, red, light brownish gray and reddish yellow clay about 63 cm (25 in) thick. Very pale brown and red sandstone is at a depth of 80 cm (32 in). Depth to bedrock ranges from 50 to 125 cm (20 to 50 in). Wildcat soils occur on hillslopes with slopes of 1 to 15 percent. These soils have high available water capacity and slow permeability. They are slightly acid to moderately alkaline and noncalcareous throughout. Runoff is medium and the hazard of erosion is slight. The series was established in Coconino County in 1960 and the name was taken from a nearby spring.

Some properties of the Wildcat soils

Horizon	Depth (cm)	O.M.%	Sand%	Silt%	Clay%	CEC meq/100g	B.S.%	pH	BD g/cm^3
E	0-18	1.79	43	46	11	11	68	6.0	1.58
Bt	18-60	1.17	21	28	51	40	79	5.0	1.93
R	60-								

Wineg (MS9, MS10). Typic Argiustolls; fine-loamy, mixed, mesic (Brown). Wineg soils are deep and well drained. They typically have a grayish brown sandy loam surface layer about 5 cm (2 in) thick. The subsoil is dark brown clay loam about 15 cm (6 in) thick and the substratum is light brown gravelly sandy clay loam about 15 cm (6 in) thick. The underlying material to a depth of 150 cm (60 in) or more is pinkish gray and pink sandy loam. Wineg soils occur on fan terraces with slopes of 1 to 15 percent. These soils have high available water capacity and moderately slow permeability. They are neutral to moderately alkaline and calcareous in the underlying material. A calcic horizon with more than 15 percent calcium carbonate is between 33 and 100 cm (13 to 40 in). Runoff is medium and the hazard of erosion is moderate. The series was established in Yavapai County in 1962.

Winkel (TA5). Typic Paleorthids; loamy-skeletal, mixed, thermic, shallow (Calcisols). Winkel soils are shallow and well drained. Typically, they have a reddish brown gravelly fine sandy loam surface layer about 15 cm (6 in) thick. The underlying material is light reddish brown very cobbly fine sandy loam about 25 cm (10 in) thick. An indurated carbonate hardpan is at a depth of 40 cm (16 in). Depth to the hardpan ranges from 28 to 48 cm (11 to 19 in). Winkel soils occur on mesas with slopes of 1 to 10 percent. These soils have very low available water capacity and moderate permeability above the hardpan. They are strongly alkaline and calcareous throughout. Runoff is slow to medium and the hazard of erosion is slight. The series was established in Washington County, Utah, in 1972.

Winona (MS2, MS5). Lithic Ustollic Calciorthids; loamy-skeletal, carbonatic, mesic (Lithosols). Winona soils are shallow and very shallow and well drained. They typically have brown, gravelly and cobbly loam surface layers about 30 cm (12 in) thick that grades to yellowish brown cobbly loam about 8 cm (3 in) thick overlying pale yellow, massive, dense limestone. Depth to bedrock ranges from 15 to 50 cm (6 to 20 in). Winona soils occur on limestone and calcareous sandstone hillslopes with slopes of 2 to 45 percent. These soils have very low available water capacity and moderate permeability. They are moderately alkaline and calcareous throughout. Runoff is medium and the hazard of erosion is slight to moderate. The series was established in Coconino County in 1967 in the Long Valley area.

Some properties of the Winona soils

Horizon	Depth (cm)	O.M.%	Sand%	Silt%	Clay%	CEC meq/100g	CaCO$_3$%	pH	BD g/cm^3
A	0-5	2.55	55	32	13	20	3	7.7	
Bk	5-38	2.84	48	29	23	23	27	7.9	
R	38-								

Zeniff (MH3). Typic Haplustalfs; fine-loamy, mixed, mesic (Reddish Chestnut). Zeniff soils are deep and well drained. Typically, they have a light brown fine sandy loam surface layer about 35 cm (14 in) thick. The subsoil is brown loam and silt loam to 150 cm (60 in). Zeniff soils occur on stream terraces with slopes of 0 to 10 percent. These soils have high available water capacity and moderate permeability. They are mildly alkaline and dominantly noncalcareous throughout. Runoff is slow and the hazard of erosion is high. The series was established in the Holbrook-Showlow area, Navajo County, in 1957.

Some properties of the Zeniff soils

Horizon	Depth (cm)	O.M.%	Sand%	Silt%	Clay%	CEC meq/100g	CaCO$_3$%	pH	BD g/cm^3
A	0-35	1.44	54	33	13	17	0	7.2	
Bt	35-150	1.28	32	46	24		0	7.3	

Ziegler (MS4, MS7). Aridic Argiustolls; clayey over fragmental, montmorillonitic, mesic (Reddish Chestnut). Ziegler soils are moderately deep and well drained. They typically have a reddish brown gravelly clay loam surface layer about 8 cm (3 in) thick. The subsoil is reddish brown gravelly clay, gravelly clay loam and very gravelly clay loam about 53 cm (21 in) thick. The underlying material to a depth of 150 cm (60 in) or more is light reddish brown and gray cinders. Depth to cinders ranges from 50 to 100 cm (20 to 40 in). Ziegler soils occur on cinder cones with slopes of 1 to 40 percent. These soils have moderate available water capacity and moderate to moderately slow permeability. They are neutral to moderately alkaline and calcareous in the lower subsoil and underlying material. Runoff is medium and the hazard of erosion is slight. The series was established in Apache County in 1966.

Some properties of the Ziegler soils

Horizon	Depth (cm)	O.M.%	Sand%	Silt%	Clay%	CEC meq/100g	B.S.%	pH	BD g/cm^3
A	0-8	3.36	18	60	22	26	78	6.5	
Bt	8-68	1.45	18	30	52	50	90	6.9	
Cr	68-90	0.29	87	9	4		92	7.8	

Appendix F
Glossary

aeolian—Pertaining to the wind, especially as it affects rocks, soils and deposits such as loess, dune sand and some volcanic tuffs the constituents of which were transported (blown) and laid down. Also pertains to sedimentary structures such as ripple made by wind or to geologic processes such as erosion and deposition accomplished by the wind.

aggregate, soil—See soil aggregate.

Albaqualfs—A great group in the suborder Aqualfs of Alfisols that characteristically exhibits somewhat poor to poor drainage.

albic horizon—A mineral soil horizon from which clay and free iron oxides have been removed or in which the oxides have been segregated to the extent that the color of the horizon is determined mostly by the color of the primary sand and silt particles rather than by the coatings on these particles.

Alfisols—The Order of mineral soils that has umbric or ochric epipedons, argillic horizons, and that holds water at less than 15 bars tension during at least three months when the soil is warm enough for plants to grow outdoors. Alfisols have a mean annual soil temperature of less than 8 C (46.4 F) or a base saturation in the part of the argillic horizon of 35 percent or more when measured at pH 8.2.

alluvium—Soil material deposited by water flowing in streams.

amphibole—A group of dark, rock-forming, ferromagnesian silicate minerals that are closely related in crystal form and composition. It is characterized by a cross-linked double chain of tetrahedra with a silicon:oxygen ratio of 4:11, by columnar or fibrous prismatic crystals and by good prismatic cleavage in two directions parallel to the crystal faces and intersecting at angles of about 56° and 124°. Amphibole varies in color from white to black and is an abundant and widely distributed constituent in igneous and metamorphic rocks.

andesite—A dark-colored, fine-grained extrusive rock.

anthropic epipedon—Said of an epipedon that is similar to a mollic epipedon but in which the content of soluble P_2O_5 is greater than 250 ppm. It develops due to long periods of cultivation and fertilization.

anticline—A fold, the core of which contains the stratigraphically older rocks. It is convex upward.

argillans—Clay skins; coatings (cutans) composed mostly of clay on the surfaces of blocky peds and of stones and lining void walls in the subsoil.

argillic horizon—A mineral soil horizon that is characterized by the illuvial accumulation of layer-lattice silicate clays. The argillic horizon has a certain minimum thickness depending on the thickness of the solum, a minimum quantity of clay in comparison with an overlying eluvial horizon depending on the clay content of the eluvial horizon, and usually has coatings of oriented clay on the surface of pores or peds or bridging sand grains.

Aridisols—The Order of mineral soils that has an aridic moisture regime, an ochric epipedon, and other pedogenic horizons but no oxic horizon.

association, soil—See soil association.

augite—A term often used as a synonym of pyroxene. It may contain titanium and ferric iron. Augite usually is black, greenish black or dark green and commonly occurs as an essential constituent in many basic igneous rocks and in certain metamorphic rocks.

authigenic—Said of constituents and minerals that have not been transported or that were derived where they are now. Also said of minerals that came into existence at the same time, or subsequently to, the formation of the rock of which they constitute a part. The term, as used, often refers to a mineral such as quartz or feldspar formed after deposition of the original sediment.

badland—A land type generally devoid of vegetation and broken by an intricate maze of narrow ravines, sharp crests and pinnacles resulting from serious erosion of soft geologic materials. Most common in arid or semiarid regions.

bar—A term used as an international unit of pressure equal to .987 atmospheres. Negative pressure or tension of soil water is measured in bars (b) and millibars (mb).

basalt—A term used to describe dark-colored, fine-grained igneous rocks that are either intrusive or extrusive.

biotic community—A plant and animal association consisting of one or more plants and animals that is distinct from other biotic communities.

biotite—A widely distributed and important rock-forming mineral of the mica group. It is generally black, dark brown or dark green and forms a constituent of crystalline rocks or a detrital constituent of sandstones and other sedimentary rocks. Also a general term to designate all ferromagnesian micas. Biotite is known also as black mica, iron mica and magnesia mica.

Boralfs—A suborder of Alfisols that has formed in cool places. Boralfs have frigid or cryic but not pergelic temperature regimes and have udic moisture regimes. Boralfs are not saturated with water for periods long enough to limit their use for most crops.

boreal—Pertaining to the northern biotic area that is characterized by tundra and taiga and by dominant coniferous forests.

breccia—A coarse-grained clastic rock composed of large (greater than sand size, or 2 mm in diameter), angular and broken rock fragments cemented together in a finer-grained matrix that may or may not be similar to the larger fragments and that can be of any composition, origin or mode of accumulation; the consolidated equivalent of rubble. Breccia is similar to conglomerate except that most of the fragments have sharp edges and unworn corners; the term formerly included conglomerate and is still sometimes so used in Europe. The rock can be formed in many ways, but chiefly by sedimentation and igneous and tectonic activities.

calcareous soil—Soil containing free lime (carbonates) that effervesces visibly when treated with diluted (1:10) hydrochloric acid.

calcic horizon—A mineral soil horizon of secondary carbonate enrichment that is more than 15 cm (6 in) thick, has a calcium carbonate equivalent of more than 15 percent and has at least 5 percent more calcium carbonate equivalent than the underlying C horizon.

cambic horizon—A mineral soil horizon that has a texture of loamy very fine sand or finer, has soil structure rather than rock structure, contains some weatherable minerals and is characterized by the alteration or removal of mineral material as indicated by mottling or gray colors, stronger chromas or redder hues than in underlying horizons, or the removal of carbonates. The cambic horizon lacks cementation or induration and has too few evidences of illuviation to meet the requirements of the argillic or spodic horizon.

carbonate—A sediment formed by the organic or inorganic precipitation from an aqueous solution of carbonates of calcium, magnesium or iron, such as limestone or dolomite.

cation—An atom, a group of atoms, or compounds that are positively charged electrically as the result of the loss of electrons.

chaparral—A thicket of shrubs and thorny bushes.

chlorosis—An abnormal condition of plants in which the green parts lose their color or turn yellow.

chronosequence—A sequence of related soils that differ from one another in certain aspects primarily as a result of time as a soil-forming factor.

cienega—A marshy area where the ground is wet due to the presence of seepage or springs, often with standing water and abundant vegetation. The term is commonly applied in arid regions such as the U.S. Southwest.

circumpolar (plant communities)—Pertains to low-lying plants that are commonly found around the poles, such as those plants that grow in both Arctic Eurasia and Arctic North America.

clastic—Pertaining to or being a rock or sediment composed principally of broken fragments that are derived from pre-existing rocks or minerals and that have been transported individually for some distance from their places of origin. Also said of the texture of such a rock. The term is often used to indicate a source from within the depositional basin, as compared with detrital. The commonest "clastics" are sandstones and shales.

clay—The smallest mineral grains, less than 0.002 mm (0.000079 in) in diameter.

cleft—An abrupt chasm, cut, breach or other sharp opening such as a craggy fissure in a rock, a wave-cut gully in a cliff, a trench in the ocean bottom, a notch in the rim of a volcanic crater or a narrow recess in a cave floor.

cliff—Any high, very steep to perpendicular or overhanging face of rock (sometimes earth or ice) occurring in the mountains or rising above the shore of a lake or river; a precipice. A cliff is usually formed by erosion, less commonly by faulting.
  a. buttressed—A cliff that has a protruding rock mass that resembles the buttress of a building; a spur running down from a steep slope.
  b. recessed—A cliff that has lost part of its face through the processes of weathering and erosion.

climax species—The most advanced plant or plants able to grow under and in dynamic equilibrium with the prevailing environment.

clod—A compact, coherent mass of soil ranging in size from 0.5 to 1.0 cm (0.2 to 0.4 in) to as much as 20 to 25 cm (8 to 10 in). It is produced artificially, usually by the activity of man by plowing, digging, etc., especially when these operations are performed on soils that are either too wet or too dry for normal tillage operations.

colluvium—A deposit of rock fragments and soil material

accumulated at the base of steep slopes as a result of gravitational action.

complex, soil—See soil complex.

coniferous—Pertaining to any of a large group of cone-bearing trees and shrubs, mostly evergreen, such as pines, spruce and cedar.

consistence, soil—See soil consistence.

cryic—A soil temperature regime that has mean annual soil temperatures of more than 0 C (32 F) but less than 8 C (46 F), more than 5 C (9 F) difference between mean summer and mean winter soil temperatures at 50 cm (20 in) and cold summer temperatures.

cuesta—A hill or ridge with a gentle slope on one side and a steep slope or cliff on the other. Cuestas are common in the U.S. Southwest.

cutan—A modification of the texture, structure or fabric of soil material along a natural surface within it and caused by a concentration of a particular soil constituent. It can be composed of any of the component substances of the soil material.

dacite—A fine-grained extrusive rock with the same general composition as andesite but having less calcic feldspar.

deciduous—Pertaining to vegetation that sheds leaves annually as opposed to plants that are evergreen.

detrital—Pertaining to or formed from detritus. Said especially of rocks, minerals and sediments. The term is often used to indicate a source from outside the depositional basin, as compared with clastic.

dioritic—Pertaining to a group of rocks intermediate in composition between acidic and basic rocks. They characteristically are composed of dark-colored amphibole, acid plagioclase, pyroxene and sometimes a small amount of quartz.

dolomite—Magnesian limestone that consists mostly of the mineral dolomite.

dolomitic—Pertaining to that which contains dolomite.

drainage, soil—See soil drainage.

druse—An irregular cavity or opening in a vein or rock, having its interior surface or walls lined (encrusted) with small projecting crystals usually of the same minerals as those of the enclosing rock, and sometimes filled with water.

duricrust—The hardened crust formed in soil and porous rock by cementation, particularly with siliceous, ferruginous or aluminous precipitates.

duripan—A horizon in a mineral soil that is characterized by cementation by silica and possibly by accessory cements.

ecotone—A transition zone between two ecologic communities. Members of both communities may compete within this zone.

eluvial horizon—A horizon that has lost bases, iron, clay, etc., through the soil-forming processes. E horizons are eluvial.

encinal—Pertaining to the live oak, especially the California live oak.

Entisols—The Order of mineral soils that has no distinct pedogenic horizons within 1 m (3.3 ft) of the soil surface.

epipedon—Soil horizon that forms at the surface of the soil.

equant—Said of a crystal in an igneous or sedimentary rock having the same or nearly the same diameters in all directions.

erosion—The wearing away of the land surface by running water, wind, ice or other geological agents, including such processes as gravitational creep, or detachment and movement of soil or rock by these processes.

   a. creep—Slow mass movement of soil and soil material down relatively steep slopes primarily under the influence of gravity, but facilitated by saturation with water and by alternate freezing and thawing.

   b. rill and gully—Rill erosion is the process in which numerous small channels several centimeters in depth are formed. This action is sometimes followed by gully erosion, a process whereby water accumulates in the narrow channels (rills) and over short periods removes the soil to considerable depths, ranging from 0.5 m (1.65 ft) to as much as 25 to 30 m (82.5 to about 100 ft).

escarpment—A long, more or less continuous cliff or relatively steep slope facing in one general direction. It breaks the general continuity of the land by separating two level or gently sloping surfaces. Erosion or faulting creates an escarpment.

evapotranspiration—The combined loss of water from a given area during a specified period of time by evaporation from the soil surface and by transpiration from plants.

exchangeable cation—A positively charged ion attached in available forms to clay and organic constituents of soils. Included are those of hydrogen and the alkali and alkaline earth metals, calcium, magnesium, potassium and sodium. These plant nutrients can be exchanged with each other and with other positively charged ions in soil solutions.

extragrade subgroup—That subgroup within a great group that has aberrant properties that do not represent intergrades to any known kind of soil identified in the hierarchical soil classification system. See intergrade subgroup and typic subgroup.

fault—A surface or zone of rock fracture along which displacement has occurred, from a few centimeters to a few kilometers.

   a. block—A type of normal faulting in which the crust is divided into structural or fault blocks of different elevations and orientations. It is the process by which block mountains are formed.

   b. thrust—A fault with a dip of 45o or less in which the hanging wall appears to have moved upward relative to the footwall. Horizontal compression rather than vertical displacement is its characteristic feature.

feldspar—A group of abundant rock-forming minerals that constitute 60 percent of the Earth's crust. They occur as components of all kinds of rocks and as fissure minerals in clefts and druse minerals in cavities. Feldspars are usually white or nearly white and clear and translucent, although they frequently are colored by impurities. On decomposition, feldspars yield a large part of the clay of soil and the mineral kaolinite.

ferromagnesian—Containing iron and magnesium; applied to mafic minerals.

field grading of soil texture—Soil surveyors routinely do field grading of soil texture by rubbing soil between finger and thumb. With experience, a person can judge by the feel of the soil how much sand, silt and clay are present. According to the proportions of these materials textural class names are given to soils. Textures intermediate between those described below can be recognized by the relative amounts of gritty, soft and sticky material in them.

1. Stones, cobbles and gravel are coarse fragments, all with diameters greater than 0.2 cm (0.079 in), can be recognized by eye and measured with a rule. Fragments of gravel size are less than 7.6 cm (3 in) in diameter, cobbles are between 7.6 and 25.4 cm (3 and 10 in) in diameter and stones or boulders are larger.
2. Sand feels gritty and harsh. Individual grains can be seen and felt. Squeezed when moist, the soil forms a fragile cast.
3. Sandy loam feels quite gritty but also somewhat loamy. Individual sand grains can be seen and felt. There is enough silt and clay to soften the feel of this soil. Squeezed when dry, the soil forms a somewhat stable cast.
4. Loam feels somewhat gritty, somewhat smooth and possibly a little sticky and plastic. Sometimes a person calls the soil a loam because it is not sandy enough to be a sandy loam, silty enough to be a silt loam, nor clayey enough to be a clay loam. Squeezed when dry, it forms a fragile cast. Squeezed when moist, it forms a stable cast.
5. Silt loam lumps and clods prove to be very fragile when dry. When rubbed, this soil feels soft like flour and forms a fairly stable cast when squeezed. When moist this soil feels smooth. The moist cast is stable. Moist soil will not form a polished ribbon when rubbed between the thumb and finger, but will appear as a somewhat rough and non-coherent coating on the thumb and finger.
6. Clay loam is hard and lumpy when dry. Moist soil is plastic, forms a very stable cast when squeezed, and when rubbed between the thumb and finger forms a thin, somewhat fragile ribbon with a somewhat polished surface. The moist soil can be kneaded in the hand into a compact mass that does not crumble readily.
7. Clay is very hard and lumpy when dry. Moist soil is very plastic and sticky, and forms a stable cast. Elongated casts may sag under their own weight. When rubbed between the thumb and finger the soil forms a long, flexible ribbon that has a good surface polish.

filamentous actinomycetes—A non-taxonomic term applied to a threadlike group of organisms with characteristics intermediate between simple bacteria and true fungi. Most of these soil organisms are unicellular and produce a slender branched mycelium. They sporulate by segmentation of the entire mycelium, or more commonly by segmentation of special hyphae.

fine earth—Soil particles less than 2.0 mm (0.08 in) in diameter. Most soil analyses are made of fine earth, sand, silt and clay, excluding coarser fragments of gravel, stones and cobbles.

fissility—A general term for the property possessed by some rocks of splitting easily into thin sheets or layers along closely spaced, roughly planar, and approximately parallel surfaces, such as along bedding planes as in shale or along cleavage planes as in schist induced by fracture of flowage. Its presence distinguishes shale from mudstone. The term is not applied to minerals, but is analogous to cleavage in minerals.

fluvial—Produced by the action of a stream or river. For instance, a fluvial deposit consists of material transported by, suspended in, or laid down by a river or stream.

foliate—Generally said of the planar arrangement of textural or structural features in any type of rock; e.g., cleavage in slate or schistosity in metamorphic rock. It is most commonly applied to a metamorphic rock.

fragipan—A natural subsurface horizon with high bulk density relative to the solum above. It is seemingly cemented when dry, but shows a moderate to weak brittleness when moist. The layer is low in organic matter, mottled, slowly or very slowly permeable to water, and usually shows occasional or frequent bleached cracks forming polygons. It may be found in profiles of either cultivated or virgin soils but not in calcareous material.

frigid soil—A soil temperature regime that has mean annual soil temperatures of more than 0 C (32 F) but less than 8 C (46 F), more than 5 C (9 F) difference between mean summer and mean winter soil temperatures at 50 cm (20 in), and warm summer temperatures. Isofrigid is the same except the summer and winter temperatures differ by less than 5 C (9 F).

gabbro—A group of dark-colored, basic intrusive igneous rocks composed principally of basic plagioclase and augite, with or without olivine and orthopyroxene. It is the approximate intrusive equivalent of basalt. Gabbro grades into monzonite with increasing alkali-feldspar content.

gangue—The valueless rock or mineral aggregates in an ore; that part of an ore that is not economically desirable.

geomorphic—Pertaining to the form of the Earth or of its surface features.

gilgai—The microrelief of soils produced by expansion and contraction with changes in moisture. Found in soils that contain large amounts of clay that swells and shrinks considerably with wetting and drying. Usually a succession of microbasins and microknolls in nearly level areas or of microvalleys and microridges parallel to the direction of the slope.

glacial till—Unsorted glacial drift transported and deposited by ice.

glaciofluvial deposits—Sediments deposited by glacial streams. These deposits are usually sandy or gravelly and typically stratified.

glaciolacustrine deposits—Sediments deposited in glacial lakes. These include fine sands, silts and clays. They may be stratified or varved.

glaebules—A three-dimensional unit, usually prolate to equant in shape, within the matrix of a soil material. It is recognizable by a greater concentration of some constituent and by its difference in fabric as compared with the enclosing soil material . . . a small clod or lump of earth.

gleying—The process of soil mottling caused by partial oxidation and reduction of its constituent ferric iron compounds due to conditions of intermittent water saturation.

gneiss—A foliated rock formed by regional metamorphism in which bands or lenticles of granular minerals alternate with bands and lenticles in which minerals having flaky or elongate prismatic habits predominate. Generally, less than 50 percent of the minerals show the preferred parallel orientation. Although a gneiss is commonly feldspar- and quartz-rich, the mineral composition is not an essential factor in its definition. Varieties are distinguished by texture, characteristic minerals or general composition and/or origin.

graben—An elongate, relatively depressed crustal unit or block that is bounded by faults on its long sides. It is a structural form that may or may not be geomorphologically expressed as a rift valley.

granodiorite—A group of coarse-grained plutonic rocks intermediate in composition between quartz diorite and quartz monzonite containing quartz, plagioclase, and potassium feldspar, with biotite, hornblende, or, more rarely, pyroxene, as the mafic components. The ratio of plagioclase to total feldspar is at least two to one but less than nine to ten. With less alkali feldspar it grades into quartz diorite, and with more alkali feldspar, into granite or quartz monzonite.

gypsic horizon—A mineral soil horizon of secondary calcium sulfate enrichment that is more than 15 cm (6 in) thick, has at least 5 percent more gypsum than the C horizon, and in which the product of the thickness in centimeters and the percent calcium sulfate is equal to or greater than 150 percent cm.

histic epipedon—A thin organic soil horizon that is saturated with water at some period of the year unless artificially drained and that is at or near the surface of a mineral soil. The histic epipedon has a maximum thickness depending on the kind of materials in the horizon and the lower limit of organic carbon is the upper limit for the mollic epipedon.

Histosols—The Order of organic soils that has organic soil materials in more than half of the upper 80 cm (32 in), or that is of any thickness if overlying rock or fragmental materials that have interstices filled with organic soil materials.

hogback—Any ridge with a sharp summit and steep slopes of nearly equal inclination on both flanks, and resembling in outline the back of a hog. More specifically, a long, narrow, sharp-crested ridge formed by the outcropping edges of very steeply inclined or highly tilted resistant rocks such as igneous dikes and produced by differential erosion. The term is usually restricted to ridges carved from beds dipping at angles greater than 20 degrees.

hogwallow—A wallow made by swine. Also, a similar depression believed to be formed by heavy rains. Or, a faintly billowing land surface characterized by many low, coalescent or rounded mounds such as mima mounds that are slightly higher than the basin-shaped depressions between them.

horizon, diagnostic—See soil horizon.

horizon, soil—See soil horizon.

horst—An elongate, relatively uplifted crustal unit or block bounded by faults along its long sides. It is a structural form that may or may not be expressed geomorphologically.

humified—Pertains to soil that has decomposed organic matter (humus) within its profile. (See humus.)

humus—That more or less stable fraction of the soil organic matter remaining after most added plant and animal residues have decomposed. Usually it is dark colored.

hygroscopic water—Water adsorbed by a dry soil from an atmosphere of high relative humidity, water remaining in the soil after "air-drying" or water held by the soil when it is in equilibrium with an atmosphere of a specified relative humidity at a specified temperature, usually 98 percent relative humidity at 25 C (77 F).

igneous rock—Rock formed from the cooling and solidification of magma, and that has not changed appreciably since its formation.

ignimbrites—The rocks formed by the deposition and consolidation of ash flows and other volcanic materials.

illuvial horizon—A soil layer or horizon in which material carried from an overlying layer has been precipitated from solution or deposited from suspension. The layer of accumulation.

Inceptisols—The Order of mineral soils that has one or more pedogenic horizons in which mineral materials other than carbonates or amorphous silica have been altered or removed but not accumulated to a significant degree. Under certain conditions, Inceptisols may have an ochric, umbric, histic, plaggen or mollic epipedon. Water is available to plants more than half of the year or more than three consecutive months during a warm season.

indurate—Describes rock or soil compacted and hardened by the action of pressure, cementation and especially heat.

intergrade subgroup—The subgroup that contains soils of one great group but that has some properties characteristic of the soils of another great group or other class. These properties are not developed or expressed enough for the soils to be included within the second great group in the hierarchical soil classification system. (See extragrade subgroup and typic subgroup.)

interstadials—Pertaining to or formed during an interstade, a warmer substage of a glacial stage marked by a temporary retreat of the ice.

interstices—Openings or spaces between one thing and another, as an opening in a rock or soil that is not occupied by solid matter.

ion—An atom, group of atoms or compounds that are electrically charged as a result of the loss of electrons (cations) or the gain of electrons (anions).

kaolinite—A common clay mineral. A two-layer hydrous aluminum silicate having the general formula $Al_2(Si_2O_5)(OH)_4$. It consists of sheets of tetrahedrally coordinated silicon joined by an oxygen shared with octahedrally coordinated aluminium.

krotovina—A former animal burrow in one soil horizon that has been filled in with organic matter or material from another horizon.

laccolithic—Pertains to a concordant igneous intrusion with a known or assumed flat floor and a postulated dikelike feeder somewhere beneath its thickest point. It is generally lenslike in form and roughly circular, less than 8 km (5 mi) in diameter, and from a meter or so to a few hundred meters in thickness.

lacustrine—Pertaining to, produced by, or formed in a lake or lakes. For example, lacustrine sands deposited on the bottom of a lake or a lacustrine terrace formed along the margin of a lake.

latite—A porphyritic extrusive rock having plagioclase and potassium feldspar present in nearly equal amounts as phenocrysts, little or no quartz and a finely crystalline to glassy groundmass that may contain obscure potassium feldspar; the extrusive equivalent of monzonite.

leaching—The removal in solution of soil constituents such as mineral salts and organic matter from an upper to a lower soil horizon by the action of percolating water, either naturally by rainwater or artificially by irrigation.

listric surface—A curvilinear, usually concave-upward surface of fracture that curves at first gently and then more steeply from a horizontal position. Listric surfaces form wedge-shaped masses that appear to be thrust against or along each other.

Lithosols—The Order of azonal soils characterized by an incomplete solum or no clearly expressed soil morphology and consisting of freshly and imperfectly weathered rock or rock fragments. Lithosols usually develop on steep slopes.

lithosphere—The solid portion of the Earth, as compared with the atmosphere and the hydrosphere; the crust of the Earth.

litter layer—The mat of dead, decomposing organic material above mineral soil.

loam—See soil texture.

loamy sand—See soil texture.

loess—Material transported and deposited by wind and consisting mostly of silt-sized particles, including those of quartz, feldspar and carbonates. Loess and loess-derived soils are generally fertile.

maar—A low-relief, coneless volcanic crater formed by a single explosive eruption. It is surrounded by a crater ring and is commonly filled with water.

mafic—Said of an igneous rock composed chiefly of one or more ferromagnesian, dark-colored minerals. Also said of those minerals.

marl—Soft and unconsolidated calcium carbonate, usually mixed with varying amounts of clay or other impurities.

mesic—A soil temperature regime that has mean annual soil temperatures of 8 C (46 F) or more but less than 15 C (59 F), and more than 5 C (9 F) difference between mean summer and mean winter soil temperatures at 50 cm (20 in). Isomesic is the same except the summer and winter temperatures differ by less than 5 C (9 F).

metamorphic rocks—Rocks derived from pre-existing rocks but that differ from them in physical, chemical and mineralogical properties as a result of natural geological processes, principally heat and pressure. The pre-existing rocks may have been igneous, sedimentary or another metamorphic rock.

mica (illite)—A mineral that consists of complex phyllosilicates with sheet-like structures that crystallize in forms apparently orthorhombic or hexagonal but really monoclinic. These structures are characterized by low hardness and by perfect basal cleavage, readily spitting into very thin, tough and somewhat elastic plates that have a splendent pearly luster on their surfaces. They commonly occur as flakes, scales or shreds and vary in color from colorless, silvery white, pale brown, or yellow to green or black. Micas are prominent rock-forming constituents of many igneous and metamorphic rocks.

mima mound—A term used in the U.S. Northwest to identify low, circular or oval domes composed of loose, unstratified gravelly silt and soil material built upon glacial outwash on a hog-wallow landscape. The basal diameter varies from 3 m (10 ft) to more than 30 m (100 ft) and the height from 30 cm (12 in) to about 2 m (7 ft). The mounds probably are built by pocket gophers.

mineral soil—A soil that is composed mainly of mineral matter but that has some organic material also.

mollic horizon—Pertains to a mineral soil horizon that is dark colored and relatively thick, contains at least 0.5 percent organic carbon, is not massive and hard or very hard when dry, has a base saturation of more than 50 percent when measured at pH 7, has less than 250 ppm of P_2O_5 soluble in 1 percent citric acid and is dominantly saturated with bivalent cations.

Mollisols—The Order of mineral soils that has a mollic epipedon overlying mineral material with a base saturation of 50 percent or more when measured at pH 7. Mollisols may have a horizon that is argillic, natric, albic, cambic, gypsic, calcic or petrocalcic, a histic epipedon or a duripan, but not an oxic or a spodic horizon.

monocline—A unit of strata that dips or flexes from the horizontal in one direction only. It is not part of an anticline or syncline. A monocline is generally a large feature of a gentle dip.

montmorillonite—A crystalline clay mineral with a 2:1 expansible layer structure; that is, with two silicon tetrahedral sheets enclosing an aluminum octahedral sheet. It swells considerably with wetting and shrinks considerably when drying.

monzonite—A group of plutonic rocks intermediate in composition between syenite and diorite, containing approximately equal amounts of orthoclase and plagioclase, little or no quartz and commonly augite as the main mafic mineral. With decrease in the alkali feldspar content, monzonite grades into diorite or grabbo, depending on the composition of the plagioclase. With an increase in alkali feldspar, it grades into syenite.

mor—A type of forest humus in which an Oa horizon is present and in which practically no mixing of surface organic matter with mineral soil occurs. The transition from the O layer to A horizon is abrupt.

morphology, soil—See soil morphology.

mull—A type of forest humus in which an Oe horizon may or may not be present and in which no Oa horizon exists. The A horizon consists of an intimate mixture of organic matter and mineral soil. The transition between the A and the horizon beneath is gradual.

muscovite—A mineral of the mica group. It is usually colorless, whitish or pale brown and is common in metamorphic rocks (gneisses and schists), in most acid igneous rocks such as granite and in many sedimentary rocks, especially sandstone.

natric horizon—A mineral soil horizon that satisfies the requirements of an argillic horizon, but that also has a prismatic, columnar or blocky structure and a subhorizon that has more than 15 percent saturation with exchangeable sodium.

nickpoint (also knickpoint)—Any interruption or break of slope. More specifically, it is a point of abrupt change or inflection in the longitudinal profile of a stream or of its valley that occurs where a new curve or erosion (graded to a new base level after a relative lowering of the former level) intersects an earlier curve. A nickpoint results from rejuvenation, glacial erosion or the outcropping of a resistant bed.

ochric horizon—Pertains to a surface horizon of mineral soil that is too light in color, too high in chroma, too low in organic carbon or too thin to be an epipedon that is plaggen, mollic, umbric, anthropic or histic, or that is both hard and massive when dry.

O horizon—See soil horizon.

organic soils—Soils that are saturated with water and have 17.4 percent or more organic carbon if the mineral fraction is 50 percent or more clay, or 11.6 percent organic carbon if the mineral fraction has no clay, or has proportional intermediate contents, or if never saturated with water, have 20.3 percent or more organic carbon.

orogeny—The process by which structures within mountain areas were formed, including thrusting, folding and faulting in the outer and higher layers, and plastic folding, metamorphism and plutonism in the inner and deeper layers.

orthoclase—A colorless, white, cream-yellow, flesh-reddish, or grayish mineral of the alkali-feldspar group. It usually contains some sodium in minor amounts. Ordinary or common orthoclase is one of the commonest rock-forming minerals. It occurs especially in granites, acid igneous rocks and crystalline schists.

oxic horizon—A mineral soil horizon that is at least 30 cm (12 in) thick and characterized by the virtual absence of weatherable primary minerals or 2:1 lattice clays, the presence of 1:1 lattice clays and highly insoluble minerals such as quartz sand, the presence of hydrated oxides of iron and aluminum, the absence of water-dispersible clay, and the presence of low cation exchange capacity and small amounts of exchangeable bases.

paleo-—A combining form denoting the attribute of great age or remoteness in regard to time, or involving ancient conditions, or of ancestral origin or dealing with fossil forms. Also a prefix indicating pre-Tertiary origin, and origin and generally altered character of a rock to the name of which it is added.

parent material—The unconsolidated material, mineral or organic, from which the solum or true soil develops.

particle-size distribution—The amounts of the various soil separates in a soil sample. They are usually expressed as weight percentages.

ped—A unit of soil structure such as an aggregate, crumb, prism, block or granule formed by natural processes. Compare with a clod, which is formed artificially.

pediment—A broad, flat or gently sloping rock-floored erosion surface or plain of low relief. It typically is developed by subaerial agents including running water in an arid or semiarid region at the base of an abrupt and receding mountain front or plateau escarpment. Bedrock and occasionally older alluvial deposits that may be varve underlie the pediment. It is often partly mantled with a thin, discontinuous veneer of alluvium derived from the upland masses and in transit across the surface. The longitudinal profile of a pediment is normally slightly concave upward. Its outward form may resemble a piedmont slope, which continues the forward inclination of a pediment.

pedogenesis—See soil genesis.

pedologist—A soil scientist.

pedon—The smallest unit or volume of soil that represents or exemplifies all the horizons of a soil profile. It is usually a horizontal, more or less hexagonal area of about one square meter, but may be larger.

pedosphere—That shell or layer of the Earth in which soil-forming processes occur.

pedoturbation—Biologic, physical churning and cycling of soil materials that homogenizes the solum in varying degrees. Freeze-thaw and wet-dry cycles are pedoturbations.

pergelic—A soil temperature regime that has mean annual soil temperatures of less than 0 C (32 F). Permafrost is present.

periglacial—Said of the processes, conditions, areas, climates and topography at the immediate margins of former and existing glaciers and ice sheets, and influenced by the cold temperature of the ice. By extension, said of an environment in which frost action is an important factor, or of phenomena induced by a periglacial climate beyond the periphery of the ice.

petrocalcic horizon—A continuous, indurated calcic horizon that is cemented by calcium carbonate and in some places with magnesium carbonate. It cannot be penetrated with a spade or auger when dry, dry fragments do not slake in water and it is impenetrable to roots.

pH, soil—A notation used to designate the acidity or alkalinity of a soil. A pH of 7.0 indicates neutrality. Lower values indicate acidity and higher values indicate alkalinity.

phenocryst—A term first suggested by J.P. Iddings and widely used for a relatively large, conspicuous crystal in a porphyritic rock. The term "inset" has been suggested as a replacement.

phyllosilicate—A class or structural type of silicate characterized by the sharing of three of the four oxygens in each tetrahedron with neighboring tetrahedra to form flat sheets. The silicon to oxygen ratio is 2:5. Micas are an example of phyllosilicate.

phytochronosequence—The succession of plants over time in a given area that continues until the climax species is established.

piping—Erosion by percolating water in a layer of subsoil that causes caving and the formation of narrow conduits, tunnels or "pipes" through which soluble or granular soil material is removed.

plaggen epipedon—A man-made surface horizon more than 50 cm (20 in) thick that is formed by long-continued manuring and mixing.

plagioclase—One of the commonest rock-forming minerals, it has characteristic twinning and commonly displays zoning: sodium-calcium feldspar.

plastic soil—A soil capable of being continuously and permanently molded or deformed into various shapes by relatively moderate pressure.

plutonism—A general term for the phenomena associated with the formation of plutons, igneous intrusions. Also the conception of the formation of the Earth by solidification of a molten mass.

pluvial—Said of a geologic episode, change, process, deposit or feature resulting from the action or effects of rain. For example, pluvial denudation, a landslide, or gully erosion and the consequent spreading out of the eroded material below. The term sometimes includes the fluvial action of rainwater flowing in a stream channel, especially in the channel of an ephemeral stream.

polypedon—An assemblage of contiguous like pedons on a landscape. See soil body.

porphyritic—Said of the texture of an igneous rock in which larger crystals (phenocrysts) are set in a finer groundmass that may be crystalline or glassy or both. Also, said of a rock with such texture or of the mineral forming the phenocrysts. The term is sometimes restricted to cases in which the phenocrysts and groundmass formed during two different crystallization generations.

profile, soil—See soil profile.

prolate—Extended or elongated in the direction of a line joining the poles.

pyroclastic—Pertaining to clastic rock material formed by volcanic explosion or aerial expulsion from a volcanic vent; also, pertaining to rock texture of explosive origin. It is not synonymous with the adjective "volcanic."

pyroxene—A group of dark, rock-forming silicate minerals closely related in crystal form and composition. Pyroxenes are a common constituent of igneous rocks and are analogous in chemical composition to the amphiboles, except that they lack hydroxyls.

quartz—Crystalline silica, an important rock-forming mineral. It is, next to feldspar, the commonest mineral. It occurs either in colorless and transparent hexagonal crystals (sometimes colored yellow, brown, purple, red, green, blue or black by impurities) or in crystalline or cryptocrystalline masses. Quartz is the commonest gangue mineral of ore deposits, forms the major proportion of most sands and has widespread distribution in igneous, metamorphic and sedimentary rocks.

radiometric dating—Calculating an age in years for geologic materials by measuring the presence of a short-life radioactive element, such as carbon-14. Calculations are made also by measuring the presence of a long-life radioactive element plus its decay product, such as potassium-40/argon-40. The term applies to all methods of age determination based on nuclear decay of natural elements.

rhyodacite—A group of extrusive porphyritic igneous rocks intermediate in composition between dacite and rhyolite. Quartz, plagioclase and biotite are the main phenocryst minerals. These rocks have a fine-grained to glassy groundmass composed of alkali feldspar and silica minerals. Rhyodacite is the extrusive equivalent of granodiorite or quartz monzonite.

rhyolitic—Pertains to a group of extrusive igneous rocks. They are generally porphyritic and exhibit flow texture. These rocks have phenocrysts of quartz and alkali feldspar in a glassy to crytocrystalline groundmass. Rhyolite grades into rhyodacite with decreasing alkali feldspar content.

rift—A narrow cleft, fissure or other opening in rock made by cracking or splitting, as in limestone.

salic horizon—A mineral soil horizon of enrichment with secondary salts more soluble than gypsum in cold water. A salic horizon is 15 cm (6 in) or more in thickness and contains at least 2 percent salt. The product of the thickness in centimeters and percent salt by weight is 60 percent cm or more.

sand—A soil particle between 0.05 and 2.0 mm (0.002 and 0.08 in) in diameter.

sand (texture)—See soil texture.

sandy clay loam—See soil texture.

sandy loam—See soil texture.

saprolite—A soft, earthy, clay-rich, thoroughly decomposed rock formed in place by chemical weathering of igneous and metamorphic rocks. It often forms a thick, as much as 100 m (330 ft), layer or cover, especially in a humid and tropical or subtropical climate. The color is commonly some shade of red or brown.

scarp—See escarpment.

schist—A strongly foliated crystalline rock formed by dynamic metamorphism that can be readily split into thin flakes or slabs due to the well-developed parallelism of more than 50 percent of the minerals present, particularly those of lamellar or elongate prismatic habit.

schistose—Said of a rock displaying schistosity, the foliation in schist or other coarse-grained, crystalline rock due to the parallel, planar arrangement of mineral grains of the platy, prismatic or ellipsoidal types, usually mica.

sclerenchyma—A tissue in higher plants composed of cells that have thickened and lignified cell walls and often mineralized. They are usually without living protoplasm and incapable of further growth when mature. Sclerenchyma is supporting or protective tissue. Its cells are of two general types, elongated, pointed cells and shorter cells of irregular size and shape.

sclerophyllous—Pertains to an excessive development of sclerenchyma in leaves. It occurs in many desert plants.

sedimentary rock—Rock formed from materials deposited from suspension or precipitated from solution and usually more or less consolidated. The principal sedimentary rocks are sandstones, shales, limestones and conglomerates.

sequence, soil—See soil sequence.

sequum—A sequence of an eluvial horizon and its related illuvial horizon.

sesquioxides—Oxides containing three atoms of oxygen combined with two of the other constituent in a molecule; ferric oxides, for instance.

shale—A fine-grained, indurated, detrital sedimentary rock formed by compression or cementation of clay, silt or mud. It is characterized by finely stratified structure and/or fissility that is approximately parallel to the bedding along which the rock breaks readily into thin layers. It is commonly most conspicuous on weathered surfaces. Shale normally contains at least 50 percent silt, with 35 percent clay or fine mica fraction and 15 percent chemical or authigenic materials. It is generally soft but sufficiently indurate so that it will not fall apart on wetting. Less firm than argillite and slate, shale commonly has a splintery fracture, a smooth feel and is easily scratched. It may be red, brown, black, gray, green or blue.

sialic—Pertains to the upper layer of the Earth's crust that is composed of rocks rich in silica and alumina. It is the source of granitic magma and is characteristic of the continental crust.

siliceous—Said of a rock containing abundant silica, especially free silica rather than silicates.

silicic—Said of a silica-rich igneous rock or magma. The amount of silica usually constitutes about 65 percent of the rock. In addition to the combined silica in feldspars, silicic rocks generally contain free silica in the form of quartz. Granite and rhyolite are typical silicic rocks.

silt—A soil consisting of particles between 0.05 and 0.002 mm (0.002 and 0.00008 in) in equivalent diameter.

silt loam—See soil texture.

silty clay—See soil texture.

silty clay loam—See soil texture.

skeletal—Pertains to a skeletan. See skeletan.

skeletan—A cutan consisting of skeleton grains adhering to the surface, e.g., bleached sand and silt grains high in quartz and low in feldspar.

skeleton grains—Relatively stable and not readily translocated grains of soil material, concentrated or reorganized by soil-forming processes, e.g., a mineral grain, or a resistant siliceous or organic body larger than colloidal size.

slickensides—Polished and grooved surfaces produced by one mass sliding past another. Slickensides are common in Vertisols.

slope of soil—The slope of soil is reported as gradient, i.e., the number of meters (feet) of rise or fall per 30 m (100 ft) horizontal distance. If the soil surface rises 3 m (10 ft) vertically over a horizontal distance of 30 m (100 ft) the slope gradient is 10 percent.

soil aggregate—A cluster of soil particles. See ped.

soil association—A group of defined and named taxonomic soil units occurring together in an individual and characteristic pattern over a geographic region. It is comparable with plant associations in many ways. The term also describes a mapping unit used on general soil maps in which two or more defined taxonomic units occurring together in a characteristic pattern are combined because the scale of the map or the purpose for which it is being made does not require delineation of the individual soils.

soil body—A single soil individual on the landscape. A unit of the soilscape.

soil complex—A mapping unit used in detailed soil surveys where two or more defined taxonomic units are so intimately intermixed geographically that it is undesirable or impractical to separate them because of the scale being used. A more intimate mixing of smaller areas of individual taxonomic units than that described under soil association.

soil consistency—The degree of cohesion or adhesion of the soil mass. Terms used for describing consistency at various soil moisture contents are below.
  a. wet soil — nonsticky, slightly sticky, sticky, very sticky, nonplastic, slightly plastic, plastic and very plastic;
  b. moist soil — loose, very friable, friable, firm, very firm and extremely firm;
  c. dry soil — loose, soft, slightly hard, hard, very hard and extremely hard; and
  d. cementation — weakly cemented, strongly cemented and indurated.

soil drainage—Natural soil drainage refers to the speed with which water is removed from the soil surface and through the soil itself. Seven classes are recognized: excessive, somewhat excessive, well, moderately well, somewhat poor (imperfect), poor and very poor. Artificial drainage refers to removal of water by ditching, tilling and construction of surface waterways and terraces.

soil genesis—The mode of the origin of the soil, with special reference to the processes of soil-forming factors responsible for the development of the solum, or true soil, from unconsolidated parent material.

soil horizon—A layer of soil or soil material approximately parallel to the land surface and differing from adjacent genetically related layers in physical, chemical and biological properties or characteristics such as color, structure, texture, consistency, kinds and numbers of organisms present, and degree of acidity or alkalinity. See the Soil Horizons section of the first chapter.

soil morphology—The physical constitution of a soil profile as exhibited by the kinds, thickness and arrangement of the horizons in the profile and by the texture, structure, consistency and porosity of each horizon. Also, the structural characteristics of the soil or any of its parts.

soil profile—A vertical section of the soil through all its horizons into the parent material.

soilscape—The soil portion of a landscape, bounded by the surface of organic litter above and the lower boundary of the rooting zone of perennial plants.

soil sequence—An arrangement of soils along a continuum.

soil series—The basic unit of soil classification. A subdivision of a soil family that consists of soils that are essentially alike in all major profile characteristics except the texture of the A horizon.

soil structure—The combination or arrangement of primary soil particles into secondary particles, units or peds. These secondary units may be arranged in the profile in such a manner as to give a distinctive characteristic pattern, but usually are not. The secondary units are characterized and classified on the basis of size, shape and degree of distinctness into classes, types and grades, respectively. See soil structure classes, soil structure grades and soil structure types.

soil structure classes—A grouping of soil structural units or peds on the basis of size. See soil structure and soil structure types.

soil structure grades—A grouping or classification of soil structure on the basis of inter- and intra-aggregate adhesion, cohesion or stability within the profile. Four grades of structure designated from 0 to 3 are described below.
  a. structureless (0) — no observable aggregation or no definite and orderly arrangement of natural lines of weakness. Massive, if coherent; single-grain, if non-coherent.
  b. weak (1) — poorly formed indistinct peds, barely observable in place.
  c. moderate (2) — well-formed, distinct peds, moderately durable and evident, but not distinct in undisturbed soil.
  d. strong (3) — durable peds that are quite evident in undisturbed soil, adhere weakly to one another, withstand displacement and become separated when the soil is disturbed.

soil structure types—A classification of soil structure based on the shape of the aggregates or peds and their arrangement in the profile.

soil texture—The relative proportions of the various soil separates in a soil. The textural classes may be modified by the addition of suitable adjectives when coarse fragments are present in substantial amounts; for example, "stony silt loam," or "silt loam, stony phase." The sand, loamy sand and sandy loam are further subdivided on the basis of the proportions of the various sand separates present. The limits of the various classes and subclasses are defined below.
  a. sand—Soil material that contains 85 percent or more of sand; percentage of silt, plus 1.5 times the percentage of clay, shall not exceed 15.
     1. coarse sand — 25 percent or more very coarse and coarse sand and less than 50 percent any other one grade of sand.
     2. sand — 25 percent or more very coarse, coarse and medium sand and less than 50 percent fine or very fine sand.
     3. fine sand — 50 percent or more fine sand or less than 25 percent very coarse, coarse and medium sand and less than 50 percent very fine sand.
     4. very fine sand — 50 percent or more very fine sand.
  b. loamy sand — Soil material that contains at the upper limit 85 percent to 90 percent sand; percentage of silt plus 1.5 times the percentage of clay is not less than 15. At the lower limit it contains not less than 70 percent to 85 percent sand; percentage of silt plus twice the percentage of clay does not exceed 30.
     1. loamy coarse sand — 25 percent or more very coarse and coarse sand and less than 50 percent any other one grade of sand.
     2. loamy sand — 25 percent or more very coarse, coarse and medium sand and less than 50 percent fine or very fine sand.
     3. loamy fine sand — 50 percent or more fine sand or less than 25 percent very coarse, coarse and medium sand and less than 50 percent very fine sand.
     4. loamy very fine sand — 50 percent or more very fine sand.
  c. sandy loam — Soil material that contains either 20 percent clay or less; percentage of silt plus twice the percentage of clay exceeds 30 and 52 percent or more sand; or less than 7 percent clay, less than 50 percent silt and between 43 percent and 52 percent sand.
     1. coarse sandy loam — 25 percent or more very coarse and coarse sand and less than 50 percent any other one grade of sand.
     2. sandy loam — 30 percent or more very coarse, coarse and medium sand, but less than 25 percent very coarse sand, and less than 30 percent very fine or fine sand.
     3. fine sandy loam — 30 percent or more fine sand and less than 30 percent very fine sand or between 15 percent and 30 percent very coarse, coarse and medium sand.
     4. very fine sandy loam — 30 percent or more very fine or greater than 40 percent fine and very fine sand, at least half of which is very fine sand and less than 15 percent very coarse, coarse and medium sand.
  d. loam — Soil material that contains 7 percent to 27 percent clay, 28 percent to 50 percent silt and less than 52 percent sand.
  e. silt loam — Soil material that contains 50 percent or more silt and 12 percent to 27 percent clay or 50 percent to 80 percent silt and less than 12 percent clay.
  f. silt — Soil material that contains 80 percent or more silt and less than 12 percent clay.
  g. sandy clay loam — Soil material that contains 20 percent to 35 percent clay, less than 28 percent silt and 45 percent or more sand.
  h. clay loam — Soil material that contains 27 percent to 40 percent clay and 20 percent to 45 percent sand.
  i. silty clay loam — Soil material that contains 27 percent to 40 percent clay and less than 20 percent sand.
  j. sandy clay — Soil material that contains 35 percent or more clay and 45 percent or more sand.
  k. silty clay — Soil material that contains 40 percent or more clay and 40 percent or more silt.
  l. clay — Soil material that contains 40 percent or more clay, less than 45 percent sand and less than 40 percent silt.

solum—The upper part of a soil profile in which soil-forming processes occur. In a mature soil, the A and B horizons constitute the solum.

spodic horizon—A mineral soil horizon that is characterized by the illuvial accumulation of amorphous materials composed of aluminum and organic carbon with or without iron. The spodic horizon has a certain minimum thickness, and a minimum quantity of extractable carbon plus iron plus aluminum in relation to its content of clay.

syenite—A group of plutonic rocks containing alkali feldspar, a small amount of plagioclase (less than in monzonite), one or more mafic minerals and quartz, if present, only as an accessory. With an increase in the quartz content, syenite grades into granite. Its name is derived from Syene,

Egypt, where the rock was quarried in ancient times.

syncline—A fold, the core of which contains the stratigraphically younger rocks; it is concave upward.

taiga—A swampy area of coniferous forest sometimes lying between tundra and steppe regions.

tectonics, plate—Global tectonics based on an Earth model characterized by a small number of large, broad, thick plates that float on viscous underlayers in the mantle. The plates move more or less independently and grind against others like ice floes in a river. Much of the dynamic activity is concentrated along the periphery of the plates, which are propelled from the rear by seafloor spreading. The continents form a part of the plates and move with them, like logs frozen in ice floes.

terrace—A level, usually narrow, plain bordering a river, lake or the sea. Rivers sometimes are bordered by terraces at different levels.

texture, soil—See soil texture.

topochronosequence—A sequence of related soils that differ from one another primarily because of differences of topography and time as soil-formation factors.

toposequence—A sequence of related soils that differ from one another primarily because of different topography as a soil-formation factor.

tubules—Irregular, hollow, twig-like calcareous concretions characteristic of loess deposits.

typic subgroup—That subgroup that represents the central concept of the soils classified in a great group within the hierarchical soil classification system. (See extragrade subgroup and intergrade subgroup.)

udic—A soil moisture regime that is neither dry for as long as 90 cumulative days nor for as long as 60 consecutive days in the 90 days following the summer solstice at periods when the soil temperature at 50 cm (20 in) is above 5 C (41 F).

umbric epipedon—A surface layer of mineral soil that has the same requirements as the mollic epipedon with respect to color, thickness, organic carbon content, consistence, structure and P_2O_5 content, but that has a base saturation of less than 50 percent when measured at pH 7.

varve—A distinct band representing the annual deposit in sedimentary materials regardless of origin and usually consisting of two layers, one a thick, light-colored layer of silt and fine sand and the other a thin, dark-colored layer of clay.

vermiculite—A group of platy or micaceous clay minerals closely related to chlorite and montmorillonite. The minerals are derived generally from the alteration of micas and they vary widely in chemical composition. They are characterized by marked exfoliation when heated above 150 C (302 F). At high temperature, their granules greatly expand into long, worm-like threads that entrap air and produce a lightweight and high water-absorbent material that is used as an insulator and as an aggregate in concrete and plaster.

Vertisols—The Order of mineral soils that has 30 percent or more clay, deep, wide cracks when dry and gilgai microrelief, intersecting slickensides or wedgeshaped structural aggregates tilted at an angle from the horizontal.

volcanic tuffs—A compacted pyroclastic deposit of volcanic ash and dust that may or may not contain up to 50 percent sediments such as sand or clay.

water table—The upper surface of groundwater or that level in the ground where the water is at atmospheric pressure.

List of References

Aandahl, A.R. 1965. The first comprehensive soil classification system. Journal of Soil and Water Conservation 20:-243-246.

Allen, D.L. 1962. Our wildlife legacy. Funk & Wagnalls, New York. 422 p.

Antevs, E.V. 1955. Geologic-climatic dating in the West. American Antiquity 20:317-335.

———. 1962. Quaternary climates in Arizona. American Antiquity 28:193-198.

ARIS. 1975. Arizona precipitation. Publication No. 5 (map series). State Climatologist, The Laboratory of Climatology, Arizona State University, Tempe.

Arkley, R.J. 1963. Calculation of carbonate and water movement in soil from climatic data. Soil Science 96:239-248.

———. 1967. Climates of some Great Soil Groups of the western United States. Soil Science 103:389-400.

———. 1971. Multiple regression equations for predicting mean annual soil temperature in the western United States. Unpublished report.

Arkley, R.J. and R. Ulrich. 1962. The use of calculated actual and potential evapotranspiration for estimating potential plant growth. Hilgardia 32:443-462.

Atwater, T. 1970. Implications of plate tectonics for the Cenozoic tectonic evolution of western North America. Bulletin of the Geological Society of America 81:3513-3536.

Auk, The. 1973. Thirty-second supplement to the American Ornithologists' Union check-list of North American birds. 90:411-419.

Bahre, S. 1976. Atlas of Arizona. Arizona Information Press, Yuma. 48 p.

Baier, W. 1969. Concepts of soil moisture availability and their effect on soil moisture estimates from a meteorological budget. Agricultural Meteorology 6:165-178.

Baier, W. and G.W. Robertson. 1966. A new versatile soil moisture budget. Canadian Journal of Plant Science 46:299-315.

Baksi, A.K. 1974. K-Ar study of the S.P. flow. Canadian Journal of Earth Sciences 11:1350-1356.

Baldwin, M., C.E. Kellogg and J. Thorp. 1938. Soil classification. Pages 979-1001 in Soils and men, yearbook of agriculture. U.S. Department of Agriculture, Washington, D.C. 1232 p.

Barsch, D. and C.F. Royse. 1972. A model for development of Quaternary terraces and pediment-terraces in the southwestern United States of America. Zeitschrift fuer Geomorphologie 16:54-75.

Barsch, D. and R.G. Updike. 1971. Late Pleistocene periglacial geomorphology (rock glaciers and blockfields) at Kendrick Peak, northern Arizona. Arizona Geological Society Digest 9:225-244.

Barth, R.C. and J.O. Klemmedson. 1978. Shrub-induced spatial patterns of dry matter, nitrogen, and organic carbon. Journal of the Soil Science Society of America 42:804-809.

———. 1982. Amount and distribution of dry matter, nitrogen, and organic carbon in soil-plant systems of mesquite and palo verde. Journal of Range Management 35:412-418.

Beasley, D. and W.J. Breed. 1975. Geologic cross section of the Grand Canyon-San Francisco Peaks-Verde Valley region. Map. Zion Natural History Association. Zion National Park, Springdale, Utah. Published by Northland Press, Flagstaff, Arizona.

Beget, J.E. 1983. Radiocarbon-dated evidence of worldwide early Holocene climate change. Geology 11:389-393.

Berdanier, C.R. et al. 1979. Mollisols in arid regimes. Agronomy Abstracts 1979:187.

Beschta, R.L. 1976. Climatology of the ponderosa pine type

in central Arizona. Arizona Agricultural Experiment Station Technical Bulletin 228. University of Arizona, Tucson. 24 p.

Beus, S., R.W. Rush and D. Smouse. 1966. Geologic investigations of experimental drainage basin 7-14, Beaver Creek Watershed, Coconino County, Arizona. Unpublished report to the Rocky Mountain Forest and Range Experiment Station, Flagstaff, Arizona.

Bird, J.M. and G.D. Isache, eds. 1972. Plate tectonics, selected papers from the Journal of Geophysical Research. American Geophysical Union, Washington, D.C. 563 p.

Blagbrough, J.W. 1971. Large nivation hollows in the Chuska Mountains, Northeast Arizona. Plateau 44:52-59.

Bockheim, J.G. 1980. Solution and use of chrono-sequences in studying soil development. Geoderma 24:71-85.

Brakenridge, G.R. 1978. Evidence for a cold, dry full-glacial climate in the American Southwest. Quaternary Research 4:22-40.

Breed, C.S. 1969. A century of conjecture on the Colorado River in the Grand Canyon. Pages 63-67 in Geology and natural history of the Grand Canyon region. Guidebook of the fifth field conference, Powell centennial river expedition. Four Corners Geological Society, Durango, Colorado. 212 p.

———. 1970. Two hypotheses of the origin and geologic history of the Colorado River. Pages 31-34 in C.T. Smith, ed., Guidebook to the Four Corners, Colorado Plateau and Central Rocky Mountain Region. National Association of Geology Teachers, Southwest Section, Cedar City, Utah. 183 p.

Breed, W.J. and E.C. Roat, eds. 1974. Geology of the Grand Canyon. Museum of Northern Arizona, Flagstaff, and Grand Canyon Natural History Association, Grand Canyon, Arizona. 185 p.

Brewer, R. 1968. Clay illuviation as a factor in particle-size differentiation in soil profiles. Pages 489-499 in Transactions of the 9th International Congress of Soil Science, volume IV, Adelaide, Australia. American Elsevier Publishing Company, New York. 776 p.

Bridges, E.M. 1978. World soils. Second edition. Cambridge University Press, Massachusetts. 128 p.

Brown, D.E. 1973. The natural vegetative communities in Arizona. Map. 1:500,000 scale. Arizona Resources Information Systems, cooperative publication 1, Phoenix, Arizona.

Brown, D.E., ed. 1982. Biotic communities of the American Southwest—United States and Mexico. Desert Plants 4:3-341.

Brown, D.E., C.H. Lowe and C.P. Pase. 1977. Biotic communities of the Southwest. Map. 1:1,000,000 scale. U.S. Forest Service general technical report RM-41. Rocky Mountain Forest and Range Experiment Station, Fort Collins, Colorado.

Bryan, A.L. and R. Gruhn. 1964. Problems relating to the neothermal climatic sequence. American Antiquity 29:307-315.

Bryan, K. 1925. Date of channel trenching (arroyo cutting) in the arid Southwest. Science 82:338-344.

———. 1941. Pre-Columbian agriculture in the Southwest as conditioned by periods of alluviation. Annals of the Association of American Geographers 31:219-242.

Bryson, R.A. 1957. The annual march of precipitation in Arizona, New Mexico, and northwestern Mexico. University of Arizona Institute of Atmospheric Physics Technical Report No. 6. University of Arizona, Tucson. 22 p.

Bryson, R.A. and W.P. Lowry. 1955. Synoptic climatology of the Arizona summer precipitation singularity. Bulletin of the American Meteorological Society 36:329-339.

Budel, J. 1959. The periglacial morphologic effects of the Pleistocene climate over the entire world. International Geology Review 1:1-16.

Budyko, M.I. 1958. The heat balance of the earth's surface. Translated by N.A. Stepanov. U.S. Department of Commerce, U.S. Weather Bureau, Washington, D.C. 259 p.

Bunting, B.T. 1965. The geography of soil. Aldine Publishing Company, Chicago. 213 p.

Buol, S.W. 1964. Calculated actual and potential evapotranspiration in Arizona. Arizona Agricultural Experiment Station Technical Bulletin 162. University of Arizona, Tucson. 48 p.

Buol, S.W., F.D. Hole and R.J. McCracken. 1980. Soil genesis and classification. Iowa State University Press, Ames. 404 p.

Burges, N.A. 1969. Some systematic requirements in soil biology. Pages 69-71 in J.G. Sheals, ed., The soil ecosystem: Systematic aspects of the environment, organisms and communities. Systematics Association Publication 8, London. 247 p.

Cady, J.G. 1960. Mineral occurrence in relation to soil profile differentiation. Pages 418-424 in Transactions of the 7th International Congress of Soil Science, volume IV, Madison, Wisconsin. International Society of Soil Science, Netherlands. 562 p.

Calvo, S.S. and P.A. Pearthree. 1982. Late Quaternary faulting west of the Santa Rita Mountains south of Tucson, Arizona. Prepublication paper for M.S. degree. Department of Geosciences, University of Arizona, Tucson.

Campbell, C.J. 1970. Ecological implications of riparian vegetation management. Journal of Soil and Water Conservation 25:49-52.

Carmichael, R.S. et al. 1978. Arizona chaparral: Plant associations and ecology. U.S. Forest Service Research Paper RM-202. Rocky Mountain Forest and Range Experiment Station, Fort Collins, Colorado. 16 p.

Cheevers, C.W. and L.J. Lund. 1981. Influence of eolian material on basaltic soils of northern Arizona. Agronomy Abstracts 1981:196-197.

Chew, R.M. 1965. Water metabolism of desert-inhabiting vertebrates. Biology Review 36:1-31.

Childs, O.E. 1948. Geomorphology of the valley of the Little Colorado River. Bulletin of the Geological Society of America 59:353-388.

Christiansen, R.L. and P.W. Lipman. 1972. Cenozoic volcanism and plate-tectonic evolution of the western United States. II. Late Cenozoic. Philosophical Transactions of the Royal Society of London, Series A 271:249-284.

LIST OF REFERENCES

Colbert, E.H. 1956. Rates of erosion in the Chinle Formation. Plateau 28:73-76.

Cole, K.L. 1982. Late Quaternary zonation of vegetation in the eastern Grand Canyon. Science 217:1142-1145.

Cole, K.L. and L. Mayer. 1982. Use of packrat middens to determine rates of cliff retreat in the eastern Grand Canyon, Arizona. Geology 10:597-599.

Colman, S.M. and R.L. Pierce. 1981. Weathering rinds on andesitic and basaltic stones as a Quaternary age indication, western United States. U.S. Geological Survey Professional Paper 1210. U.S. Government Printing Office, Washington, D.C. 56 p.

Colton, H.S. 1937. The basaltic cinder cones and lava flows of the San Francisco Mountain volcanic field, Arizona. Museum of Northern Arizona Bulletin 10, Flagstaff. 50 p.

———. 1967. The basaltic cinder cones and lava flows of the San Francisco Mountain volcanic field, Arizona. Museum of Northern Arizona Bulletin 10 (revised), Flagstaff. 50 p.

Coney, P.J. 1978. The plate tectonic setting of southeastern Arizona. Pages 285-290 in New Mexico Geological Society Guidebook, 29th field conference, November 9-11, 1978. New Mexico Geological Society, Socorro. 372 p.

Cooke, H.B.S. 1973. Pleistocene chronology: Long or short? Quaternary Research 3:206-220.

Cooke, R.V. 1974. The rainfall context of arroyo initiation in southern Arizona. Zeitschrift für Geomorphologie, Supplementband 21:63-75.

Cooke, R.V. and R.W. Reeves. 1976. Arroyos and environmental change in the American Southwest. Clarendon Press, Oxford. 213 p.

Cooley, M.E. 1962. Geomorphology and the age of volcanic rocks in northeastern Arizona. Arizona Geological Society Digest 5:97-115.

Cooley, M.E. and J.P. Akers. 1961. Ancient erosion cycles of the Little Colorado River, Arizona and New Mexico. Pages 244-248 in Short papers in the geologic and hydrologic sciences. U.S. Geological Survey Professional Paper 424-C. U.S. Government Printing Office, Washington, D.C. 398 p.

Cooley, M.E. et al. 1969. Regional hydrogeology of the Navajo and Hopi Indian reservations, Arizona, New Mexico and Utah. U.S. Geological Survey Professional Paper 521-A. U.S. Government Printing Office, Washington, D.C. 61 p.

Cox, J.E. 1968. Evaluation of climate and its correlation with soil groups. New Zealand Soil Bureau Bulletin 26:33-45.

Crocker, R.L. 1960. The plant factor in soil formation. Proceedings of the Ninth Pacific Science Congress 18:84-90.

Crosswhite, F.S. and C.D. Crosswhite. 1982. The Sonoran Desert. Pages 163-319 in G.L. Bender, ed., Reference handbook on the deserts of North America. Greenwood Press, Westport, Connecticut. 594 p.

Damon, P.E. 1971. The relationship between late Cenozoic volcanism and tectonism and orogenic-epeirogenic periodicity. Pages 15-35 in K.K. Turekian, ed., The late Cenozoic glacial ages. John Wiley and Sons, New York. 606 p.

Darwin, C. 1890. The formation of vegetable mould, through the action of worms, with observations on their habits. D. Appleton and Company, New York. 326 p.

Daubenmire, R. and J.B. Daubenmire. 1968. Forest vegetation of eastern Washington and northern Idaho. Washington Agricultural Experiment Station Bulletin 60. Washington State University, Pullman. 140 p.

Daugherty, L. 1982. Soil-climate predictors for range and forest land potentials in western United States. Pages 18-24 in Western regional technical work planning conference of the National Cooperative Soil Survey, San Diego, California. 245 p.

Davis, W.M. 1903. The mountain ranges of the Great Basin. Harvard Museum of Comparative Zoology Bulletin 42:129-177.

Denton, G.H. and W. Karlen. 1973. Holocene climatic variations and their pattern and possible cause. Quaternary Research 3:155-205.

Dice, L.R. 1943. The biotic provinces of North America. University of Michigan Press, Ann Arbor. 78 p.

Dietz, R.J. 1961. Continental and ocean basin evolution by spreading of the sea floor. Nature 190:854-857.

Duncklee, J. 1978. Glaciation evidence in Abinean Canyon. Plateau 48:73-75.

Dye, A.J. and W.H. Moir. 1977. Spruce-fir forest at its southern distribution in New Mexico. American Midland Naturalist 97:133-146.

Eberly, L.D. and T.B. Stanley. 1978. Cenozoic stratigraphy and geologic history of southwestern Arizona. Bulletin of the Geological Society of America 89:921-941.

Edney, E.B. 1967. Water balance in desert arthropods. Science 156:1059-1066.

Eicher, D.L. 1976. Geologic time. Fourth edition. Prentice-Hall, Englewood Cliffs, New Jersey. 150 p.

Englemann, M.D. 1961. The role of soil arthropods in the energetics of an old field community. Ecological Monographs 31:221-238.

Ewing, J. and M. Ewing. 1967. Sediment distribution on the mid-ocean ridges with respect to spreading of the sea floor. Science 156:1590-1592.

Evans, A.C. 1948. Studies on the relationships between earthworms and soil fertility. Applied Biology 35:1-13.

Fairbridge, R.W. 1983. The Pleistocene-Holocene boundary. Quaternary Science reviews 1:215-244.

Fenneman, N.M. 1931. Physiography of western United States. McGraw-Hill Book Company, New York. 534 p.

Feth, J.H. 1961. A new map of western conterminous United States showing the maximum known or inferred extent of Pleistocene lakes. Pages 110-112 in Short papers in the geologic and hydrologic sciences. U.S. Geological Survey Professional Paper 424-B. U.S. Government Printing Office, Washington, D.C. 343 p.

Flach, K.W., W.D. Nettleton and R.E. Nelson. 1974. The micromorphology of silica-cemented soil horizons in western North America. Pages 715-729 in G.K. Rutherford, ed., Soil microscopy. Limestone Press, Kingston, Ontario, Canada. 857 p.

Flach, K.W. et al. 1969. Pedocementation: Induration by silica, carbonates, and sesquioxides in the Quaternary. Soil

Science 107:442-452.

Ford, T.D. et al. 1974. Rock movement and mass wastage in the Grand Canyon. Pages 116-128 in W.J. Breed and E.C. Roat, eds., Geology of the Grand Canyon. Museum of Northern Arizona, Flagstaff, and Grand Canyon Natural History Association, Grand Canyon, Arizona. 185 p.

Freckman, D.W. and R. Mankau. 1977. Distribution and trophic structure of nematodes in desert soils. Ecology Bulletin 25:511-514.

Galloway, R.W. 1970. The full glacial climate in the southwestern United States. Annals of the Association of American Geographers 60:245-256.

———. 1983. Full-glacial southwestern United States: Mild and wet or cold and dry? Quaternary Research 19:236-248.

Gardner, R.L. 1972. Origin of the Mormon Mesa caliche, Clark County, Nevada. Bulletin of the Geological Society of America 83:143-186.

Gay, L.W. 1981. Potential evapotranspiration for deserts. Pages 172-194 in D.D. Evans and J.L. Thames, eds., Water in desert ecosystems. Published by Dowden, Hutchinson and Ross, New York; distributed by Academic Press, New York. 280 p.

Gelderman, F.W. 1970. Soil survey of the Safford area, Arizona. 1:20,000 scale. U.S. Department of Agriculture, Soil Conservation Service, Washington, D.C. 57 p. +maps.

Gerard, B.M. 1967. Factors affecting earthworms in pastures. Journal of Animal Ecology 36:235-252.

Gilbert, G.K. 1875. Report of the geology of portions of Nevada, Utah, California and Arizona examined in the years 1871 and 1872. U.S. Geographic Surveys West of the One-Hundredth Meridian 3:17-187.

Gile, L.H. 1975. Holocene soils and soil-geomorphic relations in an arid region of southern New Mexico. Quaternary Research 5:321-360.

———. 1977. Holocene soils and soil-geomorphic relations in a semi-arid region of southern New Mexico. Quaternary Research 7:112-132.

Gile, L.H. and R.B. Grossman. 1979. The Desert Project soil monograph: Soils and landscapes of a desert region astride the Rio Grande Valley near Las Cruces, New Mexico. U.S. Department of Agriculture, Soil Conservation Service, Washington, D.C. 984 p.

Gile, L.H. and J.W. Hawley. 1968. Age and comparative development of desert soils on the Gardner Spring Radiocarbon Site, New Mexico. Proceedings of the Soil Science Society of America 32:709-716.

———. 1972. The prediction of soil occurrence in certain desert regions of the southwestern United States. Proceedings of the Soil Science Society of America 36:119-124.

Gile, L.H., F.F. Peterson and R.B. Grossman. 1966. Morphological and genetic sequences of carbonate accumulation in desert soils. Soil Science 101:347-360.

Glass, B. et al. 1967. Geomagnetic reversals and Pleistocene chronology. Nature 216:437-442.

Goudie, A. 1977. Environmental change. Oxford University Press, England. 244 p.

Gray, J. 1961. Early Pleistocene paleoclimatic record from the Sonoran Desert. Science 133:38-39.

Green, P. 1966. Mineralogical and weathering study of a red-brown earth formed on granodiorite. Australian Journal of Soil Research 4:181-191.

Greene, R.A. and G.H. Murphy. 1932. The influence of two burrowing rodents, *Dipodomys spectabilis spectabilis* (kangaroo rat) and *Neotoma albigula albigula* (pack rat) on desert soils in Arizona. Ecology 13:359-363.

Greene, R.A. and C. Reynard. 1932. The influence of two burrowing rodents, *Dipodomys spectabilis* (kangaroo rat) and *Neotoma albigula albigula* (pack rat) on desert soils in Arizona. Ecology 13:73-80.

Grinnell, J. 1923. The burrowing rodents of California as agents in soil formation. Journal of Mammalogy 4:137-149.

Grove, J. 1979. The glacial history of the Holocene. Progress in Physical Geography 3:1-54.

Guild, W.J. McL. 1955. Earthworms and soil structure. Pages 83-98 in D.K. McE. Kevan, ed., Soil zoology. Academic Press, New York. 512 p.

Hadley, R.F. 1967. Pediments and pediment-forming processes. Journal of Geological Education 15:83-89.

Hamblin, W.K. 1974. Late Cenozoic volcanism in the western Grand Canyon. Pages 142-169 in W.J. Breed and E.C. Roat, eds., Geology of the Grand Canyon. Museum of Northern Arizona, Flagstaff, and Grand Canyon Natural History Association, Grand Canyon, Arizona. 185 p.

Hamblin, W.K. and M.G. Best, eds. 1970. Guidebook to the Geology of Utah. No. 23: The western Grand Canyon district. Utah Geological Society, Salt Lake City. 156 p.

Hasse, E.F. 1972. Survey of floodplain vegetation along the lower Gila River in southwestern Arizona. Journal of the Arizona Academy of Science 7:75-81.

Hastings, J.R. 1959. Vegetation change and arroyo cutting in southeastern Arizona during the past century: An historical review. Pages 24-39 in Arid lands Colloquia 1958-59. Office of Arid Lands Studies, University of Arizona, Tucson. 92 p.

Hastings, J.R. and R.M. Turner. 1965. The changing mile. University of Arizona Press, Tucson. 317 p.

Haynes, C.V., Jr. 1968. Geochronology of late-Quaternary alluvium. Pages 591-631 in R.B. Morrison and H.E. Wright, Jr., eds., Means of correlation of Quaternary successions, Vol. 8, Proceedings VII. University of Utah Press, Salt Lake City. 631 p.

Hecht, M.E. and R.W. Reeves. 1981. The Arizona atlas. Office of Arid Lands Studies, University of Arizona, Tucson. 164 p.

Hendricks, D.M. 1974. A micromorphological study of soil formation in sandy loam alluvium in southern Arizona. Pages 408-427 in G.K. Rutherford, ed., Soil microscopy. Limestone Press, Kingston, Ontario, Canada. 857 p.

Hendricks, D.M. and S.E. Davis. 1979. Influence of slope aspect on soil properties on Greens Peak, Apache County, Arizona. Agronomy Abstracts 1979:190.

Hevly, R.H. and T.N.V. Karlstrom. 1974. Southwest paleoclimate and continental correlations. Pages 257-295 in T.N.V. Karlstrom, G.A. Swann and R.L. Eastwood, eds.,

Geology of northern Arizona with notes on archeology and paleoclimate. Part I—regional studies. Geological Society of America, Rocky Mountain Section, Flagstaff, Arizona. 405 p.

Hilden, O. 1965. Habitat selection in birds: A review. Annales Zoologici Fennici 2:53-75.

Hill, J.D. 1970. Quantitative micromorphological evidence of clay movement. Pages 33-42 in D.A. Osmond and P. Bullock, eds., Micromorphological techniques and applications. Agricultural Research Council, Soil Survey, Technical Monograph No. 2, Harpenden, England. 110 p.

Hillel, D. 1982. Introduction to soil physics. Academic Press, New York. 364 p.

Hole, F.D. 1981. Effects of animals on soil. Geoderma 25:75-112.

Holmes, R.T. and S.K. Robinson. 1981. Tree species preferences of foraging insectivorous birds in a northern hardwoods forest. Oecologia 48:31-53.

Hooper, D.J. 1969. Some problems in the systematics of soil nematodes. Pages 131-142 in J.G. Sheals, ed., The soil ecosystem: Systematic aspects of the environment, organisms, and communities. Systematics Association Publication 8, London. 247 p.

Howard, A. and R. Dolan. 1981. Geomorphology of the Colorado River in the Grand Canyon. Journal of Geology 89:269-298.

Hugie, V.K. and H.B. Passey. 1963. Cicadas and their effect upon soil genesis in certain soils in southern Idaho, northern Utah and northeastern Nevada. Proceedings of the Soil Science Society of America 27:78-83.

Humphrey, R.R. 1933. A detailed study of desert rainfall. Ecology 14:31-34.

———. 1958. The desert grassland. University of Arizona Press, Tucson. 62 p.

Hungerford, C.R. 1980. Wild horse territory, a study of the free-ranging horses on the Carson National Forest. Final Report to U.S. Forest Service, Phoenix, Arizona. 13 p. Mimeo.

Hunt, C.B. 1956. Cenozoic geology of the Colorado Plateau. U.S. Geological Survey Professional Paper 279. U.S. Government Printing Office, Washington, D.C. 99 p.

———. 1967. Physiography of the United States. W.H. Freeman and Co., San Francisco. 480 p.

———. 1969. Geologic history of the Colorado River. Pages 59-130 in The Colorado River region and John Wesley Powell. U.S. Geological Survey Professional Paper 669-C. U.S. Government Printing Office, Washington, D.C. 145 p.

———. 1974. Grand Canyon and the Colorado River, their geologic history. Pages 129-141 in W.J. Breed and E.C. Roat, eds., Geology of the Grand Canyon. Museum of Northern Arizona, Flagstaff, and Grand Canyon Natural History Association, Grand Canyon, Arizona. 185 p.

Huntington, E. 1914. The climate factor as illustrated in arid America. Carnegie Institute of Washington Publication 192, Washington, D.C. 341 p.

Huntoon, P.W. 1974a. The post-Paleozoic structural geology of the eastern Grand Canyon, Arizona. Pages 82-115 in W.J. Breed and E.C. Roat, eds., Geology of the Grand Canyon. Museum of Northern Arizona, Flagstaff, and Grand Canyon Natural History Association, Grand Canyon, Arizona. 185 p.

———. 1974b. The karstic groundwater basins of the Kaibab Plateau, Arizona. Water Resources Research 10:579-590.

———. 1975. The Surprise Valley landslide and widening of the Grand Canyon. Plateau 48:1-12.

Hurst, F.B. 1951. Climates prevailing in the yellow-gray earth and yellow-brown earth zones in New Zealand. Soil Science 72:1-19.

Jacks, G.V. and R.O. Whyte. 1939. The rape of the earth; a world survey of soil erosion. Faber and Faber Ltd., London. 312 p.

James, F.C. 1971. Ordinations of habitat relationships among breeding birds. Wilson Bulletin 83:215-236.

Jameson, D.A. 1969. Rainfall patterns of vegetation zones in northern Arizona. Plateau 41:105-111.

Jenny, H. 1941. Factors of soil formation. McGraw-Hill, New York. 281 p.

———. 1958. Role of the plant factor in the pedogenic functions. Ecology 39:5-16.

———. 1961. Derivation of state factor equations of soils and ecosystems. Proceedings of the Soil Science Society of America 25:385-388.

———. 1980. The soil resource: origin and behavior. Springer-Verlag, New York. 377 p.

Johnsgard, P.A. 1973. Grouse and quails of North America. University of Nebraska, Lincoln. 553 p.

Johnson, R.R. and D.A. Jones, coordinators. 1977. Importance, preservation, and management of riparian habitat: A symposium, Tucson, Arizona, July 9, 1977. U.S. Forest Service general technical report RM-43. Rocky Mountain Forest and Range Experiment Station, Fort Collins, Colorado. 217 p.

Johnson, W.M. 1963. The pedon and polypedon. Proceedings of the Soil Science Society of America 27:212-215.

Jurwitz, L.R. 1953. Arizona's two season rainfall pattern. Weatherwise 6:69-99.

Kangieser, P.C. 1959. Climate of the Southwest (abstract). Bulletin of the Geological Society of America 70:1727.

Kearney, T.H. and R.H. Peebles. 1969. Vegetation of Arizona. University of California Press, Berkeley. 1085 p.

Keller, G.R., L.W. Braile and P. Morgan. 1979. Crustal structure, geophysical models and contemporary tectonism of the Colorado Plateau. Tectonophysics 61:131-147.

Kelley, V.C. 1955. Monoclines of the Colorado Plateau. Bulletin of the Geological Society of America 66:789-804.

Kent, D., N.D. Opdyke and M. Ewing. 1971. Climate change in the North Pacific using ice-rafted detritus as a climatic indicator. Bulletin of the Geological Society of America 82:2741-2754.

Kesel, R.H. 1977. Some aspects of the geomorphology of inselbergs in central Arizona, USA. Zeitschrift für Geomorphologie 21:119-146.

Kevan, D.K. 1962. Soil animals. Philosophical Library, Inc., New York. 237 p.

Koford, C.B. 1958. Prairie dogs, whitefoces and blue grama. Wildlife Monographs 3:1-78.

Konstantinov, H.R. and L.V. Popovich. 1980. Method of calculation of thermal regime of soils at depths of 5, 10 and 20 cm. based on temperature and humidity of air measured at meteorological stations. Pages 150-160 in I.A. Gol'tsberg and F.F. Dairtaya, eds., Soil climate. Translated from Russian by B.B. Bhattacharya. Amerind Publication Co., New Delhi, India. 256 p.

Koons, E.D. 1945. Geology of the Uinkaret Plateau, northern Arizona. Geological Society of America, Bulletin 56:151-180.

Krauskopf, K.B. 1956. Dissolution and precipitation of silica at low temperatures. Geochimica et Cosmochimica Acta 10:1-26.

Kuchler, A.W. 1964. Potential natural vegetation of the conterminous United States. American Geographical Society of New York, Special Publication No. 36, New York. 116 p.

Kuhnelt, W. 1976. Soil biology with special reference to the animal kingdom. English edition translated from German by N. Walker. Faber and Faber, London. 483 p.

Laney, R.S., R.H. Raymond and C.C. Winikka. 1978. Maps showing water-level declines, land subsidence, and earth fissures in south-central Arizona. 1:126,720 scale. U.S. Geological Survey WRI 78-38, Tucson, Arizona.

Lange, A.L. 1956. Cave evolution in Marble Gorge of the Colorado River. Plateau 29:12-21.

Lanner, R.M. and J.R. Van Devender. 1974. Morphology of pinyon pine needles from fossil packrat middens in Arizona. Forest Science 20:107-211.

Lehr, J.H. 1978. A catalogue of the flora of Arizona. Desert Botanical Garden, Phoenix, Arizona. 203 p.

Lipman, P.W. et al. 1972. Cenozoic volcanism and plate-tectonic evolution of the western United States. I. Early and middle Cenozoic. Philosophical Transactions of the Royal Society of London, Series A 271:217-248.

Little, E.L. 1941. Alpine flora of San Francisco Mountain, Arizona. Madroño 6:65-81.

Lowe, C.H. 1959. Contemporary biota of the Sonoran Desert: Problems. Pages 54-74 in Arid Lands Colloquia 1958-59. Office of Arid Lands Studies, University of Arizona, Tucson. 92 p.

Lowe, C.H., ed. 1972. The vertebrates of Arizona. University of Arizona Press, Tucson. 270 p.

Lowe, C.H. and D.E. Brown. 1973. The natural vegetation of Arizona. Arizona Resources Information Systems Cooperative Publication 2, Phoenix, Arizona. 53 p.

Lucchitta, I. 1972. Early history of the Colorado River in the Basin and Range Province. Bulletin of the Geological Society of America 83:1933-1948.

—————. 1974. Structural evolution of northwest Arizona and its relation to adjacent Basin and Range Province structures. Pages 336-354 in T.N.V. Karlstrom, G.A. Swann and R.L. Eastwood, eds., Geology of northern Arizona with notes on archeology and paleoclimate. Part I—regional studies. Geological Society of America, Rocky Mountain Section, Flagstaff, Arizona. 405 p.

Machetti, M.N. 1978. Dating Quaternary faults in the southwestern United States by using buried calcic paleosols. Research of the U.S. Geological Survey 6:369-381.

Mammerickx, J. 1964. Quantitative observations on pediments in the Mohave and Sonoran deserts (southwestern United States). American Journal of Science 262:417-435.

Martin, P.S. 1963. The last 10,000 years, a fossil pollen record of the American Southwest. University of Arizona Press, Tucson. 87 p.

Martin, S.C. and R.M. Turner. 1977. Vegetation change in the Sonoran Desert region, Arizona and Sonora. Journal of the Arizona Academy of Science, 12:59-69.

Martin, W.P. and J.E. Fletcher. 1943. Vertical zonation of great soil groups on Mt. Graham, Arizona, as correlated with climate, vegetation, and profile characteristics. Arizona Agricultural Experiment Station Technical Bulletin 99. University of Arizona, Tucson. 65 p.

Marvin, U.B. 1973. Continental drift, the evolution of a concept. Smithsonian Institution Press, Washington, D.C. 239 p.

Mather, J.R. and G.A. Yoshioka. 1968. The role of climate in the distribution of vegetation. Annals of the Association of American Geographers, 58:28-41.

McDonald, J.E. 1956. Variability of precipitation in an arid region: A survey of characteristics for Arizona. University of Arizona Institute of Atmospheric Physics Technical Report 1. University of Arizona, Tucson. 88 p.

McFadden, L.D. 1978. Soils of the Cañada del Oro Valley, southern Arizona. M.S. thesis. University of Arizona, Tucson. 116 p.

—————. 1981. Quaternary evolution of the Cañada del Oro Valley, southeastern Arizona. Arizona Geological Society Digest 13:13-20.

McGee, W.J. 1897. Sheet-flood erosion. Bulletin of the Geological Society of America, 8:87-112.

McKee, E.D. and E.H. McKee. 1972. Pliocene uplift of the Grand Canyon region—time of drainage adjustment. Bulletin of the Geological Society of America, 83:1923-1932.

McKee, E.D. et al. 1967. Evolution of the Colorado River in Arizona. Museum of Northern Arizona, Bulletin 44, Flagstaff. 67 p.

Mehringer, P.J. 1967a. Murray Springs, an arid-postglacial pollen record from southern Arizona. American Journal of Science 265:786-777.

—————. 1967b. Pollen analysis and the alluvial chronology. The Kiva 32:96-101.

Meikle, R.W. and T.R. Treadway. 1981. A mathematical method for estimating soil temperature in Canada. Soil Science 131:320-326.

Melton, M.A. 1961. Multiple Pleistocene glaciation of the White Mountains, Apache County, Arizona. Bulletin of the Geological Society of America, 72:1279-1282.

—————. 1965. The geomorphic and paleoclimatic significance of alluvial deposits in southern Arizona. Journal of Geology, 73:1-39.

Menges, C.M. 1981. The Sonoita Creek Basin: Implications for late Cenozoic tectonic evolution of basins and ranges in southeastern Arizona. M.S. thesis. University of Arizona, Tucson. 239 p.

Menges, C.M. and L.D. McFadden. 1981. Evidence for a latest Miocene to Pliocene transition from basin-range tectonic to post-tectonic landscape evolution in southeastern Arizona. Arizona Geological Society Digest 13:151-160.

Mercer, J.H. 1972. The lower boundary of the Holocene.

LIST OF REFERENCES

Quarternary Research 2:15-24.

Merriam, C.H. 1890. Results of a biological survey of the San Francisco Mountain region and desert of the Little Colorado River in Arizona. U.S. Department of Agriculture, Bureau of Biological Survey of American Fauna 3:1-132.

Merrill, R. K. and T.L. Péwé. 1972. Late Quaternary glacial chronology of the White Mountains, east-central Arizona. Journal of Geology 80:493-50l.

———. 1977. Late Cenozoic geology of the White Mountains, Arizona. Arizona Bureau of Geology and Mining Technology Special Paper No. l. University of Arizona, Tucson. 65 p.

Moir, W.H. and J. A. Ludwig. 1979. A classification of spruce-fir and mixed conifer habitat types of Arizona and New Mexico. U.S. Forest Service Research Paper RM-207. Rocky Mountain Forest and Range Experiment Station, Fort Collins, Colorado. 47 p.

Moore, R.B., G.E. Ulrich and E.W. Wolfe. 1974. Field guide to the geology of the eastern and northern San Francisco volcanic field, Arizona. Pages 495-520 in T.N.V. Karlstrom, G.A. Swann and R.L. Eastwood, eds., Geology of northern Arizona with notes on archeology and paleoclimate. Part II — area studies and field guides. Geological Society of America, Rocky Mountain Section, Flagstaff, Arizona. 397 p.

Moore, R.B., E.W. Wolfe and G.E. Ulrich. 1976. Volcanic rocks of the eastern and northern parts of the San Francisco volcanic field, Arizona. Journal of Research of the U.S. Geological Survey 4(5):549-560.

Moore, T.C. 1965. Origin and disjunction of the alpine tundra flora on San Francisco Mountain, Arizona. Ecology 46:860-864.

Morner, N.A. 1972. When will the present interglacial end? Quaternary Research 2:341-349.

Morrison, R.B. 1965. Geologic map of the Duncan and Candor Peak Quadrangle, Arizona and New Mexico. 1:48,000 scale. Miscellaneous Geologic Investigation Map I-442. U.S. Geological Survey, Washington, D.C. text 5 p.

Morrison, R.B., C.M. Menges and L.K. Lepley. l981. Neotectonic maps of Arizona. Arizona Geological Society Digest 13:179-184.

Museum of Northern Arizona. 1976. Geologic map of the Grand Canyon National Park. 1:1,000,000 scale. Museum of Northern Arizona, Flagstaff.

Nettleton, W.D. et al. 1975. Genesis of argillic horizons in soils of desert areas of the southwestern United States. Proceedings of the Soil Science Society of America 39:919-926.

Newhall, F. 1980. Calculation of soil moisture regime from the climatic record. USDA Soil Conservation Service. Mimeo.

Nichol, A.A. 1937. The natural vegetation of Arizona. Arizona Agricultural Experiment Station Technical Bulletin 6:181-222. University of Arizona, Tucson.

———. 1952. The natural vegetation of Arizona. Arizona Agricultural Experiment Station Technical Bulletin 127:185-230. University of Arizona, Tucson.

Nikiforoff, C.C. 1937. General trends of desert type of soil formation. Soil Science 43:105-131.

Oertel, A.C. 1968. Some observations incompatible with clay illuviation. Pages 481-488 in Transactions of the 9th International Congress of Soil Science, volume IV, Adelaide, Australia. American Elsevier Publishing Company, New York. 776 p.

Oertel, A.C. and J.B. Giles. 1967. Development of a red-brown earth profile. Australian Journal of Soil Research 5:133-147.

Oppenheimer, J.M. and J.S. Sumner. 1980. Depth-to-bedrock map, Basin and Range Province, Arizona. 1:500,000 scale. Laboratory of Geophysics, University of Arizona, Tucson.

———. 1981. Gravity modeling of the basins in the Basin and Range Province, Arizona. Arizona Geological Society Digest 13:111-116.

Osborn, H.B. 1983. Timing and duration of high rainfall rates in the southwestern United States. Water Resources Research 19:1036-1042.

Ouellet, C.E. 1973. Macroclimatic model for estimating monthly soil temperatures under short-grass cover in Canada. Canadian Journal of Soil Science 53:263-274.

Pakiser, L.C. 1963. Structure of the crust and upper mantle in the western United States. Journal of Geophysical Research 68:5747-5758.

Palmer, A.R. 1983. The decade of North American geology — 1983 Geologic time scale. Geology 11:503-504.

Palmer, W.C. and A.V. Havens. 1958. A graphical technique for determining evapotranspiration by the Thornthwaite Method. Monthly weather review 86:123-128.

Parkes, K.C. 1978. A guide to forming and capitalizing compound names of birds in English. The Auk 95:324-326.

Patton, D.R. 1978. RUN WILD: A storage and retrieval system for wildlife habitat information. U.S. Forest Service General Technical Report RM-51. Rocky Mountain Forest and Range Experiment Station, Fort Collins, Colorado. 8 p.

———. 1979. How to use RUN WILD data files stored on microfiche. U.S. Forest Service Research Note RM-377. Rocky Mountain Forest and Range Experiment Station, Fort Collins, Colorado. 2 p.

Peirce, H.W., P.E. Damon and M. Shafiqullah. 1979. An Oligocene (?) Colorado Plateau edge in Arizona. Tectonophysics 61:1-24.

Penman, H.L. 1956. Evaporation: An introductory survey. Netherlands Journal of Agricultural Science 4:9-29.

Peterson, F.F. 1981. Landforms of the Basin and Range Province defined for soil survey. Nevada Agricultural Experiment Station Technical Bulletin 28. University of Nevada, Reno. 52 p.

Péwé, T.L. 1978. Terraces of the Lower Salt River Valley in relation to the late Cenozoic history of the Phoenix Basin, Arizona. Pages 1-13 in D.M. Burt and T.L. Péwé, eds., Guidebook to the geology of central Arizona. Special Paper No. 2. Arizona Bureau of Geology and Mining Technology, University of Arizona, Tucson. 176 p.

Péwé, T.L. and R.G. Updike. 1970. Guidebook to the geology of the San Francisco Peaks, Arizona. Plateau 43:45-102.

Potts, A.S. 1970. Frost action in rocks: Some experimental data. Transactions of the Institute of British Geographers 49:109-123.

Powell, J.W. 1875. Exploration of the Colorado River and its tributaries. Smithsonian Institution, Washington, D.C. 291 p.

Rahn, P.H. 1966. Inselbergs and nickpoints in southwestern Arizona. Zeitschrift fuer Geomorphologie 10:217-225.

———. 1967. Sheetfloods, streamfloods, and the formation of pediments. Annals of the Association of American Geographers 57:593-604.

Reynolds, S.J. 1983. Generalized geologic map of Arizona. Unpublished. 1:2,500,000 scale. Arizona Bureau of Geology and Mineral Technology, University of Arizona, Tucson.

Rice, R.J. 1976. Terraces and abandoned channels of the Little Colorado River between Leupp and Cameron, Arizona. Plateau 46:102-119.

Richmond, G.M. 1972. Appraisal of the future climate of the Holocene in the Rocky Mountains. Quaternary Research 2:315-322.

Ricklefs, R.E. 1979. Ecology. Second edition. Chiron Press, Portland, Oregon. 966 p.

Rigby, J.K. 1977. Southern Colorado Plateau: Field guide. Kendall/Hunt Publishing Company, Dubuque, Iowa. 148 p.

Robbie, W.A., D.G. Brewer and R.K. Jorgensen. 1982. Soil temperatures for plant communities in northern Arizona. Agronomy Abstracts 1982:271.

Robinson, H.H. 1913. The San Francisco volcanic field, Arizona. U.S. Geological Survey Professional Paper 76. U.S. Government Printing Office, Washington, D.C. 213 p.

Rowley, P.D. et al. 1978. Age of structural differentiation between the Colorado plateaus and basin and range provinces in southwestern Utah. Geology 6:51-55.

Royse, C.F. and D. Barsch. 1971. Terraces and pediment-terraces in the Southwest. Bulletin of the Geological Society of America 82:3177-3182.

Ruhe, R.V. 1967. Geomorphic surfaces and surficial deposits in southern New Mexico. New Mexico Bureau of Mines and Mineral Resources Memoir 18, Socorro, New Mexico. 61 p.

Satchell, J.E. 1967. Lumbricidae. Pages 259-322 in A. Burges and F. Raw, eds., Soil biology. Academic Press, London. 532 p.

Scarborough, R.B. and H.W. Peirce. 1978. Late Cenozoic basins of Arizona. Pages 253-259 in J.F. Callendar et al, eds., Land of Cochise. New Mexico Geological Society Guidebook, 29th field conference, November 9-11, 1978. New Mexico Geological Society, Socorro. 372 p.

Schaefer, D.A. and W.G. Whitford. 1981. Nutrient cycling by the subterraneaan termite *Gnathamitermes tubiformans* in a Chihuahuan Desert ecosystem. Oecologia 48:277-283.

Schaller, F. 1968. Soil animals. University of Michigan Press, Ann Arbor. 144 p.

Schmidt, R.H. 1979. A climatic delineation of the "real" Chihuahuan Desert. Journal of Arid Environments 2:243-250.

Schmidt-Nielson, K. 1964. Desert animals: Physiological problems of heat and water. Oxford University Press, New York. 277 p.

Scholtz, J.F. 1969. The Beaver Creek volcanics: A new formation, Coconino-Yavapai counties, Arizona. Bulletin of the Geological Society of America 80:2637.

Schuchert, C. and C.O. Dunbar. 1964. Outlines of historical geology. Fourth edition. John Wiley and Sons, New York. 268 p.

Schumm, S.A. and R.J. Chorley. 1966. Talus weathering and scarp recession in the Colorado Plateau. Zeitschrift für Geomorphologie 10:11-36.

Segota, T. 1967. Paleotemperature changes in upper and middle Pleistocene. Eiszeitalter und Gegenwart 18:127-141.

Sellers, W.B. 1972. The climate. Pages 78-83 in Arizona: Its people and resources. University of Arizona Press, Tucson. 385 p.

Sellers, W.D. and R.H. Hill. 1974. Arizona climate 1931-1972. Revised second edition. University of Arizona Press, Tucson. 616 p.

Shafiqullah, M. et al. 1980. K-Ar geochronology and geologic history of southwestern Arizona and adjacent areas. Arizona Geological Society Digest 12:201-260.

Sharp, R.P. 1942. Multiple Pleistocene glaciation on San Francisco Mountain, Arizona. Journal of Geology 60:481-503.

Shlemon, R.J. 1978. Quaternary soil-geomorphic relationships, southeastern Mojave Desert, California and Arizona. Pages 187-207 in W.C. Mahaney, ed., Quaternary soils, 3rd symposium on Quaternary research, Toronto, Canada, May 21-23, 1976. Geo Abstracts, Inc., Norwich, England. 508 p.

Shreve, F. 1942. The vegetation of Arizona. Pages 10-23 in T. Kearney and R. Peebles, eds., Flowering plants and ferns of Arizona. U.S. Department of Agriculture Miscellaneous Publication 423. U.S. Government Printing Office, Washington, D.C. 1069 p.

Shreve, F. and I.L. Wiggins. 1951. Vegetation and flora of the Sonoran Desert. Carnegie Institute Publication No. 591, Washington, D.C. Various pagings.

———. 1964. Vegetation and flora of the Sonoran Desert. Stanford University Press, Stanford, California. Two volumes, 1740 p.

Simonson, R. 1959. Outline of a generalized theory of soil genesis. Proceedings of the Soil Science Society of America 23:152-156.

Simonson, R.W. 1962. Soil classification in the United States. Science 137:1027-1034.

———. 1964. The soil series as used in the USA. Soil Science 5:17-24.

———. 1971. Soil association maps and proposed nomenclature. Proceedings of the Soil Science Society of America 35:959-965.

———. 1978. A multiple-process model of soil genesis. Pages 1-25 in W.C. Mahaney, ed., Quaternary Soils, 3rd symposium on Quaternary research, Toronto, Canada, May 21-23, 1976. Geo Abstracts, Norwich, England. 508 p.

Sims, R.W. 1969. A numerical classification of megascolecoid earthworms. Pages 143-153 in J.G. Sheals, ed., The soil ecosystem: Systematic aspects of the environment, organisms and communities. Systematics Associa-

LIST OF REFERENCES

tion Publication 8, London. 247 p.

Singer, M.J. and P. Nkedi-Kizza. 1980. Properties of an exhumed Tertiary Oxisol in California. Journal of the Soil Science Society of America 44:587-590.

Smith, G.D. 1963. Objectives and basic assumptions of the new soil classification system. Soil Science 96:6-16.

———. 1965. Lectures on soil classification. Pedologie 4:1-134.

Smith, G.D., F. Newhall and L.H. Robinson. 1964. Soil-temperature regimes—their characteristics and predictability. Soil Conservation Service Technical Publication 144. U.S. Department of Agriculture, Washington, D.C. Various pagings.

Smith, H.V. 1956. The climate of Arizona. Arizona Agricultural Experiment Station Bulletin 279. University of Arizona, Tucson. 99 p.

Soil Survey Staff. 1975. Soil taxonomy: A basic system of soil classification for making and interpreting soil surveys. U.S. Soil Conservation Service Agricultural Handbook No. 436. U.S. Department of Agriculture, Washington, D.C. 754 p.

———. 1981. Soil survey manual. Revised May 1981. (Unpublished)

Spencer, J.S. Jr. 1966. Arizona forests. U.S. Forest Service Resource Bulletin INT-6. Intermountain Forest and Range Experiment Station, Ogden, Utah. 56 p.

Steila, D. 1972. Drought in Arizona: A drought identification methodology and analysis. Division of Economic and Business Research, University of Arizona, Tucson. 65 p.

Stevens, P.R. and J.W. Walker. 1970. The chronosequence concept and soil formation. Quarterly Review of Biology 45:333-350.

Stewart, J.H. 1978. Basin-range structure in western North America: A review. Pages 1-32 in R.B. Smith and G.P. Easton, eds., Cenozoic tectonics and regional geophysics of the western Cordillera. Geological Society of America Memoir 152, Boulder, Colorado. 388 p.

———. 1980. Regional tilt patterns of late Cenozoic basin-range fault blocks, western United States. Bulletin of the Geological Society of America Part I 91:460-464.

Stokes, W.L. 1973. Geomorphology of the Navajo country. Pages 61-67 in H.L. James, ed., Guidebook of Monument Valley and vicinity: Arizona and Utah, twenty-fourth field conference, October 4-6, 1973. New Mexico Geological Society, Socorro. 206 p.

Stout, J.D. and O.W. Heal. 1967. Protozoa. Pages 149-195 in A. Burges and F. Raw, eds., Soil biology. Academic Press, London. 532 p.

Strahler, A.N. 1944. Valleys and parts of the Kaibab and Coconino plateaus, Arizona. Journal of Geology 52:361-367.

———. 1948. Geomorphology and structure of the West Kaibab Fault Zone and Kaibab Plateau, Arizona. Bulletin of the Geological Society of America 59:513-540.

Suggate, R.P. 1965. The definition of "interglacial." Journal of Geology 74:619-626.

———. 1974. When did the last interglacial end? Quaternary research 4:246-252.

Sutton, R.L. 1974. The geology of Hopi Buttes, Arizona. Pages 647-671 in T.N.V. Karlstrom, G.A. Swann and R.L. Eastwood, eds., Geology of Northern Arizona with notes on archeology and paleoclimate. Part II—area studies and field guides. Geological Society of America, Rocky Mountain Section, Flagstaff, Arizona. 397 p.

Tarling, D.H. and D.H. Runcorn, eds. 1972. Implications of continental drift to the earth sciences. Academic Press, London. Various pagings.

Taylor, D.R. 1982. Soil survey of Coconino County, Arizona, central part. U.S. Department of Agriculture Soil Conservation Service in cooperation with Arizona Agricultural Experiment Station. 212 p.

Thomas, J.W. et al. 1979. Deer and elk. Pages 104-127 in J.W. Thomas, ed., Wildlife habitat in managed forests: The Blue Mountains of Oregon and Washington. U.S. Department of Agriculture Forest Service Agricultural Handbook No. 553, Washington, D.C. 512 p.

Thompson, G.A. and M.L. Zoback. 1979. Regional geophysics of the Colorado Plateau. Tectonophysics 61:149-181.

Thornbury, W.D. 1965. Regional geomorphology of the United States. John Wiley and Sons, New York. 609 p.

Thornthwaite, C.W. 1948. An approach toward a rational classification of climate. Geographical Review 38:55-94.

Thornthwaite, C.W. and F.K. Hare. 1955. Climatic classification in forestry. Unasylva 9:50-59.

Thornthwaite, C.W. and J.R. Mather. 1955. The water balance. Publications in Climatology Volume VIII, Number 1. Drexel Institute of Technology, Centerton, New Jersey. 86 p.

———. 1957. Instructions and tables for computing potential evapotranspiration and the water balance. Publications in Climatology Volume X, Number 3. Drexel Institute of Technology, Centerton, New Jersey. 125 p.

Thorp, J. 1949. Effects of certain animals that live in soils. Science Monthly 68:180-191.

Thorud, D.B. and P.F. Ffolliott. 1973. A comprehensive analysis of a major storm and associated flooding in Arizona. Arizona Agricultural Experiment Station Technical Bulletin 202. University of Arizona, Tucson. 30 p.

Tiedemann, A.R. and J.O. Klemmedson. 1973. Effect of mesquite on physical and chemical properties of the soil. Journal of Range Management 28:27-29.

———. 1977. Effect of mesquite trees on vegetation and soils in the desert grassland. Range Management, Journal 30:361-367.

Turner, C. and R.G. West. 1968. The subdivision and zonation of interglacial periods. Eiszeitalter und Gegenwart 19:93-101.

U.S. Department of Agriculture Soil Conservation Service. 1976. Unpublished report of progress: Field review of Navajo County area, central part.

U.S. Forest Service. Unpublished reports.

University of Arizona Press. 1972. Arizona: Its people and resources. Revised second edition. University of Arizona, Tucson. 411 p.

Updike, R.G. and T.L. Péwé. 1974. Glacial and pre-glacial

deposits in the San Francisco Mountain area, northern Arizona. Pages 557-566 in T.N.V. Karlstrom, G.A. Swann and R.L. Eastwood, eds., Geology of northern Arizona with notes on archeology and paleoclimate. Part II—area studies and field guides. Geological Society of America, Rocky Mountain Section, Flagstaff, Arizona. 397 p.

———. 1976. San Francisco Peaks—a guidebook to the geology. Second edition. Museum of Northern Arizona, Flagstaff. 80 p.

Van Devender, T.R. 1973. Late Pleistocene plants and animals of the Sonoran Desert: A survey of ancient packrat middens in southwestern Arizona. Ph.D. dissertation. University of Arizona, Tucson. 179 p.

———. 1977. Holocene woodlands in the southwestern deserts. Science 198:189-192.

Van Devender, T.R. and W.G. Spaulding. 1979. Development of vegetation and climate in the southwestern United States. Science 204:701-710.

Van Dork, J. 1976. O^{18} record of the Atlantic Ocean for the entire Pleistocene Epoch. Pages 147-164 in R.M. Cline and J.D. Hays, eds., Investigation of late Quaternary paleoceanography and paleoclimatology. Geological Society of America Memoir 145, Boulder, Colorado. 464 p.

Vine, F.J. 1966. Spreading of the ocean floor, new evidence. Science 154:1405-1415.

Vorhies, C.T. and W.P. Taylor. 1933. The life histories and ecology of jack rabbits, *Lepus alleni* and *Lepus californicus* spp., in relation to grazing in Arizona. Arizona Agricultural Experiment Station Technical Bulletin 49:478-587. University of Arizona, Tucson.

Wallace, H.R. 1956. The seasonal emergence of larvae from cysts of the beet eelworm, *Heterodera schachtii* Schmidt. Arizona Agricultural Experiment Station Technical Bulletin 49:227-238. University of Arizona, Tucson.

Wallwork, J.A. 1969. Some basic principles underlying the classification and identification of cryptostigmatid mites. Pages 155-168 in J.G. Sheals, ed., The soil ecosystem: Systematic aspects of the environment, organisms and communities. Systematics Association Publication 8, London. 247 p.

———. 1970. Ecology of soil animals. McGraw-Hill, New York. 283 p.

———. 1976. The distribution and diversity of soil fauna. Academic Press, New York. 355 p.

———. 1982. Desert soil fauna. Praeger Publishers, New York. 296 p.

Welch, T.G. and J.O. Klemmedson. 1975. Influence of the biotic factor and parent material on distribution of nitrogen and carbon in ponderosa pine ecosystems. Pages 137-158 in B. Bernier and C.H. Winget, eds., Forest soils and forest land management, proceedings of the 4th North American Forest Soils Conference, Laval University, Quebec. Les Presses de l'Universite Laval, Quebec, Canada. 675 p.

Wells, P.V. 1976. Macrofossil analysis of wood rat (Neotoma) middens as a key to the Quaternary vegetational history of arid America. Quaternary Research 6:223-248.

———. 1979. An equable glaciopluvial in the West: Pleniglacial evidence of increased precipitation on a gradient from the Great Basin to the Sonoran and Chihuahuan deserts. Quaternary Research 12:563-579.

Whittaker, R.H. et al. 1968. A soil and vegetation pattern in the Santa Catalina Mountains, Arizona. Soil Science 105:440-450.

Williams, S.T., F.L. Davies and D.M. Hall. 1969. A practical approach to the taxonomy of actinomycetes isolated from soil. Pages 101-117 in J.G. Sheals, ed., The soil ecosystem: Systematic aspects of the environment, organisms and communities. Systematics Association Publication 8, London. 247 p.

Wilson, E.D. and R.T. Moore. 1959. Structure of Basin and Range Province in Arizona. Pages 89-105 in L.A. Heindl, ed., Southern Arizona guidebook II. Arizona Geological Society, University of Arizona, Tucson. 290 p.

Wilson, J.T. 1965. A new class of faults and their bearing on continental drift. Nature 207:343-347.

Wooten, R.C., Jr., and C.S. Crawford. 1975. Food, ingestion rates and assimilation in the desert millipede *Orthoporus ornatus* (Girard) (Diplopoda). Oecologia 20:231-236.

Wright, H.E. 1972. Interglacial and post glacial climate: The pollen record. Quaternary Research 2:274-282.

Yaalon, D.H. 1971. Soil forming processes in time and space. Pages 29-40 in D.H. Yaalon, ed., Paleopedology: Origin, nature and dating of paleosols. International Society of Soil Science and Israel Universities Press, Jerusalem. 350 p.

Zagwijn, W.H. 1974. The Pliocene-Pleistocene boundary in western and southern Europe. Boreus 3:75-97.

Zinke, P.J. 1962. The pattern of individual forest trees on soil properties. Ecology 43:130-133.

Zinke, P.J. and R.L. Crocker. 1962. The influence of giant sequoia on soil properties. Forest Science 8:2-11.

Index

Abruptic Durargids, 65
acacia, 50, 57
aeolian materials, 12, 31, 117, 213
aerobic decomposition, 62
agave, 111, 155
Albaqualfs, 159, 213
albic horizon, 64, 213
alfalfa, 75, 76, 79, 82, 84, 95, 96, 109, 125
Alfisols, 66, 67, 68, 159, 160, 213
algae, 57
algerita, 155
allophane, 27
allthorn, 53, 155
almond, desert, 155
alluvial fan, 17, 25, 26, 29, 43, 75, 78, 82, 94, 101, 102, 105, 107, 112, 118, 119, 131, 135
alluvial surfaces, 24, 25, 26, 44, 88, 98, 102, 105, 107, 109, 131, 135
alluviation, 21
alluvium, 12, 17, 21, 22, 25, 26, 69, 76, 77, 79, 82, 88, 94, 96, 98, 101, 107, 108, 112, 113, 115, 116, 118, 120, 125, 126, 130, 132, 148
alpine glaciation, 43
alpine meadows, 47
alpine rock field, 49
alpine tundra, 48, 49
altithermal time, 45
alumina sheet, 4
aluminum oxide minerals, 7
American Kestrel, 56
ammonium sulphate fertilizer, 4
amphibole, 213
anaerobic decomposition, 62
anaerobic soil environment, 10
Anasazi, 1, 2
anathermal time, 45
Andepts, 68, 159
andesite, 11, 23, 121, 213
anhydrite, 16
animals' role in soil community, 56

antelope, 130, 135, 140, 157
anthropic epipedons, 26, 213
anticline, 26, 30, 213
antlions, 57
ants, 57, 58, 59, 60
Apachian Biotic Province, 55
Aquic Haplustolls, 135
Aquic Natrustalfs, 104
Aquolls, 69, 160
arachnids, 58, 59
Archeozoic Period, 151, 153
Arctic-alpine disjuncts, 49
Arcto-Tertiary flora, 53
Argic Cryoborolls, 150
Argic Pachic Cryoborolls, 144, 150
Argids, 65, 66, 68, 160
argillic horizon, 12, 13, 24, 25, 30, 31, 60, 64, 65, 68, 213
arid-adapted crops, 77
Aridic Argiustolls, 127, 129, 130, 137, 138
Aridic Calciustolls, 97, 98
Aridic Paleustolls, 110
Aridisols, 65, 66, 67, 68, 160, 213
Arizona-Sonora Desert Museum, 48
Arizona Upland Subdivision, Sonoran Desert, 52, 53
Arivaipa Lake, 99
arrowweed, 75, 155
arroyo cutting and filling, 37, 45
arthropods, 57, 59
ash, 50, 155
aspen, quaking, 155
augite, 213
authigenic, mineral origin, 214
available water capacity, 61
axial stream, 21
axial valley troughs, 21

badgers, 56, 60, 79, 150, 157
Badland, 29, 30, 113, 115, 117, 119, 164-165, 214

235

basalt, 11, 17, 23, 24, 28, 31, 69, 121, 130, 141, 214
basaltic plains, 129
base saturation, 64, 161
basic volcanic rocks, 11, 17, 23, 24, 28, 31, 69, 121, 130, 141
Basin and Range Province, 16, 17, 18-26, 34, 142
basin fill, 16, 17, 21
bass, 135
bats, 60, 157
bear, black, 139, 143, 144, 145, 147, 148, 157
beargrass, 97, 98, 127, 155
beaver, 56
Beaver Creek, 28
bedded tuffs, 28
bees, 57
beetles, 51, 60
Bermuda high pressure cell, 34
Bidahochi Formation, 29
Big Lake, 150
biological activity, 7
biotic communities, 43, 214
biotic provinces, 55
biotite, 4, 214
birds, 36, 56, 60, 62, 79, 136, 137, 138, 139, 140, 143, 144, 145, 146, 147, 148, 149, 150, 158
biseasonal precipitation pattern, 34
blackbrush, 52, 89, 91, 106, 117, 155
Black Mesa, 2, 153
Black Mountains, 85
Black Point erosion cycle, 32
Blue-gray Gnatcatcher, 56, 158
Blue Grosbeak, 56, 156
blue paloverde, 52, 158
Boralfs, 68, 159, 214
Borolls, 68, 69, 160
box canyon, 29
brackenfern, 144, 155
brush control, 131, 132
bubbler irrigation, 84
burroweed, 96, 100, 105, 106, 155
burrowing animals, 56, 57, 61
bursage, 75, 76, 77, 78, 81, 82, 83, 84, 90, 91, 102, 112, 155

cactus
    cholla, 52, 83, 101, 105, 112, 155
    organpipe, 52, 156
    pricklypear, 101, 108, 112, 156
    saguaro, 52, 112, 156.
calcareous parent material, 25, 31, 109
calcic horizon, 24, 25, 64, 68, 69, 214
Calciorthids, 91, 160
calcite, 5
calcium, exchangeable, 4
calliandra, 98, 100, 101, 103, 110, 111
cambic horizon, 64, 68, 214
Camborthids, 87, 113, 120, 168
Cambrian rocks, 151, 152
Canadian Life Zone, 46, 48, 56
canopy drip, 9
Canyon de Chelly, 15, 32, 128
Canyon Land Section, 26
capillary withdrawal, 13
carbon dioxide, 5, 6
carbonate accretion, 12, 13, 24, 68
carbonates, 5, 9, 11, 13, 24, 28, 68, 70, 161, 214
carbonatic family, 70
carnivores, 57
Carrizo-Chuska Mountains, 30, 31, 43, 49, 149
catclaw, 75, 81, 88, 95, 96, 101, 108, 110, 111, 155
cation exchange, 4
cation exchange capacity, 4, 9, 161
cations, exchangeable, 9, 12, 215

cattle, 61, 62, 104, 109
ceanothus, 124, 128, 132, 138
Cedar Ranch age, 28
Cenozoic Era, 17, 23, 151, 152, 153
Cenozoic sediments, 17
central-type volcanoes, 17
Cerbat Mountains, 85
chalazite, 16
Chaco Canyon, 2
channel cutting, 21, 30
chaparral, 47, 48, 49, 51, 214
Chihuahuan Desert, 23, 43, 48, 53, 57
Chihuahuan desert scrub, 53, 56
Chihuahuan grassland, 52
Chinle Formation, 15, 29, 30
Chino Valley, 132
Chiricahua Mountains, 49
Chiricahua National Monument, 48
Chuska Formation, 30, 31
Cibola National Wildlife Refuge, 83
cienagas, 21, 26, 34
cinder cones, 10, 28, 31, 121, 141
cinders, 22, 28, 68, 69, 121, 126, 141
citrus, 75, 84
clay, 3, 4, 5, 10, 13, 25, 40, 64, 65, 69, 214
clay accumulation, 24, 25
clay destruction, 24
clay dispersion, 13
clay minerals, 4, 5, 10, 65, 69
cliff retreat, 30, 214
cliffrose, 123, 124, 136, 137, 138, 155
climate, 33-45
    change, 45
    zones, 35-38
climax vegetation, 49, 214
coatimundi, 135, 157
Cochise people, 1
Coconino Plateau, 31, 121
cold highlands, 35
cold steppes, 35
colluvium, 43, 141
Colorado Plateau Province, 11, 16, 19, 26-32, 49, 114, 142
Colorado River, 28, 29, 30, 33, 48, 52, 75, 87
composites, 47
compressional shearing, 23
continental drift, 22
continental glaciers, 45
convective activity, 34
Copper Mountains, 73
corn, 2, 109, 118, 120, 125, 130, 132, 137, 149
cotton, 75, 76
cottontail rabbits, 60, 157
cottonwood, 50, 75, 88, 118
coyotes, 60, 79, 157
creosotebush, 35, 52, 53, 57, 75, 76, 77, 78, 79, 80, 81, 82, 84, 88, 89, 90, 91, 94, 95, 102, 105, 106, 107, 108, 155
Cryorthents, 142, 147, 176
Cumulic Haplustolls, 131, 132
curleymesquite, 135, 155

dacite, 141, 215
dates, 84
Datil Section, 26
De Chelly Sandstone, 29, 30
deer, 135, 137, 138, 139, 143, 144, 145, 146, 147, 148, 149, 150
    blacktail, 136, 157
    mule, 56, 61, 62, 146, 157
Defiance Plateau, 30
Defiance Uplift, 30, 31
delineated soil body, 71
depositional events, 32

INDEX

Desert Botanical Garden, 48
desert climatic zone, 35, 36
desert dweller domains, 2
desert pavement, 12
desert region, 17
desert scrub, 48, 52, 53, 56
desert shrubs, 87
desert tortoises, 60
desert varnish, 24
desertwillow, 75, 95, 119, 155
Devonian Period, 152, 153
diorite, 30, 215
diurnal freeze-thaw cycles, 38, 39
diurnal soil dwellers, 57
doves, 79, 140, 143, 158
drip irrigation, 84
dryland farming, 2
dunes, 30, 117
Durargids, 65, 66, 160
duripans, 25, 64, 65, 68, 78, 104, 215

earthworms, 56, 57, 59
East Pacific Rise, 23
elk, 61, 139, 140, 144, 145, 148, 157
eluvial horizons, 7, 25, 215
encinal woodland, 51, 56, 215
energy loss, 8
Entisols, 66, 67, 68
Eocene Epoch, 153
eolian, see aeolian
ephemeral plant species, 53
ephemeral streams, 21
epipedons, 64
epigeon, 57
eras, 151
erosion, 8, 10, 16, 23, 28, 29, 30, 69, 215
    bank cutting, 75, 79
    creep, 28, 30
    gully, 12, 31, 75
    rill, 12
    wind, 12, 28, 34
erosional cycles, 30, 32
evaporation, 8, 9, 10, 11, 37, 41
evapotranspiration, 37, 41, 42, 43, 215
evergreen shrubs, 51
Eutroboralfs, 148, 149, 159, 179
extinct megafauna, 1
extragrade subgroup class, soil taxonomy, 65, 215

fan alluvium, 26, 43
fan terraces, 17, 21, 22, 24, 25, 44
faulting, 21, 22, 27, 31, 215
faults
    block, 16, 17, 21
    listric, 21, 23, 218
    normal, 23, 27
    San Andreas, 23
    tilted block, 21, 23
fir
    corkbark, 49
    Douglas, 49, 50, 143, 144, 145, 146, 147, 148
    white, 49, 50, 155
filaree, 81, 155
Flandrian Interglacial, 45
flooding, 68
flooding potential, 75
floodplain alluvium, 21
floodplains, 12, 21, 22, 25, 29, 50, 75, 88, 94, 95, 96, 110, 116, 119, 131
fluvents, 60, 68
fluvial terraces, 21, 22, 24, 26, 29, 32

Fort Apache Indian Reservation, 33
formative elements, soil nomenclature, 65, 67
fossils, 15, 16
foxes, 57, 60, 79, 157
Frigid Subhumid Zone, 38, 40, 50, 56, 141-150
frogs, 60, 158
frontal storms, 34, 44
frost alternation days, 38-39

geochronology, 28, 29
geologic time scale, 151-154
geomorphic processes, 21, 216
geomorphic surface-soil relationships, 23, 30
Gila River, 21, 33, 34, 73
glacial maximum, 43
glaciation, 43, 44, 45, 154
glaciopluvial climate 44, 216
Glen Canyon group, 30
gneiss, 17, 217
gophers, 13, 60, 157
graben, 22, 23, 217
grain size distribution, 4, 69
Grand Canyon, 15, 26, 28, 29, 44, 52, 86, 87, 121, 124, 151, 153
Grand Canyon Section, 26, 28, 31
granitic rocks, 11, 23, 24, 72
granular disintegration, 23, 28
grass
    alkali sacaton, 104, 106, 115
    Arizona cottontop, 99, 101, 103
    Arizona fescue, 140, 143, 144, 145, 146, 147, 148, 149, 150
    big galleta, 80, 84, 91, 155
    black grama, 52, 80, 87, 98, 99, 102, 107, 116, 118, 119, 120, 123, 124, 125, 126, 127, 130, 131, 155
    blue grama, 52, 94, 97, 98, 99, 115, 116, 118, 119, 120, 123, 124, 125, 127, 129, 131, 132, 140, 145, 155
    bluebunch wheatgrass 150, 156
    bluestem, 128, 156
    bullgrass, 103, 155
    bush muhly, 80, 89, 94, 95, 101, 102, 105, 110, 156
    cane bluestem, 99, 108, 130, 155, 156
    fluffgrass, 106, 107, 155
    Indian ricegrass, 117, 120
    junegrass, 145, 146, 147, 148, 156
    meadow hourgrass, 156
    mountain bluegrass, 51, 135
    mountain brome, 143, 148, 155
    mountain muhly, 139, 143, 144, 146, 147, 148, 149, 150, 156
    mountain timothy, 51, 156
    needlegrass, 156
    pine dropseed, 127, 149, 155
    plains lovegrass, 99, 101, 156
    ring muhly, 127, 129, 130, 156
    Rorthrock grama, 80, 94, 155
    saltgrass, 118, 119, 156
    sand dropseed, 116, 120, 155
    sideoats grama, 52, 89, 97, 98, 99, 123, 126, 127, 129, 130, 131, 132, 140, 155
    slender grama, 97, 98, 99
    spike muhly, 123, 124, 156
    tanglehead, 136, 156
    Texas bluestem, 8, 99, 135, 136, 155
    threeawn, 95, 100, 102, 112, 117, 118, 120, 128, 131, 132, 156
    tabosa, 96, 104, 105, 108, 109, 110, 111, 140, 156
    western wheatgrass, 128, 149, 156
grasslands, 47, 48, 51-52
gravel size particles, 4, 17, 72
Great Basin Desert, 52, 57
Great Basin Desert flora, 47, 48, 52
great group category, soil taxonomy, 65, 71, 72
Green's Peak, 10, 11, 60, 150
groundwater recharge, 37

237

Gulf of Mexico, 34
gypsic horizon, 64, 68, 217
gypsum, 7, 17, 64, 68, 70, 161

hardpan, 78
Hawley Lake, 33
helium gas field, 16
hemiedophon, 57
herbivores, 61
high mountain soils, 24, 149
Histosols, 66, 67, 68, 217
hogbacks, 30, 217
Hohokam people, 1, 2
Holocene cinders, 22
Holocene Epoch, 29, 44, 45, 152, 153, 154
Hopi Buttes cycle, 32
Hopi Buttes surface, 32
Hopi Buttes volcanic field, 28
Hopi Indian Reservation, 2, 128
Hopi Mesa, 2
horses, 62
horsts, 21, 22, 23, 217
Huachuca Mountains, 133
Hualapai Mountains, 48
humidity, 8
humus, 7, 58, 217
hurricanes, 34, 44
Hyperthermic Arid Region, 37, 38, 40, 41, 53, 74

ignimbrites, 17, 217
Illinoian Glacial Stage, 154
illuvial clay, 7, 25, 31, 64, 65, 66
illuviation, 7, 8, 12, 13, 25, 217
Inceptisols, 66, 67, 68, 217
index species, 40
insects, 13, 56
interglacial time, 44, 45, 154
intergrade subgroup class, soil taxonomy, 65, 217
intermontane basins, 21, 22, 83, 88
interstadial time, 44, 154, 217
invertebrates, 55, 57
iron oxides, 5, 68
iron reduction, 10
ironwood, 78, 79, 96, 156
irrigation, 26, 84
isopods, 56

jackrabbits, 61, 157
Joshua tree, 52, 88
juniper, 50, 51, 67, 87, 99, 103, 109, 120, 123, 124, 127, 128, 129, 130, 132, 135, 136, 137, 138, 139, 140, 145, 149, 156
Jurassic Period, 152, 153

Kaibab Limestone, 31, 145
Kaibab Plateau, 31
Kanab Plateau, 31, 34, 49
Kansan Glacial Stage, 154
kaolinite, 4, 5, 217
Kayenta Formation, 15
Keet Seel Ruins, 117
Kirkland Formation, 29
Kofa National Wildlife Refuge, 83

laccolithic sills, 30
lacustrine sediments, 17, 104, 106, 218
lagomorphs, 60, 61
Lake Mary, 145
Lake Mead 87, 91
Lake Powell, 87
landforms, evolution, 30
Laramide orogeny, 27

Late Cenozoic volcanism, 27-28
lava-capped mesas, 28
leaf litter, 5, 9
leaching, 9, 12, 13, 33, 41, 43, 76
legumes, 47
lice, wood, 59
lichens, 57
limnic layers, 7
Lithic Argiustolls, 126, 129
Lithic Calciustolls, 126
Lithic Camborthids, 73, 180
Lithic Cryoborolls, 31, 146, 188
Lithic Cryorthents, 141
Lithic Haplargids, 73, 80, 98, 128, 188
Lithic Haplustolls, 99, 123, 133, 136, 138
Lithic Torriorthents, 89, 99, 116, 124, 128, 188-189
Lithic Ustollic Haplargids, 124, 127
Lithic Ustorthents, 139
lithosphere, 22, 218
Little Colorado River, 34, 35, 113, 118
lizards, 60, 158
Lukachukai Mountains, 31
Luke salt body, 16

maar crater, 28, 218
macrofauna, 56, 59
Madro-Tertiary flora, 53
mammals, 55, 57, 66
Mancos Shale, 29
manzanita, 135, 136, 138, 139, 156
Marble Canyon, 26
mariola, 53, 156
marmot, 149
medithermal time, 45
Merriam age group basalt, 29
Mesa Verde, 2
Mesic Arid Zone, 38, 40, 52, 56, 113, 114-120
Mesic Semiarid Zone, 38, 40, 52, 121-123
Mesic Subhumid Zone, 38, 40, 51, 133-140
mesofauna, 56, 59
Mesozoic Era, 152, 153
mesquite, 1, 9, 10, 35, 50, 51, 52, 75, 76, 77, 78, 79, 82, 89, 94, 95, 96, 97, 98, 100, 102, 106, 107, 108, 109, 110, 156
metamorphic core complexes, 15
Meteor Crater, 15, 124
mica, 4, 218
microbial activity, 6
microfauna, 56, 59
microhabitats, 57
microphytic-feeders, 57
millipedes, 56, 59
mineral alteration, 7, 70, 191
mineral family class, 64, 65, 69
Miocene Epoch, 152, 153
Mississippian Period, 28, 152, 153
mites, 56, 57
Mogollon people, 2
Mogollon Plateau, 35, 157
Mogollon Rim, 2, 34, 47, 51
Mohave Biotic Province, 55
Mohave Desert, 43, 44, 48, 52, 55, 57
Mohave desert scrub, 32, 56
Mohave yucca, 91, 94, 107
mollic epipedon, 64, 69, 218
Mollic Eutroboralfs, 143, 145
Mollisols, 66, 67, 68, 69, 150, 160, 218
molluscs, 59
monoclines, 26, 31, 218
montane conifer forest, 49-50
montmorillonite, 4, 5, 69, 218
monsoon, 34

INDEX

Monument Valley, 15, 115, 117
Mormon Lake, 145
Mormon Mountain volcanic field, 28, 32
Mormon-tea, 91, 115, 116, 117, 126
Mount Baldy, 28, 43, 141
Mount Floyd volcanic field, 28, 32
Mount Ord, 28
mountain dandelion, 51, 155
mountain lion, 56, 139, 143, 145, 146, 148, 149, 157
mountainmahongany, 123, 137, 138, 139, 156
mountain meadow grasslands, 31, 51, 141, 150
mountain region, 17
mountain soils, 23, 24, 25, 149
mouse, 13
    cactus, 56, 157
    pocket, 60, 157
Muggins Mountains, 73
myriopods, 59

National Cooperative Soil Survey, 6
national forests, 14
national monuments, 14
national parks, 14
national recreation areas, 14
natric horizon, 64, 219
Navahonian Biotic Province, 55
Navajo Indian Reservation, 119, 128, 149, 153
Navajo Sandstone, 29
Navajo Section, 26, 29-31
Nebraskan Glacial Age, 44, 154
nematodes, 56, 57, 58, 62
neotectonism, 21, 22
nitrogen, 9, 10, 59
North American Plate, 23
nutrient cycling, 13

oak
    Arizona, 51
    Emory, 51, 103, 135, 156
    Gambel, 143, 144, 145, 146, 147, 149, 156
    shrub live, 103, 111, 123, 126, 128, 131, 137, 156
    silverleaf, 51, 156
Ochrepts, 68, 160
ochric epipedon, 64, 68, 219
ocotillo, 52, 53, 80, 89, 108, 111, 156
Oligocene Epoch, 152, 153
omnivorous soil animals, 59
Ordovician Period, 152, 153
Organ Pipe Cactus National Monument, 48, 78
organic carbon, 10
organic litter, 10, 12, 156
organic matter, 3, 5, 7, 9, 10, 12, 26, 31, 40, 59, 60, 161
organic mineral complexes, 59
Orthents, 68
Orthids, 68
Oxisols, 68, 159
oxygen, 5, 6

Pachic Argiustolls, 140, 195
pack rat middens, 28, 43, 44
Painted Desert, 30, 115
Paleocene Epoch, 152, 153
paleoindians, 1
paleoplant communities, 43
Paleorthids, 91, 160
Paleozoic Era, 16, 151, 152, 153
paloverde, 1, 9, 10, 50, 52, 75, 76, 77, 81, 82, 88, 89, 90, 96, 99, 102, 107, 112, 156
Parker Canyon, 99
Parker Canyon Lake, 135
particle size differentiation, 24, 219

Patagonia Lake, 99
peds, 5, 219
pediments, 17, 21, 22, 24, 219
pedon, 66, 67, 219
pedoturbation, 32, 219
Pena Blanca Lake, 99
Pennsylvania Period, 152, 153
periglacial environment, 43, 219
periodic vertebrates, 60
Permian Period, 31, 36, 152, 153
Petrified Forest, 15, 115
petrocalcic horizons, 12, 24, 25, 64, 68, 219
Phoenix Basin, 21
physiographic provinces, cross sections, 18, 19
phytophages, 57
Picacho Basin, 16
piedmont slopes, 17, 21, 22, 25, 26
Pigeon, Band-tailed, 139, 143, 144, 145, 146, 147, 148, 158
Pinaleno Mountains, 9, 24, 43, 48
pine
    bristlecone, 49, 156
    Chihuahua, 51
    limber, 49, 156
    pinyon, 50, 123, 124, 125, 126, 127, 128, 129, 132, 137, 138, 139, 140, 148, 149, 156
    ponderosa, 11, 49, 50, 126, 136, 137, 138, 140, 141, 143, 144, 145, 146, 147, 148, 149, 156
Pinta Dome, 16
pinyon-juniper woodlands, 50-51
piping, 95
Piute Canyon, 117
plains grassland, 51-52
plant communities, 40, 55
plant debris, 5
plate tectonics, 22, 23
playas, 16, 17, 104
Pleistocene climatic change, 30
Pleistocene deposits, 32
Pleistocene Epoch, 22, 29, 30, 32, 43, 44, 49, 152, 153, 154
Pleistocene glaciation, 43
Pleistocene surfaces, 22
Pliocene Epoch, 28, 29, 32
plutonism, 23
pluvial climate, 43
pollen stratigraphy, 43, 45
polypedon, 66, 67, 71, 220
porcupine, 145, 146, 147, 148, 157
porphyry copper deposits, 16, 220
pot worms, 56
potential evapotranspiration, 41, 42, 43, 44
pottery designs, 2
prairie dogs, 60, 61, 157
Precambrian Era, 26, 151
precipitation, 8, 12, 33, 34-35, 37, 38, 41, 42, 44, 45, 47
prehistoric agriculture, 1
prehistoric man in Arizona, 1
primary minerals, 4
profile development, 24
Proterozoic Era, 151
protozoa, 56, 57, 59
Psamments, 68

Quail
    Bobwhite, 158
    Gambel's, 62, 147, 158
quartz, 4, 7, 220
Quaternary Period, 28, 29, 32, 152, 154

rabbitbrush, 117, 120, 125, 127, 156
rabbits, 60, 61, 136, 137, 138, 140, 157
radiometric dating, 12, 43, 151, 220

rainshadow, 34
rat
    kangaroo, 60, 61, 157
    pack, 60, 157
    wood, 60
recessed cliffs, 30
Red Lake, 17, 88
regional uplift, 27
relict soils, 26
reptiles, 16, 158
rhyodacite, 23, 220
rhyolite, 17, 121, 141, 220
riparian deciduous woodland, 50
riparian vegetation, 50, 75, 119
riverwash, 119
Roadrunner, 60, 158
rock fall, 28
rock fragments, 4
rock outcrops, 29, 78, 97, 99, 111, 115, 116, 117, 120, 123, 124, 139, 144, 146, 147, 200
rodents, 13, 57, 60, 61, 79, 136, 157

SP Crater, 31
Sacramento Valley, 15
sagebrush, 116, 117, 124, 125, 126, 127, 150, 156
Saguaro National Monument, 48
salic horizons, 65, 200
saltbush, 57, 75, 78, 79, 82, 83, 94, 95, 156
    fourwing, 106, 109, 115, 116, 156
saltcedar, 50, 156
salt deposits, 16, 17
Salt River, 21, 33
salt tolerant plants, 10
San Andreas Fault, 23
sand, 3, 4, 5, 10, 64, 72, 220
sandpaper bush, 53, 156
sand sagebrush, 120, 156
sandstone, 30, 113
San Francisco Peaks, 27, 28, 34, 48, 49, 55, 121, 153
Sangamon Intreglacial Age, 154
San Pedro Valley, 15, 75, 92
San Simon Valley, 34, 68, 106
Santa Catalina Mountains, 9, 15, 24, 48
Santa Cruz River, 75, 92, 133
Santa Cruz River Valley, 21
Santa Rita Experimental Range, 9, 61
saprolite, 23, 220
saprophages, 57
scarp retreat, 30, 220
schistose rocks, 23, 24, 220
scorpions, 57
sea-floor spreading, 22, 23
sedges, 51, 150, 156
sedimentary rocks, 17
sedimentation, 23
self-swallowing soils, 13, 32, 66, 67, 69
semiarid region, 68
semidesert grasslands, 48
shadscale, 115, 118, 156
shale, 30, 113, 220
sheep, bighorn, 62, 136, 157
Shivwits Plateau, 31, 32
short grass plains, 48
shrubby senna, 53, 156
sialic crust, 27, 221
silica cementation, 25
silica sheet, 4
silica solubility, 25
siliceous mineral family, 70, 221
silt, 3, 4, 5, 10, 72, 221
site disturbance index, 61, 62

skeletal particle size families, 69, 221
skunks, 60, 157
slab-failure, 28
slope aspect, 10, 50
slugs, 57, 59
small-scale wastage landslides, 28
snails, 57, 59
snakes, 13, 57, 58, 59, 60
snakeweed, 90, 109, 112, 131, 132, 156
snow, 35
snowline lowering, 43
soil
    acidity, 6, 12
    aeration, 62
    aggregate, 213
    alkaline, 8
    association, see soil association
    atmosphere, 5, 62
    bulk density, 161
    chronosequences, 30
    classes, 63
    classification, 38, 40, 63
    composition diagram, 3
    depth, 40
    fertility, 11
    formation, see soil formation
    horizons, 6-7, 12, 24, 25, 63, 64, 65, 68, 69, 200, 213, 214, 215, 217, 219, 221
    individual, 66
    minerals, 4, 5, 40, 64, 65, 69-70
    moisture, see soil moisture
    nitrogen, 9, 10, 59
    organic carbon, 10
    organic matter, 3, 5, 9, 40, 56, 59, 60, 68
    particle size, 4, 69
    pH, 10, 62, 219
    phase, 64
    profile, 6, 221
    saline, 76, 104, 106, 118
    saturated, 6, 10, 40
    shrinking and swelling, 13, 32
    sodic, 6, 25, 76, 104, 106
    specific surface area, 4
    structure, see soil structure
    surveys, 63
    taxa, 63
    temperature, see soil temperature
    texture, see soil texture
    type, 64
    voids, 6, 62
soil associations, 71, 213, 221
    Anthony-Sonoita, 26, 93, 112
    Anthony-Vinton-Aqua, 26, 85, 86, 88
    Badland-Torriorthents-Torrifluvents, 31, 114, 115
    Bonita-Graham-Rimrock, 93, 108
    Cabezon-Thunderbird-Springerville, 122, 129
    Caralampi-Hathaway, 93, 98
    Caralampi-White House, 93, 101
    Casa Grande-Mohall-La Palma, 76
    Casto-Martinez-Canelo, 135
    Chiricahua-Cellar, 23, 93, 103
    Continental-Latene-Pinaleno, 93, 105
    Cryorthents-Eutroboralfs, 30, 142, 147
    Eutroboralfs-Mirabal, 142, 148
    Fruitland-Camborthids-Torrifluvents, 114, 120
    Gordo-Tatiyee, 32, 51
    Gothard-Crot-Stewart, 25, 93
    Graham-Lampshire-House Mountain, 93, 111
    Gunsight-Rillito-Pinal, 74, 78
    Harqua-Perryville-Gunsight, 74, 83
    Karro-Gothard, 106

INDEX

Latene-Anthony-Tres Hermanos, 26, 86, 90
Laveen-Carrizo-Antho, 74, 81
Laveen-Rillito, 25, 74, 79
Lithic Camborthids-Rock Outcrop-Lithic Haplargids, 24, 25, 74, 80
Lithic Haplustolls-Lithic Argiustolls-Rock Outcrop, 24, 134, 136
Lithic Torriorthents-Lithic Haplargids-Rock Outcrop, 122, 128
Lithic Torriorthents-Lithic Haplustolls-Rock Outcrop, 24, 93, 99
Lithic Torriorthents-Rock Outcrop-Lithic Haplargids, 24, 86, 89
Lonti-Balon-Lynx, 122, 131
Mirabal-Baldy-Rock Outcrop, 31, 147, 152
Mirabal-Dandrea-Brolliar, 142
Moenkopie-Shalet-Tours, 114, 116
Mohall-Vecont-Pinamt, 25, 74
Nickel-Latene-Cave, 93, 107
Overgaard-Elledge-Telephone, 134, 139
Pachic Argiustolls-Lynx, 134, 140
Paleorthids-Calciorthids-Torriorthents, 25, 86, 91
Palma-Clovis-Trail, 122, 125
Pastura-Abra-Lynx, 122, 132
Pastura-Poley-Partri, 122, 130
Penthouse-Latene-Cornville, 93, 109
Roundtop-Boysag, 31, 122, 127
Roundtop-Tortuga-Jacks, 137, 138
Sheppard-Fruitland-Rock Outcrop, 30, 114, 117
Showlow-Disterheff-Cibeque, 134, 137
Signal-Grabe, 93, 110
Soldier-Hogg-McVickers, 142, 145
Soldier-Lithic Cryoborolls, 31, 49, 51, 142, 146
Sponseller-Ess-Gordo, 43, 49, 141, 144
Superstition-Rositas, 74
Torrifluvents, 26, 50, 74, 93, 95, 100
Torrifluvents-Torripsamments, 93, 94
Torriorthents-Camborthids-Rock Outcrop, 29, 52, 86, 87
Torriorthents-Torrifluvents, 114, 119
Tortugas-Purner-Jacks, 122, 123
Tours-Navajo, 114, 118
Tremant-Coolidge-Mohall, 74, 82
Tubac-Sonoita-Grabe, 93, 96
White House-Bernardino-Hathaway, 93, 97
White House-Caralampi, 93, 100
Winona-Boysag-Rock Outcrop, 31, 122, 124
soil formation factors
  climate, 7, 8
  organisms, 7, 9-10
  parent materials, 3, 7, 10-11, 23, 25, 32, 62, 69, 219
  relief, 7, 10-11, 40
  time, 7, 11-12
  topography, see relief
soil formation processes
  additions, 7, 12
  losses, 7, 12
  transformations, 7, 12
  translocations, 7, 12-13
soil formation rates, 12
soil moisture, 6, 9, 10
  control section, 40
  deficit, 42, 44
  recharge, 42
  regimes: aquic, 40; aridic, 40; torric, 40; udic, 40; ustic, 40; xeric, 40
  tension, 6
  utilization, 42, 43
soil series, 221
  Abra, 130, 131, 132, 162
  Agua, 88, 162
  Agualt, 75, 162
  Ajo, 78, 82, 162
  Amos, 139, 162
  Anklam, 99, 162-163
  Antho, 26, 75, 81, 163
  Anthony, 85, 88, 90, 95, 96, 102, 105, 112, 163
  Anway, 96, 163
  Apache, 126, 129, 163
  Arada, 91, 163
  Arizo, 91, 95, 164
  Arp, 164
  Artesia, 106
  Atacosa, 99, 164
  Avondale, 75, 164
  Baldy, 141, 144, 147, 165
  Balon, 128, 131, 165
  Bandera-like, 121, 126, 165
  Barkerville, 136, 138, 165
  Bernardino, 97, 98, 100, 108, 166
  Bitter Spring, 91, 166
  Bonita, 97, 108, 110, 111, 166
  Boysag, 31, 121, 123, 124, 127, 167-168
  Brazito, 95
  Brios, 75, 81
  Brolliar, 143, 144
  Bucklebar, 112
  Bushvalley, 144, 168
  Cabezon, 126, 136, 140, 168
  Cambern, 144, 168
  Canelo, 135
  Caralampi, 97, 98, 99, 100, 101, 169
  Carrizo, 75, 81, 169
  Casa Grande, 76, 81, 169
  Casto, 133, 135, 170
  Cave, 85, 90, 91, 105, 106, 107
  Cavelt, 78, 81
  Cellar, 23, 85, 89, 99, 103, 170-171
  Cherioni, 83, 171
  Chevelon, 137, 171
  Chilson, 171
  Chiminea, 99, 171
  Chiricahua, 23, 89, 97, 99, 103, 171-172
  Cibeque, 137, 172
  Cipriano, 78, 172
  Clay Springs, 113, 115, 116, 119, 172-173
  Clover Springs, 143, 144, 145, 146, 150, 173
  Clovis, 115, 116, 125
  Cogswell, 104, 173
  Comoro, 95, 96, 101, 102, 103, 135, 173-174
  Contine, 77, 174
  Continental, 89, 90, 96, 105, 107, 109, 174
  Coolidge, 79, 82, 83, 84, 174
  Cordes, 128, 129, 130, 131, 136, 174-175
  Cornville, 109, 175
  Cristobal, 78, 80, 83, 161, 175
  Cross, 126, 129, 136, 175
  Crot, 104, 175-176
  Courthouse, 99, 175
  Dandrea, 143, 176
  Deloro, 99, 161, 176
  Disterheff, 137, 176
  Dona Ana, 102, 105, 106, 107, 177
  Dry Lake, 104
  Duncan, 104
  Dye, 123, 138, 178
  Eba, 96, 105, 178
  Ebon, 77, 178
  Elfrida, 104, 106, 178
  Elledge, 137, 138, 178
  Ess, 144, 179
  Estrella, 75, 179
  Faraway, 136, 140, 179
  Flattop, 91
  Forest, 96, 107, 179
  Fruitland, 113, 115, 117, 120, 180
  Gachado, 80, 180

soil series (continued)
  Gadsden, 75, 180
  Gila, 88, 90, 95, 105, 181
  Gilman, 75, 76, 79, 81, 181
  Glenbar, 75, 181
  Glendale, 88, 95, 106, 181
  Gordo, 141, 144, 150, 181
  Gothard, 93, 104, 106, 181-182
  Grabe, 92, 95, 96, 106, 107, 110, 135, 182
  Graham, 93, 99, 108, 110, 111, 182
  Guest, 95, 96, 97, 104, 107, 108, 110, 111, 182
  Gunsight, 73, 78, 80, 83, 183
  Hantz, 106, 183
  Harqua, 73, 76, 78, 80, 83, 84, 183
  Harrisburg, 88, 183
  Hathaway, 92, 97, 98, 100, 101, 133, 183-184
  Hogg, 145, 154
  Holtville, 75, 184
  House Mountain, 99, 111, 184
  Hubert, 125, 184
  Indio, 75, 184
  Ives, 116, 118, 119, 120, 125, 185
  Jacks, 123, 127, 128, 138, 139
  Jacques, 129, 137, 138
  Jocity, 116, 118, 119, 125, 139, 185
  Karro, 104, 106, 105, 186
  Kimbrough, 107, 186
  Kofa, 75
  Krentz, 108, 186
  Lagunita, 75, 81, 186
  Lampshire, 89, 92, 99, 111, 187
  La Palma, 76
  Latene, 90, 102, 105, 107, 109, 187
  Laveen, 77, 78, 79, 81, 187
  Lehmans, 85, 89, 188
  Ligurta, 80, 81, 83, 161, 188
  Limpia, 111, 188
  Lomitas, 80, 83, 189
  Lonti, 128, 131, 132, 189
  Luth, 143, 144, 145, 146, 150, 189
  Luzena, 136, 189-190
  Lynx, 123, 124, 127, 128, 129, 130, 131, 132, 136, 137, 138, 139, 140, 190
  Mabray, 92, 99, 190
  Maripo, 75, 190
  Martinez, 133, 135, 190-191
  McAllister, 96, 106, 191
  McVickers, 145, 191
  Millard, 137, 191
  Millet, 125, 191-192
  Mirabal, 143, 144, 147, 148, 192
  Moano, 89, 136, 192
  Moenkopie, 66, 113, 115, 116, 117, 123, 124, 193
  Mohall, 76, 77, 79, 82, 193
  Mohave, 66, 90, 105, 193
  Mokiak, 136, 194
  Mormon Mesa, 91, 194
  Navajo, 118, 119, 194
  Nickel, 85, 88, 90, 97, 99, 107, 109, 194
  Nolan, 98, 100, 194-195
  Oracle, 99, 101, 103, 195
  Overgaard, 139, 145, 195
  Palma, 116, 125, 195
  Palomino, 145, 146, 196
  Palos Verde, 100, 102, 112, 196
  Pantano, 99, 196
  Partri, 130, 196
  Pastura, 121, 123, 126, 128, 130, 132, 196-197
  Penthouse, 109, 197
  Perryville, 73, 78, 79, 80, 197
  Pima, 92, 95, 96, 106, 107, 108, 110, 111, 197
  Pimer, 75, 197
  Pinal, 78, 198
  Pinaleno, 89, 90, 96, 97, 99, 102, 105, 107, 198
  Pinamt, 77, 80, 82, 198
  Poley, 124, 128, 132, 198
  Purgatory, 116, 199
  Purner, 123, 199
  Retriever, 99, 199
  Rillino, 102, 107, 109, 199
  Rillito, 78, 80, 81, 82, 88, 199-200
  Rimrock, 108, 111, 200
  Romero, 99, 103, 200
  Rond, 127, 138, 200
  Roundtop, 31, 127, 138, 139, 201
  Rositas, 84, 200
  Rune, 128, 130, 201
  Saint Thomas, 91, 99, 201
  Sanchez, 145, 201-202
  Schrap, 99, 202
  Shalet, 111, 115, 117, 125, 202
  Sheppard, 115, 117, 125, 202
  Siesta, 144, 202-203
  Signal, 98, 110, 203
  Sizer, 144, 203
  Soldier, 31, 145, 146, 203, 220
  Sonoita, 96, 112, 204
  Sontag, 108, 204
  Sponseller, 142, 144, 204
  Springerville, 123, 126, 129, 131, 204-205
  Stellar, 107, 205
  Stewart, 104, 205
  Suncity, 66
  Superstition, 84, 205-206
  Tatiyee, 141, 150, 206
  Telephone, 139, 206
  Thunderbird, 123, 126, 129, 140
  Tobler, 80, 206
  Toltec, 76, 207
  Tonopah, 91, 207
  Tortugas, 123, 124, 136, 138
  Tours, 113, 115, 116, 118, 119, 120, 124, 137, 138, 139, 207
  Trail, 116, 118, 119, 125, 207-208
  Tremont, 73, 76, 77, 79, 82, 208
  Tres Hermanos, 90, 96, 102, 105, 107, 112, 208
  Trix, 75, 208
  Tubac, 92, 96, 112, 208-209
  Tusayan, 124, 209
  Valencia, 112, 209
  Vecont, 76, 77, 209
  Venzia, 129, 209-210
  Vint, 75, 81, 210
  Vinton, 88, 95, 210
  Waldroup, 129, 210
  Webring, 124, 210
  White House, 97, 98, 100, 101, 108, 133, 135, 210
  Whitlock, 102, 105, 211
  Wilcoxson, 143
  Wildcat, 145, 211
  Wineg, 132, 211
  Winkel, 91, 211
  Winona, 31, 121, 127, 211
  Zeniff, 137, 212
  Ziegler, 129, 212
soil structure, 5
  angular blocky, 5, 13
  columnar, 5
  granular, 5, 7
  prismatic, 5, 7
  subangular blocky, 5
Soil Taxonomy, 38, 40, 63-70
soil temperature families, 69

INDEX

soil temperature regimes
  cryic, 38, 215
  frigid, 69, 72, 216
  hyperthermic, 38, 69
  isofrigid, 69
  isohyperthermic, 69
  isomesic, 69
  isothermic, 69
  mesic, 38, 72, 218
  pergellic, 49, 219
  thermic, 69, 72
soil texture, 4, 5
  clay, 3, 4, 5, 10, 13, 25, 40, 64, 65, 69, 214
  clay loam, 4
  loam, 4, 5, 72, 218
  sand, 3, 4, 5, 20, 64, 72, 220
  silt, 3, 4, 5, 10, 72, 221
soil texture triangle, 4
soil water
  capillary, 13
  field capacity 6, 41
  hygroscopic, 6, 217
  noncapillary, 6, 13
  oven-dry, 6
  permanent wilting point, 6
soilscape, 67, 221
solar radiation, 10
soluble salts, 9, 10, 12
Sonoran Biotic Province, 55
Sonoran Desert, 43, 48, 52, 53
sotol, 111
spadefoot toad, 60
Sparrow, Chipping, 56, 158
spruce
  blue, 49, 156
  Engelman, 10, 49, 156
spruce-alpine fir forests, 49, 56, 148
squirreltail, 129, 130, 131, 132, 138, 143, 144, 145
stadial time, 154
steppe climates, 36
steppes, 35
stratification, 26, 72, 82
stratigraphic column, 151-153
stream competence, 21
structureless soils, 5
subduction, 22
subgroup category, soil taxonomy, 65, 66, 71, 72
Subhumid Region, 68
suborder category formative elements, soil taxonomy, 65
suborder, 65, 66, 71, 72
subsidence, 16
subsurface diagnostic horizons, 64
Sulphur Springs Valley, 96
Sunset Crater, 15
surface diagnostic horizons, 64
surface water runoff, 10
surplus water, 43
syncline, 26, 223

talus, 28, 29, 30, 144
tamarisk, 75, 88, 118, 119, 156
Tappan age basalt, 28, 29, 31
telescoping effect, 44
temperature-precipitation zones, 72
termites, 57, 58, 60
terrace deposits, 22, 29, 223
Tertiary Period, 29, 32, 152, 153
Tertiary rocks, 17, 26, 28, 154
Thermic Arid Zone, 38, 40, 85, 86
Thermic Semiarid Zone, 3, 38, 40, 52, 53, 92, 93
thistle, Russian, 117, 156

time zero, 12
toads, 60, 158
Torrifluvents, 77, 78, 79, 94, 98, 102, 103, 105, 109, 115, 119, 120, 127, 160, 207
Torriorthentic Haplustolls, 126
Torriorthents, 80, 87, 91, 98, 115, 119, 120, 160
Torripsamments, 94, 160, 207
transition zone, 17
transpiration, 8, 33, 41
tree-ring dating, 28
trenches, deep sea, 22
Triassic Period, 152, 153
trout stocked lakes, 135, 140, 150
Tucson Mountain Park, 48
tumbleweed, see thistle, Russian
turkshead, 84, 156
turpentine brush, 103, 156
Typic Argiborolls, 143, 144
Typic Argiustolls, 108
Typic Calciorthids, 78, 79, 81, 82, 83, 84, 90, 102, 105, 107, 109
Typic Chromusterts, 108, 129
Typic Cryorthents, 147
Typic Durargids, 66, 76
Typic Durorthids, 78, 80, 104
Typic Eutroboralfs, 145, 148
Typic Glossoboralfs, 145, 146
Typic Haplargids, 76, 77, 82, 83, 90, 96, 102, 105, 109, 112
Typic Haplustalfs, 137, 139
Typic Natrargids, 76, 104, 106
Typic Paleargids, 96
Typic Paleorthids, 78, 107
Typic Torrifluvents, 75, 81, 88, 90, 95, 110, 112, 115, 116, 118, 119, 120, 125
Typic Torriorthents, 81, 115, 116, 119, 120
Typic Torripsamments, 84
Typic Ustorthents, 147, 148
Udic Haplustalfs, 123, 135, 138, 139
Udic Paleustalfs, 135
Ultisols, 66, 67, 68, 160
unconformity, 17, 151
Usterts, 69, 160
Ustollic Calciorthids, 132, 137
Ustollic Haplargids, 97, 98, 100, 101, 103, 109, 125, 130, 131
Ustollic Paleorthids, 130, 132
Ustorthents, 87, 150
V-shaped canyon, 29
valley axial troughs, 21
vegetation zones, 44, 48, 49, 55, 56
vertebrates, 56, 60-61
Vertisols, 13, 32, 66, 67, 69, 160, 223
Vishnu Schist, 151, 153
voids, 3, 5, 6, 40, 62
volcanic ash, 25, 31, 68, 69, 108
volcanic rocks, 11, 17, 19, 22, 23, 24, 26, 27, 28, 29, 31-32, 72, 141
Walnut Canyon, 124
warm highlands, 35
warm steppes, 35
water-adapted vegetation, 6
water budget, 41, 42
water cycle, 37
water, erosion, 12
water, forms in soil, 6
water harvesting, 2
water, need, 40
water, saturation, 6, 10
water table, 20, 223
water vapor, 5, 6
Wasatch Mountains, 35
weathered rocks, 8
weathering, 3, 4, 23, 24, 25, 28, 30, 31, 68, 70
  chemical, 31

weathering (continued)
  differential, 29
  hydrolytic, 68
  physical, 31, 43
welded ash flow, 17
Western Pacific Plate, 23
White Mountain volcanic field, 28
White Mountains, 2, 5, 33, 34, 35, 38, 43, 44, 55
whitebrush, 53, 156
whitethorn, 53, 101, 102, 107, 112
wild burros, 60
wild daisy, 51, 155
Wild Turkey, 137, 143, 144, 145, 147, 148, 149, 150, 158
Willcox Playa, 16, 17, 104
willow, 50
wilting point, 41
wind deposition, 30
wind direction, 10
winterfat, 52, 130, 156
Wisconsin Glaciation, 44, 154
wolfberry, 96, 156
wolftail, 99, 103, 109, 123, 130, 131, 132, 156
Woodhouse basalt age group, 28, 29
worms, 13, 56, 59
Wren, Cactus, 56, 158
Wupatki National Monument, 124
Yarmouth Interglacial Age, 44
Yuma soil moisture budget, 47
zeolite, 16